LONDON MATHEMATICAL SOCIETY LECTURE NOTE SERIES

Managing Editor: Professor M. Reid, Mathematics Institute,
University of Warwick, Coventry CV4 7AL, United Kingdom

The titles below are available from booksellers, or from Cambridge University Press at www.cambridge.org/mathematics

London Mathematical Society Lecture Notes Series: 383

Motivic Integration and its Interactions with Model Theory and Non-Archimedean Geometry

Volume I

Edited by

RAF CLUCKERS
Université de Lille 1, France

JOHANNES NICAISE
Katholieke Universiteit Leuven, Belgium

JULIEN SEBAG
Université de Rennes 1, France

CAMBRIDGE
UNIVERSITY PRESS

CAMBRIDGE
UNIVERSITY PRESS

University Printing House, Cambridge CB2 8BS, United Kingdom

One Liberty Plaza, 20th Floor, New York, NY 10006, USA

477 Williamstown Road, Port Melbourne, VIC 3207, Australia

314-321, 3rd Floor, Plot 3, Splendor Forum, Jasola District Centre, New Delhi - 110025, India

79 Anson Road, #06-04/06, Singapore 079906

Cambridge University Press is part of the University of Cambridge.

It furthers the University's mission by disseminating knowledge in the pursuit of education, learning and research at the highest international levels of excellence.

www.cambridge.org
Information on this title: www.cambridge.org/9780521149761

© Cambridge University Press 2011

First published 2011

A catalogue record for this publication is available from the British Library

Library of Congress Cataloging in Publication data
Motivic integration and its interactions with model theory and non-Archimedean geometry / edited by Raf Cluckers, Johannes Nicaise, Julien Sebag.
p. cm. – (London Mathematical Society lecture note series ; 383)
ISBN 978-0-521-14976-1 (pbk.)
1. Model theory.   2. Valued fields.   3. Analytic spaces.   4. Geometry, Algebraic.
I. Cluckers, Raf.   II. Nicaise, Johannes.   III. Sebag, Julien.
QA9.7.M68   2011
511.3´4 – dc23   2011021254

ISBN 978-0-521-14976-1 Paperback

Table of Contents for Volume I

Table of Contents for Volume II

Contributors

Manuel Blickle,   *Johannes Gutenberg-Universität Mainz, Institut für Mathematik, 55099 Mainz, Germany;* email*: manuel.blickle@gmail.com*

Siegfried Bosch,   *Mathematisches Institüt, Fachbereich Mathematik und Informatik der Universität Münster, Einsteinstrasse 62, 48149 Münster, Germany;* email*: bosch@math.uni-muenster.de*

Antoine Chambert–Loir,   *Université Rennes 1, Unité de Formation et de Recherche Mathématiques, Institut de Recherche Mathématique de Rennes (IRMAR), 263 Avenue du Général Leclerc, CS 74205, 35042 Rennes Cedex, France;* email*: Antoine.Chambert-Loir@univ-rennes1.fr*

Zoé Chatzidakis,   *Université Paris 7, Unité de Formation et de Recherche Mathématiques, Case 7012, Site Chevaleret, 75205 Paris Cedex 13, France;* email*: zoe@logique.jussieu.fr*

Raf Cluckers,   *Université Lille 1, Laboratoire Painlevé, CNRS - UMR 8524, Cité Scientifique, 59655 Villeneuve d'Ascq Cedex, France and Katholieke Universiteit Leuven, Department of Mathematics, Celestijnenlaan 200B, 3001 Heverlee, Belgium;* email*: Raf.Cluckers@wis.kuleuven.be*

Françoise Delon,   *Université Paris 7, Unité de Formation et de Recherche Mathématiques, Case 7012, Site Chevaleret, 75205 Paris Cedex 13, France;* email*: delon@logique.jussieu.fr*

Immanuel Halupczok,   *Münster Universität, Mathematisches Institut und Institut für Mathematische Logik und Grundlagenforschung, Einsteinstrasse 62, 48149 Münster, Germany;* email*: math@karimmi.de*

Florian Ivorra,   *Université Rennes 1, Unité de Formation et de Recherche Mathématiques, Institut de Recherche Mathématique de Rennes (IRMAR), 263 Avenue du Général Leclerc, CS 74205, 35042 Rennes Cedex, France;* email*: Florian.Ivorra@univ-rennes1.fr*

Fumiharu Kato,   *Kyoto University, Department of Mathematics, Faculty of Science, Kyoto 606-8502, Japan;* email*: kato@kusm.kyoto-u.ac.jp*

Emmanuel Kowalski, *Eidgenössische Technische Hochschule Zürich, Departement Mathematik, HG G 64.1, Rämistrasse 101, 8092 Zürich, Switzerland;* email: *emmanuel.kowalski@math.ethz.ch*

François Loeser, *École Normale Supérieure, 4, rue d'Ulm, 75230 Paris Cedex 05, France;* email: *Francois.Loeser@ens.fr*

Johannes Nicaise, *Katholieke Universiteit Leuven, Department of Mathematics, Celestijnenlaan 200B, 3001 Heverlee, Belgium;* email: *Johannes.Nicaise@wis.kuleuven.be*

Karl Rökaeus, *Horteweg de Vries Instituut voor Wiskunde, Universiteit van Amsterdam, P.O. Box 94248, 1090 Ge AMSTERDAM, The Netherlands;* email: *S.K.F.Rokaeus@uva.nl*

Julien Sebag, *Université Rennes 1, Unité de Formation et de Recherche Mathématiques, Institut de Recherche Mathématique de Rennes (IRMAR), 263 Avenue du Général Leclerc, CS 74205, 35042 Rennes Cedex, France;* email: *Julien.Sebag@univ-rennes1.fr*

Michael Temkin, *Hebrew University of Jerusalem, The Hebrew University, The Edmond J. Safra Campus–Givat Ram, Jerusalem 91904, Israel;* email: *temkin@math.huji.ac.il*

1

Introduction

Since its creation by Kontsevich in 1995, motivic integration has developed quickly in several directions and has found applications in various domains, such as singularity theory and the Langlands program. In its development, it incorporated tools from model theory and non-archimedean geometry. The aim of the present book is to give an introduction to different theories of motivic integration and related topics.

Motivic integration is a theory of integration for various classes of geometric objects. One term in this "trade-name" seems unusual: *motivic*. What is a motive and in what sense is this theory of integration motivic? These questions lie at the heart of the theory. We will try to answer them in Section 4. First, we briefly introduce the two other protagonists of the present book: *model theory* and *non-archimedean geometry*. We will explain below how they interact with the theory of motivic integration.

1 Model theory

A central notion in model theory is *language*. A language $\mathscr{L}$ is a collection of symbols, divided into three types: function symbols, relation symbols, and constant symbols. For every function symbol and relation symbol, one specifies the number of arguments. A formula in the language $\mathscr{L}$ is built from these symbols, variables, Boolean combinations and quantifiers.

Motivic Integration and its Interactions with Model Theory and Non-Archimedean Geometry (Volume I), ed. Raf Cluckers, Johannes Nicaise, and Julien Sebag. Published by Cambridge University Press. © Cambridge University Press 2011.

A structure for the language $\mathscr{L}$ is a set M together with a concrete interpretation of the symbols in the language; for instance, for every function symbol f with n arguments one specifies a function

$$f : M^n \to M.$$

In this way, one can interpret formulas in the language $\mathscr{L}$ in the structure M, and any $\mathscr{L}$-formula φ determines a subset $\varphi(M)$ of a suitable Cartesian power of M. Such subsets $\varphi(M)$ are called $\mathscr{L}$-*definable sets* in the structure M.

Example 1.1   The language of rings $\mathscr{L}_{\text{ring}}$ is the language with two function symbols " $+$ " and " $\cdot$ ", each with two arguments, no relation symbols, and two constant symbols "0" and "1". If M is a ring, then we can interpret these symbols in A in the obvious way, so that M becomes a structure for the language $\mathscr{L}_{\text{ring}}$.

The expression φ given by

$$(\exists x)(x^2 = y) \wedge z = 1$$

is a formula in the language $\mathscr{L}_{\text{ring}}$. It has three variables: x, y and z. The variable x is *bounded* by the quantifier $(\exists x)$. The variables y and z are *free*. For any ring M, the formula φ defines a subset $\varphi(M)$ of M^2, namely the set of all couples $(y, 1)$ with y a square in M.

A *theory* in the language $\mathscr{L}$ is a collection $\mathscr{T}$ of formulas without free variables. This means in particular that, once interpreted in an $\mathscr{L}$-structure, the formulas in $\mathscr{T}$ are either true or false. A *model* for $\mathscr{T}$ is an $\mathscr{L}$-structure that satisfies every formula in $\mathscr{T}$.

Example 1.2   In the language $\mathscr{L}_{\text{ring}}$ of rings, we can express the axioms saying that a structure M is a field. For instance, commutativity of the addition is expressed by the formula

$$(\forall x)(\forall y)(x + y = y + x)$$

and the existence of a multiplicative inverse is expressed by the formula

$$(\forall x)(x = 0 \vee (\exists y)(x \cdot y = 1)).$$

The set of these axioms is a theory, and its models are precisely the fields.

Two of the main objectives of model theory are the following:

- given a language $\mathscr{L}$ and a theory $\mathscr{T}$, classify the models of $\mathscr{T}$;
- given a language $\mathscr{L}$ and a structure M, analyze the definable sets of M.

We will focus here on the second objective. A crucial idea is that the complexity of the definable sets depends both on the specific structure M and on the language $\mathscr{L}$. For instance, if $\mathscr{L}$ is the language of rings, then the definable sets for a field K are easy to describe when K is algebraically closed (they are precisely the *constructible subsets* of K^m, with $m \in \mathbb{N}$) but their structure can be much more delicate when K is an arbitrary field. For the structure $\mathbb{Z}$, the definable sets are easy to describe when we only allow addition and the natural order in our language, but they become extremely complicated when we also consider multiplication; for example, we can define the set of prime numbers. There exists no algorithm that, with input a formula φ in the language of rings, decides whether $\varphi(\mathbb{Z})$ is empty or not.

A property that can greatly reduce the complexity of definable sets is *quantifier elimination*. This property also plays an important role in the theory of motivic integration, since it facilitates the analysis of definable sets in valued fields. We say that a structure M for a language $\mathscr{L}$ has quantifier elimination if every definable set can be described by a formula in $\mathscr{L}$ *without* quantifiers. Classical examples are algebraically closed fields in the language of rings (this is essentially Chevalley's theorem) and real closed fields in the language of rings with the additional relation $\leq$ (Tarski's theorem). Again, this property depends both on the structure and the language. For instance, $\mathbb{C}$ has quantifier elimination in the language of rings; $\mathbb{R}$ does not have quantifier elimination in the language of rings, since the set of nonnegative real numbers can be defined by the formula

$$(\exists x)(y = x^2)$$

but not by a formula without quantifiers; $\mathbb{R}$ acquires quantifier elimination if we add the relation $\leq$ to the language.

In the theory of motivic integration, one uses various languages to describe the structure of valued fields. One of the most prominent ones is the language $\mathscr{L}_{\mathrm{DP}}$ of Denef-Pas. It is a bit more complicated than the type of language we have considered above: it describes simultaneously three *sorts* of sets, corresponding to the valued field K, its value group Γ with ∞ added, and its residue field k. On K and k one considers the language of rings, and on Γ the language of ordered groups. The three sorts are related in the language $\mathscr{L}_{\mathrm{DP}}$ by the valuation map

$$v_K : K \to \Gamma \cup \{\infty\}$$

and a certain multiplicative map, called the angular component map,

$$\overline{ac} : K \to k.$$

We will mostly be dealing with Henselian discretely valued fields. In that case, $\Gamma = \mathbb{Z}$ and the angular component map can be defined by choosing a uniformizer π in the valuation ring $R = \mathscr{O}_K$ and putting

$$\overline{ac}(x) = \pi^{-v_K(x)} x \mod \pi R \quad \in k$$

for $x \neq 0$, and $\overline{ac}(0) = 0$. In this language of Denef-Pas, as well as in related languages, the analysis of definable sets in Henselian valued fields is highly developed and is based on specific quantifier elimination and cell decomposition results. It are these definable sets that can be measured by various theories of motivic integration.

For a more thorough introduction to model theory, in particular the model theory of valued fields, we refer to Chapter 2.

2 Non-archimedean geometry

The aim of non-archimedean geometry is to develop a formalism of analytic geometry over a non-archimedean field K (a field endowed with an absolute value that satisfies the ultrametric property), such as the field $\mathbb{Q}_p$ of p-adic numbers. In the naïve approach, analytic functions on open parts of K^n (w.r.t. the metric topology) are defined to be K-valued functions that can locally be written as converging power series, just like in the complex case. However, this does not lead to a satisfactory theory with the good properties of complex analytic geometry[1]. Since the topology on K is totally disconnected, there are too many analytic functions (in particular, all locally constant functions satisfy the naïve definition) and very few K-analytic manifolds (every compact p-adic analytic manifold is isomorphic to a finite disjoint union of closed unit balls; see Chapter 7). Several theories of non-archimedean geometry deal with this problem in various ways.

---

[1] Nevertheless, it should be noted that, since the foundational work by Denef and van den Dries [15], this naïve approach has led to interesting geometrical results on non-archimedean local fields, for example in analogy to real subanalytic geometry after Hironaka and Łojasiewicz.

Historically, the first of these theories was *rigid geometry*. Rigid geometry was created by Tate in the sixties [37] and further developed by Grauert, Remmert, Kiehl and others. In rigid geometry, as in Grothendieck's scheme theory, one starts from the "correct" K-algebra of analytic functions, and one constructs from this algebra a space on which these functions live in a natural way. For instance, the analytic functions on the closed unit disc D in K are the formal power series over K that converge *globally* on D, and the associated space is the maximal spectrum of this algebra. This spectrum is endowed with a certain *Grothendieck topology* that turns it into a connected space. In this Grothendieck topology only certain types of open covers are allowed; this *rigidifies* the topology on a K-analytic variety and explains the name of the theory (Tate called the naïve K-analytic varieties *wobbly*). The foundations of rigid geometry are explained in Chapter 3.

In the early seventies, Raynaud showed in [33] how a certain class of rigid K-varieties can be realized as generic fibers of formal schemes over the ring of integers R of K, and how different formal models of a fixed rigid variety can be related via so-called *admissible blow-ups* (formal blow-ups whose centers are contained in the special fiber of the formal scheme). This result makes it possible to deduce many fundamental properties of rigid varieties from the corresponding properties of formal schemes. Raynaud's theory was systematically developed by Bosch, Lütkebohmert and Raynaud in the nineties in a series of papers [6, 7, 8, 9]. The foundations of the theory were recently written down and expanded in [1]. The philosophy behind Raynaud's theory is explained in Chapter 4.

At the end of the eighties, Berkovich developed a new approach to non-archimedean geometry, generalizing the notion of point on an analytic space [5]. In Tate's approach, the points on a rigid variety correspond to maximal ideals of a so-called *Tate algebra A* (a quotient of a K-algebra of converging power series). The residue fields at these maximal ideals can be shown to be finite extensions of K, so that they carry a unique extension of the absolute value $|\cdot|_K$ on K. A point of the Berkovich spectrum associated to A corresponds to a *prime* ideal $\mathscr{P}$ of A *plus* an extension of $|\cdot|_K$ to the residue field $Frac(A/\mathscr{P})$. An advantage of Berkovich's theory is that his spaces carry a natural topology with very good properties. For instance, the closed unit disc becomes Hausdorff, compact and arc-connected. However, for the construction of a sufficiently rich category of non-archimedean analytic spaces, the introduction of a Grothendieck topology is still necessary. A surprising and quite useful feature of Berkovich's theory is that it also works well over fields with the trivial absolute value.

A defect of Tate's rigid spaces is that they do not contain enough points to detect the exactness of a sequence of abelian sheaves by looking at the stalks. There are two other approaches to non-archimedean geometry where the topological space associated to a rigid K-variety X is the one predicted by topos theory (a sober topological space whose category of sheaves is equivalent to the category of sheaves w.r.t. the rigid Grothendieck topology on X; such a space is unique if it exists [2, IV.4.2.4]). The first approach was developed by Huber, the second by Fujiwara and Kato.

Huber's theory of *adic spaces* is explained in [22]. It is based on valuation spectra of a large class of topological rings. An important difference with Berkovich's theory is that points correspond to valuations of arbitrary rank, and not only of rank one. For separated rigid spaces, the associated Berkovich space can be realized as the biggest Hausdorff quotient of Huber's adic space. For more background on the comparison between different topological spaces appearing in non-archimedean geometry, we refer to [34].

The approach of Fujiwara and Kato is based on Raynaud's theory of formal models. In order to construct the generic fiber of a formal R-scheme, they take the projective limit over all the admissible blow-ups of the formal scheme in the category of locally ringed spaces, and they tensor the structure sheaf with K. The underlying topological space of this projective limit is the so-called *Zariski Riemann space*. Fujiwara and Kato's theory is explained in Chapter 4. This construction can be set up in a quite general framework, leading to an interesting blend of rigid geometry and birational geometry.

3 Motives and the Grothendieck ring of varieties

In a broad sense, a *motive* is an object that captures the geometric essence of a certain given structure and explains the fundamental common ground underlying various manifestations of this structure. More precisely, motives in algebraic geometry are supposed to provide a "universal cohomology theory". They should give a deep geometric explanation for the similarities in the behaviour of several cohomology theories (Betti, de Rham, ℓ-adic, crystalline,...).

Motives were introduced by Grothendieck. In his letter to Serre dated 16–8–1964 (see [13]), Grothendieck wrote:

J'appelle «motif» sur k quelque chose comme un groupe de cohomologie ℓ-adique d'un schéma algébrique sur k, mais considéré comme indépendant de ℓ, et avec sa structure «entière», ou disons pour l'instant «sur ℚ», déduite de la théorie des cycles algébriques. La triste vérité, c'est que pour le moment je ne sais pas définir la catégorie abélienne des motifs [...][2]

Since then, several categories of motives have been proposed, but a large part of the theory remains conjectural. Grothendieck defined the category of *Chow motives*, which should be direct sums of "pure" motives. The good properties of this category are predicted by Grothendieck's *standard conjectures*, which are still open today. It is also conjectured that there should exist a larger category of "mixed" motives, but this object is still out of reach. Instead, people have been trying to construct the so-called derived category of mixed motives. We refer to Chapter 12 for more background.

Very roughly, Chow motives over a field k can be thought of as "pieces" of smooth and projective algebraic k-varieties that capture certain parts of the graded cohomology space of the variety. This philosophy can be implemented at a more elementary level, leading to the so-called *Grothendieck ring of varieties* $K_0(\mathrm{Var}_k)$. This ring was already considered by Grothendieck, in the same letter to Serre [13]:

Soit k un corps [...] et soit L(k) le «groupe K» défini par les schémas de type fini sur k, avec comme relations celles qui proviennent d'un découpage en morceaux [...][3]

The elements of the Grothendieck ring are *virtual varieties*. More precisely, $K_0(\mathrm{Var}_k)$ is the free abelian group on isomorphism classes $[X]$ of k-varieties X, modulo the relations

$$[X] = [Y] + [X \setminus Y]$$

---

[2] I call a "motive" over k something like an ℓ-adic cohomology group of an algebraic scheme over k, but considered as independent of ℓ, and with its "integral" structure, or, let us say for the moment "over ℚ" structure, deduced from the theory of algebraic cycles. The sad truth is that for the moment I do not know how to define the abelian category of motives [...]

[3] Let k be a field [...] and let $L(k)$ be the "K-group" defined by schemes of finite type over k together with relations coming from decomposition into pieces [...]

for every k-variety X and every closed subvariety Y of X. These relations are called *scissor relations*, because they allow to cut up a variety into subvarieties. The ring product in $K_0(\mathrm{Var}_k)$ is induced by the product of k-varieties. The Grothendieck ring of varieties is studied in detail in Chapter 5. Its structure is still quite mysterious, and related to many fundamental questions in algebraic geometry.

Taking the class of a k-variety in the Grothendieck ring $K_0(\mathrm{Var}_k)$ is, in a sense, the most general way to "measure the size" of the variety. More precisely, it is the universal additive and multiplicative invariant: every invariant of k-varieties with values in a ring A that is additive (i.e., respects the scissor relations) and multiplicative w.r.t. the product of varieties factors through a unique realization morphism $K_0(\mathrm{Var}_k) \to A$. This applies, in particular, to many cohomological invariants, such as the Euler characteristic, the Poincaré polynomial and (in characteristic zero) the Hodge-Deligne polynomial. If k is finite, then counting the number of rational points on a k-variety is clearly an additive and multiplicative operation, so we obtain the *point counting realization*

$$\sharp : K_0(\mathrm{Var}_k) \to \mathbb{Z} : [X] \mapsto |X(k)|.$$

More generally, for any finite field extension k' of k, we have a ring morphism

$$\sharp_{k'} : K_0(\mathrm{Var}_k) \to \mathbb{Z} : [X] \mapsto |X(k')|.$$

We denote by $\mathbb{L}$ the class $[\mathbb{A}^1_k]$ of the affine line in the Grothendieck ring $K_0(\mathrm{Var}_k)$, and by $\mathscr{M}_k$ the localization of $K_0(\mathrm{Var}_k)$ with respect to $\mathbb{L}$. If k is finite, then $\sharp\mathbb{L} = |k|$, and the point counting realization localizes to a ring morphism

$$\sharp : \mathscr{M}_k \to \mathbb{Z}[|k|^{-1}].$$

Drinfeld called the elements in $K_0(\mathrm{Var}_k)$ the "poor man's motives". If k has characteristic zero, there is a natural realization morphism from $K_0(\mathrm{Var}_k)$ to the Grothendieck ring of the additive category of Chow motives. In his letter to Serre, Grothendieck already conjectured the existence of such a morphism, and he asked whether it is very far from being bijective. We will discuss this issue in Chapter 5.

4 Why is motivic integration called motivic?

There are at least two explanations for the adjective "motivic" in "motivic integration", and they correspond to two types of applications of the theory.

4.1 Motivic invariants of varieties and singularities

A first explanation is that motivic integrals take their values in a localized Grothendieck ring of geometric objects (or an appropriate completion) where these geometric objects can be varieties, motives, or definable sets. In this way, motivic integration produces motivic invariants of algebraic varieties, rigid varieties and formal singularities. These invariants can be specialized to various cohomological invariants.

The central tool in this context is the *change of variables formula* for motivic integrals, which enables you to compute the motivic integral on a new space where the geometry of the data is easier, or where different motivic integrals can be compared. In this way, for instance, Kontsevich proved his famous theorem on the equality of Hodge numbers of birationally equivalent complex Calabi-Yau varieties, which was the start of the theory of motivic integration (see Section 6.1 and Chapter 6). This approach has also led to several new invariants of algebraic singularities that can be defined in terms of a motivic integral and computed explicitly on a resolution of singularities. We refer to Chapter 6 and to [39] for a survey.

4.2 Geometric nature of p-adic integrals

A second explanation is that motivic integrals can be specialized to p-adic integrals in a precise sense. They reflect the geometric structure behind a large class of p-adic integrals, and explain the uniform behaviour of these p-adic integrals for sufficiently large primes p. An example of this phenomenon is Denef and Loeser's *motivic zeta function* that appears as a universal geometric object floating above Igusa's p-adic zeta functions associated to a polynomial with integer coefficients for different values of the prime p (see [29] for a survey). More recently, Cluckers, Hales and Loeser have applied these ideas to the study of p-adic orbital integrals appearing in the Langlands program [10]. They use Cluckers and Loeser's *transfer principle* for motivic integrals [12] to transfer equalities of integrals between the mixed characteristic case (*p*-adic fields) and the equal characteristic case (Laurent series over a finite field) for sufficiently large residual characteristic. They show that this principle applies to the integrals appearing in the Fundamental Lemma.

4.3 Complex geometry versus *p*-adic geometry

These two types of applications are certainly not disjoint. For instance, Denef and Loeser's motivic zeta function is not only a universal *p*-adic zeta function, but also a very rich invariant of complex hypersurface singularities; see Chapter 7. Moreover, before the discovery of motivic integration, *p*-adic integrals had been used to prove results about complex algebraic varieties. The immediate precursor of Kontsevich's theorem, namely Batyrev's theorem stating that birationally equivalent Calabi-Yau varieties have equal Betti numbers [4], was proved using *p*-adic integration and the Weil Conjectures (see Section 6.1). Earlier, Denef and Loeser had already used a similar technique to define the topological zeta function of a complex hypersurface singularity, by considering *p*-adic zeta functions over rings of Witt vectors $W(\mathbb{F}_{p^e})$ for all $e \in \mathbb{Z}_{>0}$, interpolating for real $e > 0$, and formally taking a limit for $e \to 0$ [17]. Nowadays, the topological zeta function can be realized as a specialization of the motivic zeta function. We refer to [24] for a survey on the *p*-adic prehistory of motivic integration.

5 Motivic versus *p*-adic integration

The easiest way to understand motivic integration is to view it as a geometric form of *p*-adic integration. Let us try to make this more precise.

Let R be a complete discrete valuation ring with residue field k and quotient field K. We fix a uniformizer π in R. For every integer $m \geq 0$, we have the projection map

$$\theta_m : R^n \to (R/\pi^{m+1})^n.$$

We denote by v_K the discrete valuation on R, and we define an absolute value on K by putting $|x|_K = \epsilon^{v_K(x)}$ for every $x \in K^*$, where ϵ is any element in $]0, 1[$. If k is finite, we take $\epsilon = |k|^{-1}$. The absolute value $|\cdot|_K$ is non-archimedean and defines a totally disconnected topology on K.

5.1 p-adic integration

Assume that k is finite. Then K is either a finite extension of $\mathbb{Q}_p$ or (once we fix a section of the projection $R \to k$) the field of Laurent series $k((\pi))$. We denote by q the cardinality of k. A good example to keep in mind is the case $R = \mathbb{Z}_p$, $k = \mathbb{F}_p$, $K = \mathbb{Q}_p$ and $\pi = p = q$. Finiteness of k implies that K is a locally compact group. Hence, K carries a Haar measure μ_{Haar}, which is unique if we normalize it by putting $\mu_{\text{Haar}}(R) = 1$. We denote again by μ_{Haar} the product measure on K^n, for every $n > 0$ (it is also a Haar measure on the locally compact group K^n). With slight abuse of notation, by p-adic integration, we mean integration w.r.t. the Haar measure on K, also in the case where K is a Laurent series field.

A *cylinder* is a subset C of R^n of the form $\theta_m^{-1}(C_m)$, for some $m \geq 0$ and some subset C_m of $(R/\pi^{m+1})^n$. It is not difficult to compute the measure of such a cylinder: by additivity and translation invariance of μ_{Haar}, it is given by

$$\mu_{\text{Haar}}(C) = |C_m| q^{-(m+1)n}. \tag{1.1}$$

To prove this, it is enough to consider the case where C_m is a point; then R^n is a disjoint union of $q^{(m+1)n}$ translates of C, so the normalization $\mu_{\text{Haar}}(R^n) = 1$ implies (1.1). The cylinders form a basis for the topology on R^n, and the measure of more general measurable subsets of R^n can be computed *via* approximation by cylinders.

Now we want to extend this construction to an *arbitrary* complete discrete valuation ring R (if R has mixed characteristic, we usually need to assume that k is perfect for technical reasons). If k is infinite, then K is no longer locally compact, so that we do not have a Haar measure, and we have to find another way to measure subsets of K^n. This is precisely what motivic integration does.

In the theory of motivic integration, one can roughly distinguish two types of approaches: the ones based on geometry, and the ones based on model theory. In fact, both approaches are geometric in nature, but they differ in the fact that the first approach uses properties from algebraic geometry, where the second approach uses model theory, to analyze the shapes, sizes, and general properties of the objects that are to be measured. Both approaches are linked by various comparison and specialization results, see below and Chapter 8.

5.2 The geometric approach to motivic integration

Here the idea is to parameterize the elements of $(R/\pi^{m+1})^n$ by k-varieties, and to measure the size of a k-variety by taking its class in the Grothendieck ring $K_0(\mathrm{Var}_k)$.

The most basic example is the case where R has equal characteristic. For simplicity, we assume that k is algebraically closed. If we fix a section of the projection $R \to k$, then we can identify R with $k[[\pi]]$, and an element of $(R/\pi^{m+1})^n$ is simply an n-tuple

$$(a_{1,0} + \ldots + a_{1,m}\pi^m, \ldots, a_{n,0} + \ldots + a_{n,m}\pi^m)$$

of truncated power series in $k[[\pi]]/\pi^{m+1}$. Interpreting the coefficients $a_{i,j}$ as affine coordinates, we can identify $(R/\pi^{m+1})^n$ with the set of closed points of the affine space $\mathbb{A}_k^{(m+1)n}$. If R has mixed characteristic, we can make a similar construction using Witt vector coordinates.

A *cylinder* in R^n is a subset C of the form

$$C = \theta_m^{-1}(C_m) \tag{1.2}$$

for some $m \in \mathbb{N}$ and some *constructible* subset C_m of

$$(R/\pi^{m+1})^n = \mathbb{A}_k^{(m+1)n}(k).$$

We define the motivic measure of a cylinder by mimicking the formula (1.1), as follows.

Every constructible subset Y of an algebraic k-variety X can be written as a disjoint union

$$Y = Y_1 \sqcup \ldots \sqcup Y_r$$

of subvarieties Y_i of X, and defines a class

$$[Y] := [Y_1] + \cdots + [Y_r]$$

in $K_0(\mathrm{Var}_k)$ that is independent of the chosen decomposition. The object $\mathbb{L} = [\mathbb{A}_k^1]$ in $K_0(\mathrm{Var}_k)$ plays the role of the cardinality of the residue field k. This is justified by the fact that, if F is a finite field, the cardinality $|F|$ equals the number of F-rational points on $\mathbb{A}_F^1$. The motivic measure of the cylinder C in (1.2) is then defined by

$$\mu_{\mathrm{mot}}(C) = [C_m]\mathbb{L}^{-mn} \quad \in \mathcal{M}_k. \tag{1.3}$$

This definition does not depend on the choice of m, since for every $i \in \mathbb{N}$, the truncation map

$$(R/\pi^{m+i+1})^n \to (R/\pi^{m+1})^n$$

is a trivial fibration with fiber $\mathbb{A}_k^{in}$.

The chosen normalization in (1.3) is different from the one in (1.1); in geometric applications, it is more natural to give R measure $\mathbb{L}$ (the class of the reduction modulo π), rather than measure 1.

The most basic class of integrable functions on R^n are the functions of the form $\mathbb{L}^\alpha$, with α a function from R^n to $\mathbb{Z} \cup \{\infty\}$ that takes only finitely many values and whose fibers over elements of $\mathbb{Z}$ are cylinders. The motivic integral of such a function is defined by

$$\int_{R^n} \mathbb{L}^\alpha = \sum_{i \in \mathbb{Z}} \mu_{\mathrm{mot}}(\alpha^{-1}(i))\mathbb{L}^i \quad \in \mathscr{M}_k.$$

The motivic measure of more complicated subsets of R^n is defined by approximation by cylinders. In this procedure, it is necessary to replace $\mathscr{M}_k$ by a certain *dimensional completion*; see Chapter 6. Via this completion, we can also integrate a larger class of functions. Taking the dimensional completion of $\mathscr{M}_k$ corresponds to replacing $\mathbb{Q}$ by $\mathbb{R}$ as value ring in the theory of p-adic integration.

It is also possible to generalize the measure space R^n, by considering spaces $X(R)$ of R-sections on (formal) R-schemes X. This leads to the notion of *arc scheme* (Chapter 6) and *Greenberg scheme* (Chapter 7).

Let us return to formula (1.3). If k is an algebraic closure of a finite field $\mathbb{F}_q$, and the constructible set C_m is defined over $\mathbb{F}_q$, then we can take the motivic measure of C in the Grothendieck ring $\mathscr{M}_{\mathbb{F}_q}$. Applying the point counting realization

$$\sharp : \mathscr{M}_{\mathbb{F}_q} \to \mathbb{Z}[q^{-1}]$$

we find (modulo normalization) the Haar measure of the cylinder

$$C(\mathbb{F}_q) := \theta_m^{-1}(C_m(\mathbb{F}_q)) \subset \mathbb{F}_q[[t]]^n.$$

Explicitly, we have

$$q^n \mu_{\mathrm{Haar}}(C(\mathbb{F}_q)) = \sharp \mu_{\mathrm{mot}}(C).$$

In this way, the motivic integral of an appropriate measurable function $\mathbb{L}^\alpha$ on R^n can be specialized to the p-adic integral of the function q^α on $\mathbb{F}_q[[t]]^n$. More generally, by counting points over finite extensions of $\mathbb{F}_q$, we can specialize the motivic integral to the corresponding p-adic integral over unramified extensions of $\mathbb{F}_q[[t]]$. Similar results hold when R has mixed characteristic. For details, see Chapters 7 and 14.

Example 5.1 Assume that $R = k[[t]]$, with k an algebraically closed field, and let us compute the motivic measure of the group of units R^* inside R. The set R^* is a cylinder in R: it is the inverse image of k^* under the projection

$$\theta_0 : R \to k.$$

Hence, we find

$$\mu_{\text{mot}}(R^*) = [\mathbb{A}_k^1 \setminus \{0\}] = \mathbb{L} - 1 \quad \in \mathcal{M}_k.$$

If $R = \mathbb{F}_q[[t]]$, then the Haar measure of R^* is given by

$$\mu_{\text{Haar}}(R^*) = q^{-1}(q - 1).$$

We see that

$$q \cdot \mu_{\text{Haar}}(R^*) = \sharp\mu_{\text{mot}}(R^*)$$

so that the p-adic measure is obtained from the motivic measure by counting points.

5.3 The model theoretic approach to motivic integration

One of the underlying motivations of the model theoretic approach to motivic integration was to tighten and generalize the link between p-adic integrals and motivic integrals. Another goal was to obtain Fubini theorems for motivic integrals. These goals have been achieved by broadening the geometric tools: instead of only using equations, one also uses quantifiers and other symbols to describe sets uniformly. As value ring for motivic integrals, one replaces the Grothendieck ring of varieties by a Grothendieck ring of definable sets over the residue field. This Grothendieck ring of definable sets is better suited to keep track of rational points uniformly in the residue field, so that it is a richer and finer assessment of size, see Example 5.4 below. Likewise, the sets that one wants to measure are allowed to be more general than cylinders over constructible sets, and they are given in a uniform way (that is, independent of the Henselian discretely valued field) by formulas in the language $\mathscr{L}_{\text{DP}}$ of Denef-Pas, where the quantifiers over the residue field can play an important role. Very roughly, the idea is to define the measure of a definable set over the valued field by realizing it as a family of balls parameterized by definable sets over the residue field and the value group. Important tools to achieve this are quantifier elimination and *cell decomposition*, which allows to break up a definable set into more elementary pieces, called *cells*. The idea to use cell decompsotion in the context of p-adic integrals is due to Denef [14].

One of the strong points of model theory is that it can handle families very naturally, namely definable families, giving rise Fubini theorems for iterated integrals. A class of functions that is rich enough to have statements of Fubini type, and such that the functions can still be described in a uniform, field independent way, is the class of constructible motivic functions of [11] in equicharacteristic zero, and of Chapter 8 in mixed characteristic.

Example 5.2 The formula φ given by $\infty > v(x) = v(y) \geq 0$ gives a uniform description, for any discretely valued Henselian field K, of the set

$$\varphi(K) := \{(x, y) \in K^2 \mid \infty > v_K(x) = v_K(y) \geq 0\}$$

where v_K denotes the discrete valuation on K. We view $\varphi(K)$ as a family of sets

$$\{y \in K \mid (x, y) \in \varphi(K)\}$$

parameterized by $x \in R \setminus \{0\}$, with R the value ring in K. It is intuitively reasonable to assign to this family, uniformly in K, the motivic measure

$$I(x) = (\mathbb{L} - 1)\mathbb{L}^{-v(x)-1}.$$

as a function of x (one can think of the members of this family as cylinders in R; we used the normalization of (1.1), not the one of (1.3)). Integrating $I(x)$ over the x-space amounts to summing the geometric power series

$$(\mathbb{L} - 1)^2 \sum_{i \in \mathbb{N}} \mathbb{L}^{-2i-2}.$$

Next consider the formula ψ given by

$$\infty > v(x) > v(y) \geq 0,$$

so that

$$\psi(K) = \{(x, y) \in K^2 \mid \infty > v_K(x) > v_K(y) \geq 0\}.$$

Suppose one would like to integrate the function $\mathbb{L}^{v_K(y)+1}$, uniformly in K, over the set

$$\{y \in K \mid (x, y) \in \psi(K)\}$$

and obtain a function in x. The most reasonable sense one can give here is

$$J(x) = \int_{y,\ \infty > v(x) > v(y) \geq 0} \mathbb{L}^{v(y)+1}$$

$$= \sum_{i=0}^{v(x)-1} (\mathbb{L} - 1)\mathbb{L}^{-i-1}\mathbb{L}^{i+1}$$

$$= v(x)(\mathbb{L} - 1).$$

Functions of the forms $I(x)$ and $J(x)$ are examples of *constructible motivic functions*. They come up naturally when considering iterated integrals, but one must include much more general functions to obtain general Fubini statements.

Another strong point of model theory is that, although one typically uses the theory of Henselian valued fields of equicharacteristic zero, one can often deduce results over $\mathbb{Q}_p$ and $\mathbb{F}_p((t))$ (and their finite field extensions) as soon as p is large enough. A basic philosophy is that if a property holds motivically, then it holds over $\mathbb{Q}_p$ and over $\mathbb{F}_p((t))$ as soon as p is large enough. A finer variant of this philosophy is that, even when something is not known motivically, then still, for large p, whether it holds over a non-archimedean local field only depends on the residue field, and hence, it will hold over $\mathbb{Q}_p$ if and only if it holds over $\mathbb{F}_p((t))$. The idea behind this philosophy is that motivic integrals can interpolate p-adic and $\mathbb{F}_p((t))$-integrals for large enough p. This property is made precise in the *transfer principle* of [12], see Section 6.2 and Examples 5.3 and 5.4.

The model theoretic approaches to motivic integration have also introduced Fourier transformation at a motivic level, [12, 19], and as such have opened the way for the further study of harmonic analysis at a motivic level. These approaches are explained in more detail in Section 6.2.

Example 5.3 Let us illustrate how one can use results in equicharacteristic zero to obtain results over non-archimedean local fields. Let φ_1 and φ_2 be formula's in the Denef-Pas language $\mathscr{L}_{\mathrm{DP}}$ (or, if you want, just in the ring language). Suppose that, for all Henselian valued fields K of equicharacteristic zero, one knows that $\varphi_1(K)$ is the same set as $\varphi_2(K)$. Then, by Gödel's completeness theorem, this property about φ_1 and φ_2 is (strictly) provable from the theory of Henselian equicharacteristic zero valued fields. On the one hand, such a proof $\mathscr{P}$ only uses finitely many axioms from the theory, and on the other hand, to say that the characteristic is zero is not possible with finitely many axioms. The finitely many axioms from the theory that suffice to prove that the definable sets $\varphi_1(K)$ and $\varphi_2(K)$ are equal for the equicharacteristic zero fields K, can at most specify that the residue field characteristic is $> N$ for some integer N. It follows that, for all non-archimedean local fields L of residue characteristic $> N$, the sets $\varphi_1(L)$ and $\varphi_2(L)$ are equal, because the proof $\mathscr{P}$ will also be valid for L.

5.4 Specializing motivic integrals to p-adic integrals

We have seen in Section 5.2 that motivic integrals over a local field specialize to p-adic integrals over unramified extensions of that field by point counting. This specialization (or interpolation) principle is generalized in the model theoretic approaches, where it is for example also possible to consider situations where the prime p varies. As we will see in Section 6.2, also Fourier transforms on $\mathbb{Q}_p^n$ and on $\mathbb{F}_p((t))^n$ can be interpolated by motivic integrals for p large enough.

Example 5.4 A *definable assignment* is a map X that associates to every characteristic zero field k a subset $X(k)$ of $k[[t]]^n$, for some fixed $n \in \mathbb{N}$, such that there exists a formula ψ in the language of Denef-Pas with $X(k) = \psi(k[[t]])$ for every k.

Let us sketch the computation of the motivic measure of the definable assignment X which to a field k of characteristic zero associates the set of nonzero squares inside $k[[t]]$. This definable assignment is the infinite union, over $z \in 2\mathbb{N}$, of the definable assignments A_z which to k associate the set

$$\{x \in k[[t]] \mid v(x) = z \text{ and } \overline{ac}(x) \text{ is a square in } k^*\}.$$

The A_z are so-called *cells*. Each cell A_z has measure $[\varphi]\mathbb{L}^{-z-1}$ in an appropriate localized Grothendieck ring of definable sets over k, where $[\varphi]$ is the class of the formula

$$\varphi = (\exists \eta)(\eta^2 = \xi \neq 0)$$

in the residue field variables ξ, η, and where $\mathbb{L}$ is the class of the affine line. Summing over $z \in 2\mathbb{N}$ yields $[\varphi]/\mathbb{L}(1 - \mathbb{L}^{-2})$ as measure of X.

For every prime p, we can also look at the set $\mathscr{S}_p$ of nonzero squares inside $\mathbb{Z}_p$. If $p > 2$, then $\mathscr{S}_p$ has measure

$$\mu_{\text{Haar}}(\mathscr{S}_p) = \sum_{z \in 2\mathbb{N}} \mu_{\text{Haar}}(\{x \in \mathbb{Z}_p \,|\, v(x) = z \text{ and } \overline{ac}(x) \text{ is a square in } \mathbb{F}_p^*\}).$$

The summand with index z equals

$$\sharp_{\mathbb{F}_p}([\varphi]\mathbb{L}^{-z-1}) = |\{\text{Squares in } \mathbb{F}_p^*\}| \cdot p^{-z-1} = \frac{p-1}{2}p^{-z-1},$$

so that

$$\mu_{\text{Haar}}(\mathscr{S}_p) = (p-1)/2p(1-p^{-2})$$
$$= \sharp_{\mathbb{F}_p}([\varphi]/\mathbb{L}(1-\mathbb{L}^{-2})).$$

Here the point counting operator $\sharp_{\mathbb{F}_p}$ is defined by

$$\sharp_{\mathbb{F}_p}([\psi]) = |\psi(\mathbb{F}_p)|$$

for every ring formula ψ (to be precise, $\sharp_{\mathbb{F}_p}([\psi])$ is only defined for $p \gg 0$). In this way, motivic integrals in equal characteristic zero can be specialized to p-adic integrals for $p \gg 0$.

A similar computation is valid over $\mathbb{F}_p((t))$ with $p > 2$. For $\mathbb{Q}_2$ and all its finite field extensions, an analogous calculation is possible, and even at a motivic level if one bounds the ramification degree, see Chapter 8. The situation for $\mathbb{F}_2((t))$ is behaved differently: the set of squares in $\mathbb{F}_2((t))$ has measure zero for μ_{Haar}.

6 The various theories of motivic integration

6.1 Approaches based on geometry

Kontsevich's theorem. The history of motivic integration starts with a famous lecture by Kontsevich in Orsay in 1995. The main result of the lecture was an improvement of a theorem of Batyrev saying that birationally equivalent complex Calabi-Yau varieties X, Y have the same Betti numbers [4]. Kontsevich showed that even the Hodge numbers are the same.

Let us briefly sketch the proof of Batyrev's theorem (our argument is slightly different from his, but follows the same lines). First, one reduces by a spreading out argument to the case where X and Y are defined over a number field, say $\mathbb{Q}$ for simplicity. We choose a prime p such that $X \times_{\mathbb{Q}} \mathbb{Q}_p$ and $Y \times_{\mathbb{Q}} \mathbb{Q}_p$ admit smooth and proper $\mathbb{Z}_p$-models $\mathscr{X}$, resp. $\mathscr{Y}$, and we choose gauge forms (i.e., nowhere vanishing differential forms of maximal degree) ω_X and ω_Y on X, resp. Y that extend to relative gauge forms on the models $\mathscr{X}$ and $\mathscr{Y}$.

The gauge form ω_X defines a measure $|\omega_X|$ on the set $X(K)$, for every finite unramified extension K of $\mathbb{Q}_p$, and the volume of $X(K)$ is equal to

$$p^{-e \cdot \dim(X)} |\mathscr{X}_s(\mathbb{F}_{p^e})| \qquad (1.4)$$

where $e = [K : \mathbb{Q}_p]$ and $\mathscr{X}_s$ is the reduction of $\mathscr{X}$ modulo p (see Chapter 7). The Weil Conjectures imply that we can recover the Betti numbers of X if we know the value $|\mathscr{X}_s(\mathbb{F}_{p^e})|$ for every integer $e > 0$. Hence, it suffices to show that

$$\int_{X(K)} |\omega_X| = \int_{Y(K)} |\omega_Y|$$

for every finite unramified extension K of $\mathbb{Q}_p$.

We choose a smooth projective $\mathbb{Q}$-variety Z with proper birational morphisms $f : Z \to X$ and $g : Z \to Y$. By the change of variables formula for p-adic integrals, we have

$$\int_{X(K)} |\omega_X| = \int_{Z(K)} |f^* \omega_X|$$

and the analogous equality for Y. We will show that $f^* \omega_X$ and $g^* \omega_Y$ define the same measure on $Z(K)$. It is enough to prove that these forms differ only by a factor in $\mathbb{Q}^*$ whose p-adic absolute value is one.

We denote by $U \subset X$ the domain of definition of the birational map $h = g \circ f^{-1}$ from X to Y. Since X is normal and Y is proper, we know that $X \setminus U$ has codimension at least two in X. Therefore, the differential form $h^* \omega_Y$ on U extends uniquely to a global section of the line bundle

$$\Omega_X^{\dim(X)} = \mathscr{O}_X \cdot \omega_X,$$

and we can write

$$h^* \omega_Y = \phi \cdot \omega_X$$

with ϕ in $\mathbb{Q}^*$. Pulling back to Z, we find

$$g^* \omega_Y = \phi \cdot f^* \omega_X$$

and, hence,

$$\int_{Y(K)} |\omega_Y| = \int_{Z(K)} |g^*\omega_Y| = |\phi|_K \int_{Z(K)} |f^*\omega_X| = |\phi|_K \int_{X(K)} |\omega_X|$$

for every finite unramified extension K of $\mathbb{Q}_p$. Note that $|\phi|_K = (|\phi|_p)^{[K:\mathbb{Q}_p]}$, where $|\phi|_p$ denotes the p-adic absolute value of ϕ. By (1.4), the analogous equality for Y and the Lang-Weil estimates, we know that the integrals

$$\int_{X(K)} |\omega_X| \quad \text{and} \quad \int_{Y(K)} |\omega_Y|$$

both tend to one when the degree $e = [K : \mathbb{Q}_p]$ tends to infinity. It follows that $|\phi|_p$ must be one, so that ϕ is a p-adic unit. This concludes the proof of Batyrev's theorem.

Kontsevich observed that Batyrev's result could be strengthened by "geometrizing" the arguments, replacing p-adic integration by motivic integration on the space $X(\mathbb{C}[[t]])$ (to be precise, the integration space is the so-called *arc scheme* of X, a complex scheme that parameterizes the elements of $X(\mathbb{C}[[t]])$).

In Kontsevich's construction, the motivic volume of $X(\mathbb{C}[[t]])$ is precisely the class $[X]$ of X in $\mathcal{M}_\mathbb{C}$, and the analogous property holds for Y. These volumes can then be compared in an appropriate completion $\widehat{\mathcal{M}}_\mathbb{C}$ of $\mathcal{M}_\mathbb{C}$ by dominating X and Y by a common birational model Z as above, and applying a *change of variables formula* for motivic integrals. In this way, one shows that X and Y define the same class in $\widehat{\mathcal{M}}_\mathbb{C}$. It follows, in particular, that their Hodge numbers are the same, because the Hodge-Deligne polynomial is an additive and multiplicative invariant of complex varieties that factors through the image of $\mathcal{M}_\mathbb{C}$ in the completion $\widehat{\mathcal{M}}_\mathbb{C}$. We refer to Chapter 6 for a detailed account.

Batyrev's argument based on p-adic integration depends on the Weil Conjectures because p-adic integration only gives you information on the number of points on the reduction modulo p of the variety X. Motivic integration produces the class in the Grothendieck ring of the reduction modulo t of the variety $X \times_\mathbb{C} \mathbb{C}[[t]]$, i.e., the class of X. Kontsevich's proof is more direct and easier than Batyrev's, since the proof of the change of variables formula for motivic integrals is much more elementary than that of the Weil Conjectures - but you need the marvellous idea to construct integrals of functions on $X(\mathbb{C}[[t]])$ with values in the Grothendieck ring of varieties.

Denef and Loeser's geometric motivic integration. Kontsevich presented his ideas in his 1995 lecture in Orsay, but he never published them. They were systematically developed and expanded by Denef and Loeser [16]. Denef and Loeser designed a theory of motivic integration for a large class of functions on the arc space of an algebraic variety over a field k of characteristic zero. An important new feature of Denef and Loeser's theory is the construction of the motivic measure on the arc spaces of *singular* algebraic varieties. This theory will be explained in detail in Chapter 6. Yasuda has generalized the Denef-Loeser formalism to algebraic stacks [38].

Denef and Loeser discovered many applications of their theory in several domains, including motivic versions of Igusa's zeta functions, motivic nearby cycles and Milnor fibers, and the McKay correspondence. Motivic zeta functions and Denef and Loeser's *motivic monodromy conjecture* will be discussed in Chapter 7.

Denef and Loeser's constructions are geometric in nature, but model theoretic tools (Pas's quantifier elimination) already come into play to construct a broad class of integrable functions and to prove rationality results of certain motivic generating series.

Sebag's motivic integration on formal schemes. In his 2000 Bourbaki report [26], Looijenga gave an account of Denef and Loeser's results. He placed himself in a more general setting, considering varieties over the ring of formal power series $k[[t]]$ instead of over the characteristic zero field k. In his PhD thesis, Sebag further developed this idea and generalized it to a very broad framework: formal schemes $\mathfrak{X}$ of finite type over a complete discrete valuation ring R with perfect residue field k [35]. The integration space in this theory is the *Greenberg scheme* of $\mathfrak{X}$, a k-scheme that parameterizes R'-sections on $\mathfrak{X}$ where R' runs through the finite unramified extensions of R. The integrals still take their values in the localized Grothendieck ring $\mathcal{M}_k$, or an appropriate dimensional completion. Denef and Loeser's geometric motivic integration fits completely in this framework by replacing a k-variety X by the formal t-adic completion of $X \times_k k[[t]]$.

If k is finite, then the point counting realization does not extend to the dimensional completion of $\mathscr{M}_k$. Recently, Rökaeus proposed a different kind of completion to solve this problem; see Chapter 14. Unfortunately, working with this completion, it is a difficult open problem to establish a rich class of measurable sets and to prove the change of variables formula in a general set-up.

Loeser and Sebag's motivic integration on rigid varieties. In [25], Loeser and Sebag define motivic integrals of differential forms of maximal degree on rigid varieties over a complete discretely valued field K with perfect residue field. This framework is the natural generalization of p-adic integration on p-adic manifolds. The main ingredients of the construction are Raynaud's theory of formal models for rigid varieties, Sebag's theory of motivic integration on formal schemes, and the theory of *weak Néron models*. The existence of weak Néron models makes it possible to define the motivic integral of a differential form on a smooth rigid K-variety in terms of the motivic integral of an appropriate function on the Greenberg scheme of a smooth formal scheme over the ring of integers in K. Loeser and Sebag's formalism is discussed in detail in Chapter 7. It was further generalized by Nicaise in [28].

Loeser and Sebag's theory yields interesting new invariants of rigid K-varieties, and hence, in particular, of degenerating one-parameter families of complex algebraic varieties (here $K = \mathbb{C}((t))$). One of these invariants is the *motivic Serre invariant*, a motivic generalization of Serre's invariant that classifies compact p-adic analytic manifolds [36]. The motivic Serre invariant of a rigid K-variety can be considered as a measure for the set of K-rational points on the variety. In [32, 28, 30, 31] it is shown that, under certain conditions, the motivic Serre invariant admits a cohomological interpretation in terms of a *trace formula* that is similar to the Grothendieck-Lefschetz-Verdier trace formula for varieties over finite fields. We refer to Chapter 7 for a detailed account.

6.2 Approaches based on model theory

Denef and Loeser's arithmetic motivic integration. Besides the theory of geometric motivic integration discussed in the previous section, Denef and Loeser also developed a theory of *arithmetic motivic integration* [17]. It is better behaved than geometric motivic integration if one wants to specialize motivic integrals to p-adic integrals, because it "remembers" that the residue field of a p-adic field is finite, of course without fixing the residue field. More precisely, one avoids Chevalley's theorem and one uses instead quantifier elimination results for pseudofinite fields.

Let k be a field of characteristic zero. An important tool in Denef and Loeser's theory is a realization morphism χ_c from the Grothendieck ring of the theory of k-fields to the Grothendieck ring of Chow motives over k tensored with $\mathbb{Q}$. It associates virtual Chow motives to formulas in the language of k-algebras. Assume that k is a number field, say $\mathbb{Q}$ for simplicity, and φ a formula in the language of k-algebras in n free variables. Then φ defines a subset $\varphi(F)$ of F^n for every finite field F of sufficiently large characteristic. The cardinality of this set is equal to the number of F-rational points on the virtual Chow motive associated to φ, where this number is defined by taking traces of Frobenius on the étale realization. So, roughly speaking, Denef and Loeser associate to a formula φ a motive that gives the correct number of F-rational points for every finite field F of sufficiently large characteristic.

For example, the motive associated to the formula

$$x \neq 0 \wedge (\exists y)(x = y^2)$$

is $([\mathbb{L}_{mot}] - 1)/2$, where $\mathbb{L}_{mot}$ is the *Lefschetz motive* over k. This reflects the fact that in finite fields of characteristic > 2, half of the units are squares. Accordingly, the arithmetic motivic measure of the set of squares in $k[[t]]^*$ is equal to $[\mathbb{L}_{mot}]^{-1}([\mathbb{L}_{mot}] - 1)/2$, for every field k of characteristic zero. Note that this construction only makes sense if we interpret it correctly: if k is algebraically closed, then every element in $k[[t]]^*$ is a square. We're actually assigning a motivic measure to the map that associates to every field extension k' of k the subset of $k'[[t]]$ that consists of the squares in $k'[[t]]^*$. This is an example of what is called a *definable assignment* in Denef and Loeser's theory. We've seen other examples of definable assignments in Example 5.4.

The main tool in Denef and Loeser's construction is quantifier elimination for pseudofinite fields in terms of Galois stratifications (Fried-Jarden-Sacerdote; see [18]). This kind of quantifier elimination respects point counting over finite fields better than Chevalley's theorem, as can be seen on the following example.

Example 6.1   Let k be a field of characteristic zero. Consider again the formula φ saying

$$x \neq 0 \wedge (\exists y)(x = y^2).$$

Geometrically, this corresponds to taking the image of the twofold torus covering

$$f : \mathbb{G}_{m,k} \to \mathbb{G}_{m,k} : y \mapsto y^2.$$

Scheme-theoretically, the image of this morphism is the entire torus $\mathbb{G}_m$, and we lose the information that in finite fields of characteristic > 2, only half of the units are squares.

However, for every pseudo-finite field L containing k, the image of the map

$$\mathbb{G}_{m,k}(L) \to \mathbb{G}_{m,k}(L)$$

induced by f consists precisely of the points of $\mathbb{G}_{m,k}(L)$ with trivial decomposition group w.r.t. the Galois covering f of group $\mathbb{Z}/2\mathbb{Z}$. This interpretation is used to assign a motivic measure to the formula φ.

For a self-contained construction of the realization morphism χ_c, including a geometric proof of the quantifier elimination theorem for pseudo-finite fields, we refer to [27]. Starting from χ_c, one defines an arithmetic motivic measure on the arc scheme of a variety in a similar way as in the theory of geometric motivic integration, using approximation by cylinders.

Cluckers and Loeser's theory. Cluckers and Loeser's approach to motivic integration, developed in [11] and [12], allows one to study motivic integrals in families with a general Fubini Theorem in equicharacteristic zero. In Chapter 8, a similar theory is developed in mixed characteristic with bounded ramification. Cluckers and Loeser's approach is sufficiently general to specialize to geometric motivic integration, to arithmetic motivic integration, and to Loeser and Sebag's motivic integration of [25]. Approximation by cylinders does not play a role anymore. Instead, in the construction of the measure on, say, $k'((t))$, for field extensions k' of a base field k, one uses very basic subassignments

$$X_{c,z} : k' \mapsto X_{c,z}(k') \subset k'((t))$$

of the form

$$X_{c,z}(k') = \{x \in k'((t)) \mid v(x - c) = z \text{ and } \overline{ac}(x - c) \in A(k')\}$$

for some $z \in \mathbb{Z}$ and $c \in k((t))$. Here $A(k') \subset k'$ is a subset of k' defined by a formula in the language of rings. The assignment $X_{c,z}$ has measure $\mathbb{L}^{-z}[A]$ in a suitable localized Grothendieck ring of definable sets. It is a basic example of what is called a *cell*. Any definable assignment Y with $Y(k') \subset k'((t))^n$ can be written as a finite disjoint union of cells, and for cells it is almost as elementary to integrate one variable out as it was to measure the basic cell $X_{c,z}$. In this process, infinite sums will appear, like $\sum_{z \in \mathbb{N}} \mathbb{L}^{-z}[A]$, but since they are all related to geometric power series, their sum (in our example $[A]/(1 - \mathbb{L}^{-1})$) exists in a certain localized Grothendieck ring without further need to complete.

The interpolation principle of p-adic integrals by motivic integrals remains valid as in arithmetic motivic integration, and is generalized to families of p-adic integrals. One of the central results of the theory is a change of variables formula in the general category of definable assignments with definable morphisms. Grothendieck semi-rings are used (instead of Grothendieck rings) to have a notion of non-negativity and to define integrability. A whole framework of motivic constructible functions, closed under integration, is developed.

Hrushovski and Kazhdan's theory. In [19], Hrushovski and Kazhdan build further in the model theoretic line of approaches to motivic integration. Where previously the objective had always been to put measures on Henselian discretely valued fields, here one takes a more geometric viewpoint and one puts a motivic measure on *algebraically closed* valued fields. By taking rational points over various subfields, one is able to recover p-adic points and p-adic measures for p large enough. A surprising result in this approach is that the motivic measure is shown to be "optimal", by proving unexpected isomorphisms between Grothendieck (semi-)rings of definable subsets of the algebraically closed valued fields on the one hand, and of definable subsets in residue field and value group on the other hand. To this end, the residue field and the value group of K are naturally bundled into the quotient of groups $K^*/1 + M_K$, with M_K the maximal ideal of the valuation ring of K. Instead of the language of Denef-Pas, one uses the language of rings for the valued field, together with the projection to $K^*/1 + M_K$, in which one has quantifier elimination results by S. Basarab [3]. This language is more natural than the Denef-Pas language, since not every valued field admits an angular component map. The achievements of the Cluckers-Loeser approach are preserved in the theory of Hrushovski-Kazhdan: one works with families of definable sets with natural Fubini results, one has a general change of variables formula, no completion of Grothendieck rings is necessary, and one constructs a framework of motivic functions that is closed under integration. A new kind of application that is made possible by the mentioned isomorphism theorems, is to perform a double integration in the case that the residue field itself carries a measure, for example for the rank two valued field $\mathbb{Q}_p((t))$ where the residue field is $\mathbb{Q}_p$. A part of the analysis of definable sets is still related to cells, like the basic assignments $X_{c,z}$ above, but another part of the analysis goes much further, especially the study how definable functions and definable sets interact in K and in the quotient $K^*/1 + M_K$. Although precise specialization morphisms linking this theory to the approach by Cluckers and Loeser are yet to be discovered, one can expect that there is such a connection. Hrushovski and Kazhdan apply their work to obtain motivic versions of Poisson summation [20]. Another achievement in this approach is that one works in an axiomatic set-up for algebraically closed valued fields, related to C-minimality, instead of working with a fixed language and a fixed theory. (For a detailed account on C-minimality, see Chapter 10.) A different axiomatic framework, for discretely valued Henselian fields, is used in Chapter 8. Axiomatic set-ups allow one for example to use richer languages, also including function symbols for converging power series as functions on the valuation ring, as in [23].

Towards motivic harmonic analysis and motivic representation theory. A joint achievement of [12] and [19] that was not yet mentioned is the general introduction of additive characters into the realm of motivic integration, opening the way to Fourier analysis on a motivic level. These motivic Fourier transforms interpolate the Fourier transforms over the p-adics and over $\mathbb{F}_p((t))$ for p large enough, where now the counting operator

$$\sharp : K_0(\mathrm{Var}_k) \to \mathbb{Z} : [X] \mapsto |X(k)|$$

for finite fields k is replaced by a summation operator which to a finite field extension k of $\mathbb{F}_p$, a definable set X and a definable function

$f : X \to \mathbb{A}^1$ associates the exponential sum

$$\sum_{x \in X(k)} \exp(2\pi i \frac{\mathrm{Trace}_{k/\mathbb{F}_p} f(x)}{p}).$$

Similar exponential sums appear in Chapter 13. With this specialization technique, associating to definable sets and functions on the residue field level exponential sums for all finite fields, it is possible to interpolate parameterized p-adic integrals for p large enough, for example, a Fourier transform:

$$\mathscr{F}(|g(x)|_p)(y) = \int_{x \in \mathbb{Q}_p^n} |g(x)|_p \psi(x \cdot y)|dx|$$

where g is a definable function and ψ a nontrivial additive character on $\mathbb{Q}_p$. The Fourier inversion theorem holds at a motivic level. It is shown in [10] that the integrals appearing in the Fundamental Lemma of the Langlands program and several of its variants (e.g. as in [40]) can be interpolated by motivic integrals with additive character. As such, they fall in the reach of the transfer principle for motivic integrals of [12], which states that two motivic integrals with characters are equal over $\mathbb{Q}_p$ if and only if they are equal over $\mathbb{F}_p((t))$, for large enough p. This can be compared to the classical Ax-Kochen-Eršov principle which states that, for a definable set X and for large enough p, $X(\mathbb{Q}_p)$ is empty if and only if $X(\mathbb{F}_p((t)))$ is empty.

Harmonic analysis, for a large part, works on all locally compact groups. Motivic integration is a natural tool to render harmonic analysis and representation theory even more uniform in the (totally disconnected) locally compact group, and to extend the theory to the non locally compact setting. Further applications of motivic integration to harmonic analysis and representation theory remain to be developed. The motivic study of multiplicative characters on the valued field is possibly an important next step.

7 Connections between motivic integration, model theory and non-archimedean geometry

It should be clear from what we have said above that model theory and non-archimedean geometry play an important role in the theory of motivic integration. Model theory provides the natural framework for several types of motivic integration and many applications related to *p*-adic integrals. In these theories, the measurable sets are described in model-theoretic terms, and the model-theoretic language allows to describe in a precise way how motivic integrals specialize to *p*-adic integrals. Fundamental results from model theory such as quantifier elimination and cell decomposition underly the construction of the motivic measure and the study of its properties.

The theory of motivic integration on rigid varieties produces many interesting new invariants of rigid varieties, such as the motivic Serre invariant. Conversely, non-archimedean geometry offers a convenient framework to interpret certain motivic invariants arising from singularity theory. In [32], Nicaise and Sebag used motivic integration on rigid varieties to give a new interpretation of Denef and Loeser's motivic zeta function in terms of the so-called *analytic Milnor fiber*, a rigid variety over the field of complex Laurent series associated to a complex hypersurface singularity (see Chapter 7). The analytic Milnor fiber contains a wealth of information about the singularity. This new interpretation of the motivic zeta function is important in the light of Denef and Loeser's motivic monodromy conjecture, which relates the poles of the zeta function to monodromy eigenvalues on the cohomology of the classical Milnor fiber. These topics will be discussed in Chapter 7.

There are also many deep connections between model theory and non-archimedean geometry that, unfortunately, are not discussed in this book. A spectacular recent illustration of these connections is Hrushovski and Loeser's description of the homotopy type of Berkovich spaces using advanced tools from model theory [21]. This fascinating subject deserves a book on its own.

8 Further reading

The following references are suitable for a first exploration of the topics under consideration.

8.1 Motivic integration

- R. Cluckers, T. Hales and F. Loeser: *Transfer Principle for the fundamental lemma.* To appear in: M. Harris (editor). *Stabilisation de la formule des traces, variétés de Shimura, et applications arithmétiques.* — This article explains how the general transfer principle of Cluckers and Loeser may be used in the study of the fundamental lemma.

- J. Denef: *Report on Igusa's local zeta function.* Séminaire Bourbaki, Vol. 1990/91. Astérisque No. 201-203 (1992), Exp. No. 741, pages 359–386. — A survey on Igusa's p-adic zeta function and related topics, such as the monodromy conjecture and the topological zeta function. It was written before the creation of motivic integration.

- J. Gordon and Y. Yaffe: *An overview of arithmetic motivic integration.* In: C. Cunnigham and M. Nevins (editors): *Ottawa Lectures on p-adic Groups,* pages 113–150. Fields Institute Monograph Series, American Mathematical Society, Providence, RI, 2009. — An elementary exposition, with examples, of the theory of motivic integration developed by R. Cluckers and F. Loeser, with the view towards applications in representation theory of p-adic groups.

- J. Igusa: *An introduction to the theory of local zeta functions.* AMS Studies in Advanced Mathematics, 14. International Press, Cambridge, MA, 2000. — A splendid account of the theory of Igusa's p-adic zeta functions, of which many aspects have been generalized to a motivic setting by several authors.

- F. Loeser: *Seattle lectures on motivic integration.* In: D. Abramovich et al. (editors): *Algebraic geometry–Seattle 2005. Part 2,* pages 745–784. Volume 80 of *Proceedings of Symposia in Pure Mathematics.* American Mathematical Society, Providence, RI, 2009. — A survey of the development of motivic integration from the p-adic prehistory over the work of Kontsevich and Denef-Loeser to the work of Cluckers-Loeser, including many references to applications.

- J. Nicaise: *An introduction to p-adic and motivic zeta functions and the monodromy conjecture.* In: G. Bhowmik, K. Matsumoto and H. Tsumura (editors). *Algebraic and analytic aspects of zeta functions and L-functions,* pages 141–166. Volume 21 of *MSJ Memoirs.* Mathematical Society of Japan, Tokyo, 2010. — An elementary introduction to Igusa's p-adic zeta functions, the monodromy conjecture, motivic integration and motivic zeta functions.

- W. Veys: *Arc spaces, motivic integration and stringy invariants.* In: *Singularity theory and its applications,* pages 529–572. Volume 43 of *Advanced Studies in Pure Mathematics.* Mathematical Society of Japan, Tokyo, 2010. — An elementary introduction to motivic integration and motivic invariants of algebraic singularities, with some p-adic motivation.

8.2 Model theory

- R. Cluckers: *An introduction to b-minimality*. In: *Logic Colloquium 2006*, pages 91–102. Lect. Notes Log., Assoc. Symbol. Logic, Chicago, IL, 2009. — The paper gives a modern account of cell decomposition in Henselian valued fields in the form of a framework called *b*-minimality, which plays a role in Chapter 8. Such cell decomposition results are useful to calculate *p*-adic and motivic integrals, and go back to work by P. Cohen and J. Denef.

- J. Denef, L. van den Dries: *p-adic and real subanalytic sets*. Ann. of Math. (2) 128(1) 79–138, 1988. — This foundational work on the analytic model theory of the *p*-adics and on analytic *p*-adic zeta functions has been a great source of inspiration for later research.

- W. Hodges: *Model theory*. Encyclopedia of Mathematics and its Applications, 42. Cambridge University Press, Cambridge, 1993. — A thorough introduction to modern model theory. A more concise version of the book by the same author and publisher is available under the title *A shorter model theory*.

- A. Macintyre: *Twenty years of p-adic model theory*. In: *Logic colloquium '84*, pages 121–153. Volume 120 of *Stud. Logic Found. Math.* North-Holland, Amsterdam, 1986. — An extensive survey of the model theory of valued fields in the twenty years following the work of Ax - Kochen and Eršov.

- D. Marker: *Model theory. An introduction*. Volume 217 of *Graduate Texts in Mathematics*. Springer-Verlag, New York, 2002. — A to the point introduction to many topics in model theory with a strong link to current research.

8.3 Non-archimedean geometry

- S. Bosch, U. Güntzer, R. Remmert: *Non-archimedean analysis*. Volume 261 of *Grundlehren der mathematischen Wissenschaften*. Springer Verlag, Berlin, 1984. — A systematic treatment of the foundations of non-archimedean analysis and rigid geometry.

- S. Bosch: *Half a century of rigid analytic spaces*. Pure Appl. Math. Q. (Special Issue: In honor of John Tate.) 5(4):1435–1467, 2009. — A survey of rigid geometry and the theory of formal models of rigid varieties.

- S. Bosch: *Lectures on formal and rigid geometry.* preprint, available at www.math1.uni-muenster.de/sfb/about/publ/heft378.pdf. — An introduction to rigid geometry and the theory of formal models, including proofs of all basic results.
- B. Conrad: *Several approaches to non-archimedean geometry.* In: *p-adic geometry, Lectures of the 2007 Arizona Winter School,* pages 9–63. Volume 45 of *University Lecture Series.* American Mathematical Society, Providence, RI, 2008. — A survey on rigid varieties, their formal models and Berkovich spaces, with numerous exercises and examples.
- A. Ducros: *Espaces analytiques p-adiques au sens de Berkovich.* Séminaire Bourbaki. Vol. 2005/2006. Astérisque No. 311 (2007), Exp. No. 958, pages 137–176. — A survey on Berkovich spaces containing some results on their homotopy types and discussing many applications to the Langlands program, nearby cycles, p-adic integration of one-forms, dynamical systems,....
- J. Fresnel and M. van der Put: *Rigid analytic geometry and its applications.* Volume 218 of *Progress in Mathematics.* Birkhäuser, Boston, MA, 2004. — An accessible introduction to rigid geometry. The book starts with some explicit rigid geometry on the projective line, and contains numerous examples and applications, such as semi-stable reduction of curves and uniformization of abelian varieties.
- J. Nicaise: *Formal and rigid geometry: an intuitive introduction and some applications.* Enseign. Math. (2) 54(3-4): 213–249, 2008. — A survey on rigid varieties and their formal models, with a brief discussion of applications to singularity theory, nearby cycles, deformation theory, semi-stable reduction and coverings of curves.

9 About this book

This book grew out of the workshop "Motivic integration and its interactions with model theory and non-archimedean geometry" that we organized with A. MacIntyre at the International Center for Mathematical Sciences in Edinburgh (May 12 – May 17, 2008). Since its creation by Kontsevich in 1995, motivic integration has developed quickly in several directions and has found applications in various domains, such as singularity theory and the Langlands program. In its development, it incorporated tools from model theory and non-archimedean geometry. The aim of the meeting was to bring together experts working in one of these fields, as well as researchers from neighbouring disciplines and PhD students. This has led to interesting interactions and collaborations; we warmly thank all speakers and participants for their contributions to the success of the workshop.

References

[1] A. Abbes. *Éléments de Géométrie Rigide. Volume I. Construction et étude géométrique des espaces rigides.* to appear in *Progress in Mathematics*, Birkhäuser.

[2] M. Artin, A. Grothendieck and J.-L. Verdier (eds.). *Séminaire de Géométrie Algébrique du Bois Marie (1963-64), Théorie des topos et cohomologie étale des schémas. Vol. 1.* Volume 269 of *Lecture notes in mathematics*. Springer-Verlag, Berlin, 1972.

[3] S. Basarab. Relative elimination of quantifiers for Henselian valued fields. *Ann. Pure Appl. Logic* 53(1):51–74, 1991.

[4] V. Batyrev. Birational Calabi-Yau n-folds have equal Betti numbers. In: *New trends in algebraic geometry (Warwick, 1996)*, pages 1–11. Volume 264 of *London Math. Soc. Lecture Note Series*. Cambridge Univ. Press, Cambridge, 1999.

[5] V. G. Berkovich. *Spectral theory and analytic geometry over non-archimedean fields.* Volume 33 of *Mathematical Surveys and Monographs*. AMS, 1990.

[6] S. Bosch and W. Lütkebohmert. Formal and rigid geometry. I. Rigid spaces. *Math. Ann.*, 295(2):291–317, 1993.

[7] S. Bosch and W. Lütkebohmert. Formal and rigid geometry. II. Flattening techniques. *Math. Ann.*, 296(3):403–429, 1993.

[8] S. Bosch, W. Lütkebohmert and M. Raynaud. Formal and rigid geometry. III. The relative maximum principle. *Math. Ann.*, 302(1):1–29, 1995.

[9] S. Bosch, W. Lütkebohmert and M. Raynaud. Formal and rigid geometry. IV. The reduced fibre theorem. *Invent. Math.*, 119(2):361–398, 1995.

[10] R. Cluckers, T. Hales and F. Loeser. Transfer Principle for the fundamental lemma. to appear in: M. Harris (ed.). *Stabilisation de la formule des traces, variétés de Shimura, et applications arithmétiques.*

[11] R. Cluckers and F. Loeser. Constructible motivic functions and motivic integration. *Invent. Math.* 173(1):23–121, 2008.

[12] R. Cluckers and F. Loeser. Constructible exponential functions, motivic Fourier transform and transfer principle. *Ann. Math.* 171:1011–1065, 2010.

[13] P. Colmez and J.-P. Serre (eds.) *Correspondance Grothendieck-Serre.* Volume 2 of *Documents Mathématiques*. Société Mathématique de France, Paris, 2001.

[14] J. Denef. The rationality of the Poincaré series associated to the p-adic points on a variety. *Invent. Math.* 77:1–23, 1984.

[15] J. Denef and L. van den Dries. *p*-adic and real subanalytic sets. *Ann. of Math. (2)* 128(1):79–138, 1988.

[16] J. Denef and F. Loeser. Germs of arcs on singular algebraic varieties and motivic integration. *Invent. Math.* 165:201–232, 1999.

[17] J. Denef and F. Loeser. Definable sets, motives and *p*-adic integrals. *J. Am. Math. Soc.*, 14(2):429–469, 2001.

[18] M. Fried and M. Jarden. *Field arithmetic, 3rd edition.* Volume 11 of *Ergebnisse der Mathematik und ihrer Grenzgebiete. 3. Folge.* Springer-Verlag, Berlin, 2008.

[19] E. Hrushovski and D. Kazhdan. Integration in valued fields. In: V. Ginzburg (ed.). *Algebraic geometry and number theory*, pages 261–405. Volume 253 of *Progress in Mathematics*, Birkhäuser, Basel, 2006.

[20] E. Hrushovski and D. Kazhdan. Motivic Poisson summation. *Mosc. Math. J.* 9(3):569–623, 2009.

[21] E. Hrushovski and F. Loeser. Non-archimedean tame topology and stably dominated types. arXiv:1009.0252.

[22] R. Huber. *Étale cohomology of rigid analytic varieties and adic spaces.* Volume E30 of *Aspects of Mathematics.* Friedr. Vieweg & Sohn, Braunschweig, 1996.

[23] L. Lipshitz Rigid subanalytic sets. *Amer. J. Math.* 115(1):77–108, 1993.

[24] F. Loeser. Seattle lectures on motivic integration. In: D. Abramovich et al. (eds.). *Algebraic geometry—Seattle 2005. Part 2,* pages 745–784. Volume 80 of *Proceedings of Symposia in Pure Mathematics.* American Mathematical Society, Providence, RI, 2009.

[25] F. Loeser and J. Sebag. Motivic integration on smooth rigid varieties and invariants of degenerations. *Duke Math. J.*, 119:315–344, 2003.

[26] E. Looijenga. Motivic measures. *Séminaire Bourbaki,* Vol. 1999/2000. *Astérisque* No. 276, pages 267–297, 2002.

[27] J. Nicaise. Relative motives and the theory of pseudofinite fields. *Int. Math. Res. Pap.* 2007:rpm 001, 69 pages, 2010.

[28] J. Nicaise. A trace formula for rigid varieties, and motivic weil generating series for formal schemes. *Math. Ann.*, 343(2):285–349, 2009.

[29] J. Nicaise. An introduction to *p*-adic and motivic zeta functions and the monodromy conjecture. In: G. Bhowmik, K. Matsumoto and H. Tsumura (eds.). *Algebraic and analytic aspects of zeta functions and L-functions.* Volume 21 of *MSJ Memoirs*, pages 115–140. Mathematical Society of Japan, Tokyo, 2010.

[30] J. Nicaise. A trace formula for varieties over a discretely valued field. *J. Reine Angew. Math.* 650:193–238, 2011.

[31] J. Nicaise. Geometric criteria for tame ramification. *preprint*, arXiv: 0910.3812.

[32] J. Nicaise and J. Sebag. The motivic Serre invariant, ramification, and the analytic Milnor fiber. *Invent. Math.*, 168(1):133–173, 2007.

[33] M. Raynaud. Géométrie analytique rigide d'après Tate, Kiehl, ... *Mémoires de la S.M.F.*, 39-40:319–327, 1974.

[34] M. van der Put and P. Schneider. Points and topologies in rigid geometry. *Math. Ann.* 302(1):81–103, 1995.

[35] J. Sebag. Intégration motivique sur les schémas formels. *Bull. Soc. Math. France*, 132(1):1–54, 2004.

[36] J.-P. Serre. Classification des variétés analytiques p-adiques compactes. *Topology*, 3:409–412, 1965.

[37] J. Tate. Rigid analytic spaces. Private notes, reproduced with(out) his permission by I.H.É.S. (1962). Reprinted in *Invent. Math.* 12:257–289, 1971.

[38] T. Yasuda. Motivic integration over Deligne-Mumford stacks. *Adv. Math.* 207(2):707–761, 2006.

[39] W. Veys. Arc spaces, motivic integration and stringy invariants. In: S. Izumiya et al. (eds.), *Singularity Theory and its applications*. Volume 43 of *Advanced Studies in Pure Mathematics*, pages 529–572. Mathematical Society of Japan, Tokyo, 2006.

[40] Z. Yun, with an appendix by J. Gordon. The fundamental lemma of Jacquet-Rallis in positive characteristics. To appear in *Duke Math. J.*, arXiv:0901.0900, appendix arXiv:1005.0610.

2
Introductory notes on the model theory of valued fields

Zoé Chatzidakis

Partially supported by MRTN-CT-2004-512234 and ANR-06-BLAN-0183

These notes will give some very basic definitions and results from model theory. They contain many examples, and in particular discuss extensively the various languages used to study valued fields. They are intended as giving the necessary background to read the papers by Cluckers-Loeser, Delon, Halupczok and Kowalski in this volume. We also mention a few recent results or directions of research in the model theory of valued fields, but omit completely those themes which will be discussed elsewhere in this volume. So for instance, we do not even mention motivic integration.

People interested in learning more model theory should consult standard model theory books. For instance: D. Marker, *Model Theory: an Introduction*, Graduate Texts in Mathematics 217, Springer-Verlag New York, 2002; C.C. Chang, H.J. Keisler, *Model Theory*, North-Holland Publishing Company, Amsterdam 1973; W. Hodges, *A shorter model theory*, Cambridge University Press, 1997.

1 Languages, structures, satisfaction

Languages and structures

1.1 Languages. A language is a collection $\mathscr{L}$, finite or infinite, of symbols. These symbols are of three kinds:

Motivic Integration and its Interactions with Model Theory and Non-Archimedean Geometry (Volume I), ed. Raf Cluckers, Johannes Nicaise, and Julien Sebag. Published by Cambridge University Press. © Cambridge University Press 2011.

– *function* symbols,
– *relation* symbols,
– *constant* symbols.

To each function symbol f is associated a number $n(f) \in \mathbb{N}^{>0}$, and to each relation symbol R a number $n(R) \in \mathbb{N}^{>0}$. The numbers $n(f)$ and $n(R)$ are called the *arities* of the function f, resp., the relation R. In addition, the language will also contain variable symbols (usually denoted by $x, y, \ldots$), the equality relation $=$, as well as parentheses and logical symbols $\wedge$, $\vee$, $\neg$, $\rightarrow$ (and, or, negation, implies), quantifiers $\exists$, $\forall$ (there exists, for all).

1.2 $\mathscr{L}$**-structures.** We fix a language $\mathscr{L} = \{f_i, R_j, c_k \mid i \in I, j \in J, k \in K\}$, where the f_i's are function symbols, the R_j's are relation symbols, and the c_k's are constant symbols.

An $\mathscr{L}$*-structure* $\mathscr{M}$ is then given by

– A set M, called the *universe of $\mathscr{M}$*,
– For each function symbol $f \in \mathscr{L}$, a function $f^{\mathscr{M}} : M^{n(f)} \rightarrow M$, called the *interpretation of f in $\mathscr{M}$*,
– For each relation symbol $R \in \mathscr{L}$, a subset $R^{\mathscr{M}}$ of $M^{n(R)}$, called the *interpretation of R in $\mathscr{M}$*,
– For each constant symbol $c \in \mathscr{L}$, an element $c^{\mathscr{M}} \in M$, called the *interpretation of c in $\mathscr{M}$*.

The structure $\mathscr{M}$ is then denoted by

$$\mathscr{M} = (M, f_i^{\mathscr{M}}, R_j^{\mathscr{M}}, c_k^{\mathscr{M}} \mid i \in I, j \in J, k \in K).$$

In fact, the superscript $\mathscr{M}$ often disappears, and the structure and its universe are denoted by the same letter. This is when **no confusion is possible**, for instance when there is only one type of structure on $\mathscr{M}$.

1.3   **Substructures.** Let M be an $\mathscr{L}$-structure. An $\mathscr{L}$*-substructure* of M, or simply a substructure of M if no confusion is likely, is an $\mathscr{L}$-structure N, with universe contained in the universe of M, and such that the interpretations of the symbols of $\mathscr{L}$ in N are restrictions of the interpretation of these symbols in M, i.e.:

– If f is a function symbol of $\mathscr{L}$, then the interpretation of f in N is the restriction of f^M to $N^{n(f)}$,
– If R is a relation symbol of $\mathscr{L}$, then $R^N = R^M \cap N^{n(R)}$,
– If c is a constant symbol of $\mathscr{L}$, then $c^M = c^N$.

Hence a subset of M is the universe of a substructure of M if and only if it contains all the (elements interpreting the) constants of $\mathscr{L}$, and is closed under the (interpretation in M of the) functions of $\mathscr{L}$. Note that **if the language has no constant symbol**, then the empty set is the universe of a substructure of M.

1.4 Morphisms, embeddings, isomorphisms, automorphisms.
Let M and N be two $\mathscr{L}$-structures. A map $s : M \to N$ is an ($\mathscr{L}$)-*morphism* if for all relation symbols $R \in \mathscr{L}$, function symbols $f \in \mathscr{L}$, constant symbols $c \in \mathscr{L}$, and tuples $\bar{a}, \bar{b}$ in M, we have:

$$\text{if } \bar{a} \in R, \text{ then } s(\bar{a}) \in R; \qquad s(f(\bar{b})) = f(s(\bar{b})); \qquad s(c) = c.$$

An *embedding* is an injective morphism $s : M \to N$, which satisfies in addition for all relation $R \in \mathscr{L}$ and tuple $\bar{a}$ in M, that

$$\bar{a} \in R \iff s(\bar{a}) \in R.$$

An *isomorphism* between M and N is a bijective morphism, whose inverse is also a morphism. Finally, an *automorphism* of M is an isomorphism $M \to M$.

1.5 Reducts and expansions Let $\mathscr{L} \subseteq \mathscr{L}'$ be languages, and M an $\mathscr{L}'$-structure. The *reduct of M to $\mathscr{L}$* is the $\mathscr{L}$-structure (denoted by $M|_{\mathscr{L}}$ with universe the same as M, in which one forgets the interpretation of symbols of $\mathscr{L}' \setminus \mathscr{L}$. For instance, the ring of real numbers $(\mathbb{R}, +, \cdot, 0, 1)$ is a reduct of the ordered ring of real numbers $(\mathbb{R}, +, -, \cdot, 0, 1, <)$, which is a reduct of the exponential ordered ring of the real numbers $(\mathbb{R}, +, -, \cdot, 0, 1, <, \exp)$. Conversely, M is an *expansion of the $\mathscr{L}$-structure $M|_{\mathscr{L}}$ to the language $\mathscr{L}'$*.

Thus taking a reduct of a structure is forgetting some of the relations, constants or function symbols, while taking an expansion of a structure means adding new relations, constants or function symbols.

1.6 Examples of languages, structures, and substructures.
The concrete structures considered in model theory all come from standard algebraic examples, and so the examples given below will be very familiar to you.

Example 1 - The language of groups (additive notation). The language of groups, $\mathscr{L}_G$, is the language $\{+, -, 0\}$, where $+$ is a 2-ary function symbol, $-$ is a unary function symbol, and 0 is a constant symbol.

Any group G has a natural $\mathscr{L}_G$-structure, obtained by interpreting $+$ as the group multiplication, $-$ as the group inverse, and 0 as the unit element of the group.

A substructure of the group G is then a subset containing 0, closed under multiplication and inverse: it is simply a subgroup of G. The notions of homomorphisms, embeddings, etc. between groups, have the usual meaning.

This is a good place to remark that the notion of substructure is sensitive to the language. While the inverse function and the identity element of the group G are retrievable (definable) from the group multiplication of G, the notion of "substructure" heavily depends on them. For instance, a $\{+, 0\}$-substructure of G is simply a submonoid of G containing 0, while a $\{+\}$-substructure of G can be empty.

If the group is not abelian, then one usually uses the multiplicative notation, i.e. one replaces $+$ by $\cdot$, $-$ by $^{-1}$ and 0 by 1. Here are some examples of $\mathscr{L}_G$-structures:

(1) $(\mathbb{Z}, +, -, 0)$, the natural structure on the additive group of the integers,

(2) $(\mathbb{R}, +, -, 0)$, the natural structure on the additive group of the reals,

(3) (multiplicative notation)
$(\mathbb{R}^{>0}, \cdot, ^{-1}, 1)$ the multiplicative group of the positive reals.

(4) (multiplicative notation), K a field, $n > 0$:
$(\mathrm{GL}_n(K), \cdot, ^{-1}, 1)$, the multiplicative group of invertible $n \times n$ matrices with coefficients in K.

(5) An $\mathscr{L}_G$-structure is not necessarily a group. E.g., define $+$ on $\mathbb{Z}$ by $a +^{\mathbb{Z}} b = 1$, $-^{\mathbb{Z}} a = 0$ for all $a, b \in \mathbb{Z}$, and $0^{\mathbb{Z}} = 2$. The resulting structure $(\mathbb{Z}, +^{\mathbb{Z}}, -^{\mathbb{Z}}, 0^{\mathbb{Z}})$ is not a classical structure.

Example 2 - The language of rings. The language of rings, $\mathscr{L}_R$, is the language $\{+, -, \cdot, 0, 1\}$, where $+$ and $\cdot$ are binary functions, $-$ is a unary function, 0 and 1 are constants.

A (unitary) ring S has a natural $\mathscr{L}_R$-structure, obtained by interpreting $+, -, \cdot$ as the usual ring operations of addition, subtraction and multiplication, 0 as the identity element of $+$, and 1 as the unit element of S.

A substructure of the $\mathscr{L}_R$-structure S is then simply a subring of S. Note that it will in particular contain the subring of S generated by 1, i.e., a copy of $\mathbb{Z}$ or of $\mathbb{Z}/n\mathbb{Z}$ for some integer n. Again homomorphisms and embeddings between rings have the usual meaning.

When one deals with fields, it is sometimes convenient to add a function symbol for the multiplicative inverse (denoted $^{-1}$). By convention $0^{-1} = 0$. Most of the time however, one studies fields in the language of rings.

Example 3 - The language of ordered groups, of ordered rings.

One simply adds to $\mathscr{L}_G$, resp. $\mathscr{L}_R$, a binary relation symbol, $\leq$ (or sometimes $<$). I will denote these languages by $\mathscr{L}_{og}$ and $\mathscr{L}_{or}$ respectively. I will also use the abbreviation $x < y$ for $x \leq y \wedge x \neq y$.

Example 4 - Valued fields. Here there are several possibilities.

Recall first that a *valued field* is a field K, with a map $v : K^\times \to \Gamma \cup \{\infty\}$, where Γ is an ordered abelian group, and satisfying the following axioms:

- $\forall x \; v(x) = \infty \iff x = 0$,
- $\forall x, y \; v(xy) = v(x) + v(y)$,
- $\forall x, y \; v(x + y) \geq \min\{v(x), v(y)\}$.

By convention, ∞ is greater than all elements of Γ. Note that we do not assume that Γ is archimedean, e.g. $\mathbb{Z} \oplus \mathbb{Q}$ with the anti-lexicographical ordering is possible. (Recall that the *anti-lexicographical ordering* on a product $A \times B$ of ordered groups is defined by: $(a_1, b_1) \leq (a_2, b_2) \iff (b_1 < b_2) \vee [(b_1 = b_2) \wedge (a_1 \leq a_2)]$).

1. Maybe the most natural language (used in the definition) is the *two-sorted* language with a sort for the valued field and one for the value group; each sort has its own language (the language of rings for the field sort, and the language of ordered abelian groups with an additional constant symbol ∞ for the group sort); there is a function v from the field sort to the group sort. Thus our structure is

$$\big((K, +, -, \cdot, 0, 1), (\Gamma \cup \{\infty\}, +, -, 0, \infty, \leq), v\big).$$

Formulas are built as in classical first-order logic, except that variables come with their sort. Thus for instance, in the three defining axioms, all variables are of the field sort. To avoid ambiguity, one sometimes write $\forall x \in K$, or $\forall x \in \Gamma$. Or one uses a different set of letters. For instance, the axiom stating that the map v is surjective will involve both sorts, and can be written

$$\forall \gamma \, (\in \Gamma) \, \exists x \, (\in K) \, v(x) = \gamma.$$

2. Another natural language is the language $\mathscr{L}_{\mathrm{div}}$ obtained by adding to the language of rings a binary relation symbol $|$, interpreted by

$$x \,|\, y \iff v(x) \leq v(y).$$

Note that the valuation ring $\mathscr{O}_K$ is quantifier-free definable, by the formula $1 \,|\, x$, and that the group Γ is isomorphic to $K^\times / \mathscr{O}_K^\times$, the ordering been given by the image of $|$. Hence the ordered abelian group Γ is *interpretable* in $(K, +, -, \cdot, 0, 1, |)$. See 1.17 for a formal definition.

3. A third possibility is to look at the field K in the language of rings augmented by a (unary) predicate for its valuation ring $\mathscr{O}_K$. The divisibility relation is then definable $(x \,|\, y \iff x \neq 0 \wedge (\forall z \; zx = 1 \to yz \in \mathscr{O}_K))$.

The following language has been used to study valued rings or fields with additional analytic structure.

4. The ring $\mathscr{O}_K$, in the language of rings augmented by a binary function Div, interpreted by:

$$\mathrm{Div}(x, y) = \begin{cases} xy^{-1} & \text{if } y \,|\, x, \\ 0 & \text{otherwise.} \end{cases}$$

In all four languages, the residue field k_K, as well as the residue map $\mathscr{O}_K \to k_K$, are interpretable: k_K is the quotient of $\mathscr{O}_K$ by the maximal ideal $\mathscr{M}_K$ of $\mathscr{O}_K$. And $\mathscr{M}_K$ is of course definable by the formula expressing that the element x is not invertible in $\mathscr{O}_K$.

5. In the same spirit as the first language, here are three more examples of natural many-sorted languages in which one can study valued fields. The first one has three sorts: the valued field, the value group and the residue field, in their natural language, together with the valuation map v, and the residue map res, which coincides with the usual residue map on the valuation ring, and 0 outside. Our valued field K will then be the structure

$$((K, +, -, \cdot, 0, 1), (\Gamma \cup \{\infty\}, +, -, 0, \infty, <), (k_K, +, -, \cdot, 0, 1), v, \mathrm{res}).$$

We will see later a variant of this language given by Denef-Pas. Another natural language is the following: given a valued field K, let $\mathrm{RV}(K) = K^\times / 1 + \mathscr{M}_K$. We then have an exact sequence

$$0 \to k_K^\times \to \mathrm{RV}(K) \overset{val_{\mathrm{r}}v}{\to} \Gamma \to 0,$$

where val_{rv} is the natural map induced by the valuation map. To K one associates the following two-sorted structure:

$$\big((K, +, -, \cdot, 0, 1), (\mathrm{RV}(K) \cup \{0\}, \cdot, /, 1, \leq, k^\times, +, 0), v, rv\big),$$

where the map $rv : K \to \mathrm{RV}(K)$ is the natural quotient map on $K^\times$ and sends 0 to 0; $\cdot$, $/$ and 1 give the group structure of $\mathrm{RV}(K)$ (multiplication or division by 0 can be defined to be 0); $k^\times$ is a unary predicate (for a subgroup of $\mathrm{RV}(K)$), and $+$ is a binary operation on $k = k^\times \cup \{0\}$; finally $\leq$ is interpreted by $x \leq y \iff val_{rv}(x) \leq val_{rv}(y)$. You then see that while the residue field k is definable in $\mathrm{RV}(K)$, the value group Γ is only interpretable in it.

Finally, in mixed characteristic, it is sometimes convenient or necessary to work with

$$\mathrm{RV}_n(K) = K^\times / 1 + n\mathscr{M}_K,$$

together with the natural maps $rv_n : K \to \mathrm{RV}_n(K) \cup \{0\}$. The language has now sorts indexed by the integers, and is similar to the one above.

1.7 Multi-sorted structures. Multi-sorted structures appeared naturally in example 4. The difference with the classical ("1-sorted") structures is that a structure will have several *sorts* or *universes*, say indexed by a set I which may be infinite. As already mentioned, each sort comes with its own language, and there may be relations or functions between different sorts or cartesian products of sorts. Some authors require that the universes of distinct sorts be disjoint, but it is not necessary.

Formulas

This subsection and the next are fairly boring, and I would recommend that the reader at first only reads paragraphs 1.10 and 1.13 which give examples. Formulas are built using some basic logical symbols (given below) and in a fashion which ensures unique readability. Satisfaction is defined in the only possible manner. We give here the formal definitions, and the idea is that the reader can come back to them when he needs a precise definition.

1.8 Terms. We can start using the symbols of $\mathscr{L}$ to express properties of a given $\mathscr{L}$-structure. In addition to the symbols of $\mathscr{L}$, we will consider a set of symbols (which we suppose disjoint from $\mathscr{L}$), called the *set of logical symbols*. It consists of

- logical connectives $\wedge$, $\vee$, $\neg$, and also (for convenience) $\rightarrow$ and $\leftrightarrow$,
- parentheses (and),
- a (binary relation) symbol $=$ for equality,
- infinitely many variable symbols, usually denoted x, y, x_i, etc ...
- the quantifiers $\forall$ (for all) and $\exists$ (there exists).

Fix a language $\mathscr{L}$. An $\mathscr{L}$-formula will then be a string of symbols from $\mathscr{L}$ and logical symbols, obeying certain rules. We start by defining $\mathscr{L}$-*terms* (or simply, terms). Roughly speaking, terms are expressions obtained from constants and variables by applying functions. In any $\mathscr{L}$-structure M, a term t will then define uniquely a function from a certain cartesian power of M to M. Terms are defined by induction, as follows:

- a variable x, or a constant c, are terms.
- if $t_1, \ldots, t_n$ are terms, and f is an n-ary function, then $f(t_1, \ldots, t_n)$ is a term.

Given a term $t(x_1, \ldots, x_m)$, the notation indicating that the variables occurring in t are among $x_1, \ldots, x_m$, and an $\mathscr{L}$-structure M, we get a function $F_t : M^m \to M$. Again this function is defined by induction on the complexity of the term:

- if c is a constant symbol, then $F_c : M^0 \to M$ is the function $\emptyset \mapsto c^M$,
- if x is a variable, then $F_x : M \to M$ is the identity,
- if $t_1, \ldots, t_n$ are terms in the variables $x_1, \ldots, x_m$ and f is an n-ary function symbol, then $F_{f(t_1, \ldots, t_n)} : (x_1, \ldots, x_m) \mapsto f(F_{t_1}(\bar{x}), \ldots, F_{t_n}(\bar{x}))$ ($\bar{x} = (x_1, \ldots, x_m)$).

1.9 Formulas. We are now ready to define formulas. Again they are defined by induction.

An *atomic formula* is a formula of the form $t_1(\bar{x}) = t_2(\bar{x})$ or $R(t_1(\bar{x}), \ldots, t_n(\bar{x}))$, where $\bar{x} = (x_1, \ldots, x_m)$ is a tuple of variables, $t_1, \ldots, t_n$ are terms (of the language $\mathscr{L}$, in the variables $\bar{x}$), and R is an n-ary relation symbol of $\mathscr{L}$.

The set of *quantifier-free formulas* is the set of *Boolean combinations* of atomic formulas, i.e., is the closure of the set of atomic formulas under the operations of $\wedge$ (and), $\vee$ (or) and $\neg$ (negation, or not). So, if $\varphi_1(\bar{x})$, $\varphi_2(\bar{x})$ are quantifier-free formulas, so are $(\varphi_1(\bar{x}) \wedge \varphi_2(\bar{x}))$, $(\varphi_1(\bar{x}) \vee \varphi_2(\bar{x}))$, and $(\neg \varphi_1(\bar{x}))$.

One often uses $(\varphi_1 \rightarrow \varphi_2)$ as an abbreviation for $(\neg \varphi_1) \vee \varphi_2$, and $(\varphi_1 \leftrightarrow \varphi_2)$ as an abbreviation for $(\varphi_1 \wedge \varphi_2) \vee [(\neg \varphi_1) \wedge (\neg \varphi_2)]$.

A *formula* ψ is then a string of symbols of the form

$$Q_1 x_1 Q_2 x_2 \ldots Q_m x_m \, \varphi(x_1, \ldots, x_n) \tag{1}$$

where $\varphi(\bar{x})$ is a quantifier-free formula, with variables among $\bar{x} = (x_1, \ldots, x_n)$, and $Q_1, \ldots, Q_m$ are quantifiers, i.e., belong to $\{\forall, \exists\}$. We may assume $m \le n$.

Important: the variables $x_1, \ldots, x_n$ are supposed distinct - $\forall x_1 \exists x_1 \ldots$ is not allowed. If $m \le n$, the variables $x_{m+1}, \ldots, x_n$ are called the *free variables of the formula* ψ. One usually writes $\psi(x_{m+1}, \ldots, x_n)$ to indicate that the free variables of ψ are among $(x_{m+1}, \ldots, x_n)$. The variables $x_1, \ldots, x_m$ are called the *bound variables of* ψ. If $n = m$, then ψ has no free variables and is called a *sentence*.

If all quantifiers $Q_1, \ldots, Q_m$ are $\exists$, then ψ is called an *existential formula*; if they are all $\forall$, then ψ is called a *universal formula*. One can define a hierarchy of complexity of formulas, by counting the number of alternations of quantifiers in the string $Q_1, \ldots, Q_n$. Let us simply say that a Π_2-*formula*, also called a $\forall\exists$-*formula*, is one in which $Q_1 \ldots Q_n$ is a block of $\forall$ followed by a block of $\exists$, that a Σ_2-*formula*, also called a $\exists\forall$-*formula*, is one in which $Q_1 \ldots Q_n$ is a block of $\exists$ followed by a block of $\forall$. In these definitions, either block is allowed to be empty, so that an existential formula is both a Π_2 and a Σ_2-formula. Let us also mention that a *positive formula* is one of the form $Q_1 x_1 \ldots Q_m x_m \varphi(x_1, \ldots, x_n)$, where $\varphi(\bar{x})$ is a finite disjunction of finite conjunctions of *atomic* formulas.

1.10   Warning. This is not the usual definition of a formula. A formula as in (1) is said to be in *prenex form*. The set of formulas in prenex form is **not closed** under Boolean operations. One has however a notion of *"logical equivalence"*, under which for instance the formulas $Q_1 x_1 \ldots Q_m x_m \, \varphi(x_1, \ldots, x_m, x_{m+1}, \ldots, x_n)$ and $Q_1 y_1 \ldots Q_m y_m$ $\varphi(y_1, \ldots, y_m, x_{m+1}, \ldots, x_n)$ are logically equivalent. It is then quite easy to see that a Boolean combination of formulas in prenex form is logically equivalent to a formula in prenex form. E.g,

$$(Q_1 x_1 \ldots Q_m x_m \, \varphi_1(x_1, \ldots, x_n)) \wedge (Q_1' x_1 \ldots Q_m' x_m \, \varphi_2(x_1, \ldots, x_n))$$

is logically equivalent to

$$Q_1 x_1 Q_1' y_1 \ldots Q_m' y_m (\varphi_1(x_1, \ldots, x_n) \wedge \varphi_2(y_1, \ldots, y_m, x_{m+1}, \ldots, x_n)).$$

If one wants to be economical about the number of quantifiers, one notes that in general $\forall x \, \varphi_1(x, \ldots) \wedge \forall x \, \varphi_2(x, \ldots)$ is logically equivalent to $\forall x \, (\varphi_1(x, \ldots) \wedge \varphi_2(x, \ldots))$, and $\exists x \, \varphi_1(x, \ldots) \vee \exists x \, \varphi_2(x, \ldots)$ is logically equivalent to $\exists x \, (\varphi_1(x, \ldots) \vee \varphi_2(x, \ldots))$. For negations, one uses the logical equivalence of $\neg(Q_1 x_1 \ldots Q_m x_m \, \varphi(x_1, \ldots, x_n))$ with

$Q'_1 x_1 \ldots Q'_m x_m \; \neg(\varphi(x_1, \ldots, x_n))$, where $Q'_i = \exists$ if $Q_i = \forall$, $Q'_i = \forall$ if $Q_i = \exists$. Thus the negation of a Π_2-formula is a Σ_2-formula, etc.

Logical equivalence can also be used to rewrite Boolean combinations, and one can show that any quantifier-free formula $\varphi(\bar{x})$ is logically equivalent to one of the form $\bigvee_i \bigwedge_j \varphi_{i,j}(\bar{x})$, where the $\varphi_{i,j}$ are atomic formulas or negations of atomic formulas.

1.11   Adding constant symbols, diagrams. Let $\mathscr{L}$ be a language, M an $\mathscr{L}$-structure, and A a subset of M. The language $\mathscr{L}(A)$ is obtained by adding to $\mathscr{L}$ a new constant symbol symbol $\underline{a}$ for each element a in A. M has then a natural (expansion to an) $\mathscr{L}(A)$-structure: interpret each $\underline{a}$ by the corresponding a. The *basic diagram*, or *atomic diagram* of A in M, $\mathrm{Diag}(A)$ (or $\mathrm{Diag}_M(A)$), is the set of quantifier-free $\mathscr{L}(A)$-sentences satisfied by M.

Example. Let $\mathscr{L} = \mathscr{L}_G$, and $M = \mathbb{Z}$ with the usual group structure, and $A = \{n \in \mathbb{Z} \mid n \geq -1\}$. Then $\mathrm{Diag}(A)$ will contain $\mathscr{L}(A)$-sentences of the following form:

$$\underline{1} + \underline{1} = \underline{2}, \quad \underline{-1} = -\underline{1}, \quad \underline{1} + \underline{4} \neq \underline{3},$$

and so on. Thus, an $\mathscr{L}_G(A)$-structure which is a group and a model of $\mathrm{Diag}(A)$ will be a group in which we have named the elements of a copy of $\{-1\} \cup \mathbb{N}$.

One can also look at more complicated formulas: the *elementary diagram* of A in M, $\mathrm{Diag}_{\mathrm{el}}^M(A)$[1], is the set of all $\mathscr{L}(A)$-sentences which are true in M. Thus for instance with A and M as above, $\mathrm{Diag}_{\mathrm{el}}(A)$ will express the fact that $\underline{1}$ is not divisible by 2 ($\forall x \; x + x \neq \underline{1}$). So, the natural expansion of the group $\mathbb{Q}$ to an $\mathscr{L}_G(A)$-structure is a model of $\mathrm{Diag}(A)$, but not of $\mathrm{Diag}_{\mathrm{el}}(A)$.

In most (all?) situations, we omit the underline on the constant symbol, i.e., denote the same way the constant and its interpretation.

1.12   Examples of formulas. The definitions given above are completely formal. When considering concrete examples, they get very much simplified, to agree with current usage. The first thing to note is that the formula $\neg(x = y)$ is abbreviated by $x \neq y$.

---

[1] When the theory T is complete, one often writes $T(A)$ instead of $\mathrm{Diag}_{\mathrm{el}}^M(A)$

Example 1. $\mathscr{L}_{og} = \{+, -, 0, <\}$. A term is built up from 0, $+$, $-$, and some variables. E.g., $+(0, -(+(x_1, -(x_1))))$ is a term, in the variable x_1. If we work in an arbitrary $\mathscr{L}_{og}$-structure, i.e., not necessarily a group, this expression cannot be simplified. If we work in a group, then we will first of all switch to the usual notation of $x + y$ instead of $+(x, y)$, $-x$ instead of $-(x)$ and $x - y$ instead of $x + (-y)$; then we allow ourselves to use the associativity of the group law to get rid of extraneous parentheses. The term above then becomes $0 - (x_1 - x_1)$, which can be further simplified to 0 (we are now using the fact that in all groups, the sentence $\forall x \; x - x = 0$ holds. I.e., this reduction is only valid because we are working *modulo the theory of groups*).

From now on, we will assume that our $\mathscr{L}_{og}$-structures are **commutative groups**. We add to the language some new symbols of constants, $c_1, \ldots, c_n$.

Here are some terms: $x + x$, $x + x + x, \ldots, nx$, $-nx$ $(n \in \mathbb{N})$, $c_1 + c_2$, $2c_3$. General form of a term $t(x_1, \ldots, x_m)$:

$$\sum_{i=1}^{m} n_i x_i + \sum_{j=1}^{n} \ell_j c_j,$$

where the n_i, ℓ_j belong to $\mathbb{Z}$. This notation can be a little dangerous, as it suggests a uniformity in the coefficients. *One should insists on the fact that if n and m are distinct integers, then the terms nx and mx are different.* [So, in general, the set of torsion elements of a group is not definable in the group G, since an element g is torsion if and only if **for some** n **in** $\mathbb{N}$, $ng = 0$. There are of course exceptions, e.g., if the order of torsion elements is bounded.]

Quantifier-free-formulas: apply relations and Boolean connectives to terms: $\bar{x} = (x_1, \ldots, x_m)$, $t_1(\bar{x}), \ldots, t_4(\bar{x})$ terms:

$$\bigl(t_1(\bar{x}) = t_2(\bar{x}) \wedge t_3(\bar{x}) < t_4(\bar{x})\bigr) \vee \; \bigl(t_1(\bar{x}) < t_2(\bar{x})\bigr).$$

Example 2. $\mathscr{L}_R = \{+, -, \cdot, 0, 1\}$. Again, terms as defined formally, are extremely ugly. But, in case all $\mathscr{L}_R$-structures considered are commutative rings, they can be rewritten in a more natural fashion. From now on, **all $\mathscr{L}_R$-structures are commutative rings**.

If $n \in \mathbb{N}^{>1}$ the term $1 + 1 + \cdots + 1$ (n times) will simply be denoted by n. Similarly $x + x + \cdots + x$ (n times) is denoted by nx, and $x \cdot \ldots \cdot x$

(n times) by x^n. An arbitrary term will then be of the form $f(x_1, \ldots, x_n)$, where $f(X_1, \ldots, X_n) \in \mathbb{Z}[X_1, \ldots, X_n]$.

Quantifier-free formulas are finite disjunctions of finite conjunctions of equations and inequations. Thus, in the ring $\mathbb{C}$, they will define the usual *constructible sets* which are defined over $\mathbb{Z}$. If we want to get all constructible sets, we should work in the language $\mathscr{L}_R(\mathbb{C})$, obtained by adding constant symbols for the elements of $\mathbb{C}$.

If one adds $\leq$ to the language, and assumes that our structures are **ordered rings**, then quantifier-free formulas can be rewritten as finite conjunctions of finite disjunctions of formulas of the form

$$f(\bar{x}) = 0, \quad g(\bar{x}) > 0, \tag{2}$$

where f, g are polynomials over $\mathbb{Z}$. Here, $x < y$ stands for $x \leq y \wedge x \neq y$, and one uses the equivalences $x \neq 0 \iff x < 0 \vee x > 0$, $x > 0 \iff (-x) < 0$. If M is an ordered ring, then $\mathscr{L}_{or}(M)$-quantifier-free formulas will be as above, except that f and g are polynomials over M. In case M is the ordered field $\mathbb{R}$, one then gets the usual *semi-algebraic sets*.

Satisfaction

1.13 **Satisfaction.** Let M be an $\mathscr{L}$-structure, $\varphi(\bar{x})$ an $\mathscr{L}$-formula, where $\bar{x} = (x_1, \ldots, x_n)$ is a tuple of variables occurring freely in φ, and $\bar{a} = (a_1, \ldots, a_n)$ an n-tuple of elements of M. We wish to define the notion M *satisfies* $\varphi(\bar{a})$, (or $\bar{a}$ *satisfies* φ *in* M, or $\varphi(\bar{a})$ *holds in* M, *is true in* M), denoted by

$$M \models \varphi(\bar{a}).$$

(The negation of $M \models \varphi(\bar{a})$ is denoted by $M \not\models \varphi(\bar{a})$, or ... by $M \models \neg\varphi(\bar{a})$.) Satisfaction is what it should be if you read the formula aloud. Here is a formal definition, by induction on the complexity of the formulas. It is fairly boring, and if you wish you can skip it. Let $\bar{a}$, $\bar{b}$ be tuples in M,

 – If $\varphi(\bar{x})$ is the formula $t_1(\bar{x}) = t_2(\bar{x})$, where t_1, t_2 are $\mathscr{L}$-terms in the variable $\bar{x}$, then

$$M \models t_1(\bar{a}) = t_2(\bar{a}) \text{ if and only if } F_{t_1}(\bar{a}) = F_{t_2}(\bar{a}).$$

 – If $\varphi(\bar{x})$ is the formula $R(t_1(\bar{x}), \ldots, t_m(\bar{x}))$, where $t_1, \ldots, t_m$ are terms and R is an m-ary relation, then

$$M \models R(t_1(\bar{a}), \ldots, t_m(\bar{a})) \text{ if and only if } (F_{t_1}(\bar{a}), \ldots, F_{t_m}(\bar{a})) \in R^M.$$

– If $\varphi(\bar{x}) = \varphi_1(\bar{x}) \vee \varphi_2(\bar{x})$, then

$$M \models \varphi(\bar{a}) \text{ if and only if } M \models \varphi_1(\bar{a}) \text{ or } M \models \varphi_2(\bar{a}).$$

– If $\varphi(\bar{x}) = \varphi_1(\bar{x}) \wedge \varphi_2(\bar{x})$, then

$$M \models \varphi(\bar{a}) \text{ if and only if } M \models \varphi_1(\bar{a}) \text{ and } M \models \varphi_2(\bar{a}).$$

– If $\varphi(\bar{x}) = \neg\varphi_1(\bar{x})$, then

$$M \models \varphi(\bar{a}) \text{ if and only if } M \not\models \varphi_1(\bar{a}).$$

– If $\varphi(\bar{x}) = \exists y\, \psi(\bar{x}, y)$, where the free variables of ψ are among $\bar{x}, y$, then

$$M \models \varphi(\bar{a}) \text{ if and only if } \textbf{there is } c \in M \text{ such that } M \models \psi(\bar{a}, c).$$

– If $\varphi(\bar{x}) = \forall y\, \psi(\bar{x}, y)$, then

$$M \models \varphi(\bar{a}) \text{ if and only if } M \models \neg(\exists y\, \neg\psi(\bar{a}, y))$$
$$\text{if and only if } \textbf{for all } c \text{ in } M, \quad M \models \varphi(\bar{a}, c).$$

Note that of course, for all $\bar{a}$ in M, one has

$$M \models \forall y\, \psi(\bar{a}, y) \text{ if and only if } M \models \neg(\exists y\, \neg\psi(\bar{a}, y)).$$

1.14 Parameters, definable sets. Let M be an $\mathscr{L}$-structure, $\varphi(\bar{x}, \bar{y})$ a formula ($\bar{x}$ an n-tuple of variables, $\bar{y}$ an m-tuple of variables), and $\bar{a} \in M^n$. Then the set $\{\bar{b} \in M^m \mid M \models \varphi(\bar{a}, \bar{b})\}$ is called a *definable set*. We also say that it is *defined over $\bar{a}$* by the formula $\varphi(\bar{a}, \bar{y})$, or that it is *$\bar{a}$-definable*. The tuple $\bar{a}$ is a *parameter of the formula* $\varphi(\bar{a}, \bar{y})$. When $\bar{a}$ varies over M^n, the sets $\{\bar{b} \in M^m \mid M \models \varphi(\bar{a}, \bar{b})\}$, which are sometimes denoted by $\varphi(\bar{a}, M^m)$ or by $\varphi(\bar{a}, M)$, form a *family of uniformly definable sets*.

Let M be an $\mathscr{L}$-structure. The set of $\mathscr{L}(M)$-definable subsets of M^n is clearly closed under unions, intersections and complements (corresponding to the use of the logical connectives $\vee$, $\wedge$ and $\neg$). If $S \subseteq M^{n+1}$ is defined by the formula $\varphi(\bar{x}, \bar{a})$, $\bar{x} = (x_1, \ldots, x_{n+1})$, and $\pi : M^{n+1} \to M$ is the projection on the first n coordinates, then $\pi(S)$ is defined by the formula $\exists x_{n+1}\, \varphi(\bar{x}, \bar{a})$, and the complement of $\pi(S)$ by the formula $\forall x_{n+1}\, \neg\varphi(\bar{x}, \bar{a})$.

Thus an alternate definition of $\mathscr{L}$-subsets of M is as follows: it is the smallest collection $\mathscr{S} = (\mathscr{S}_n)_{n \in \mathbb{N}}$, where each $\mathscr{S}_n$ is a set of subsets of M^n, which satisfies the following conditions:

- $\mathscr{S}_1$ contains all singletons of constants; if $f \in \mathscr{L}$ is an n-ary function symbol, then the graph of f is in $\mathscr{S}_{n+1}$; if R is an n-ary function symbol, then the interpretation of R is in $\mathscr{S}_n$; $\mathscr{S}_2$ contains the diagonal.
- Each $\mathscr{S}_n$ is closed under Boolean operations $\cup$, $\cap$, and complement.
- $\mathscr{S}$ is closed under (finite) cartesian products.
- If $\pi : M^{n+1} \to M^n$ is a projection on an n-subset of the coordinates, and $S \in \mathscr{S}_{n+1}$, then $\pi(S) \in \mathscr{S}_n$.

1.15 An example. Consider the $\mathscr{L}_{og}$-formula

$$\varphi(x, y) := x < y \land (\forall z\, x < z \to z = y \lor y < z),$$

where as usual $x < y$ is an abbreviation for $x \leq y \land x \neq y$. In an ordered group G, this formula expresses that y is an immediate successor of x. Thus, in $(\mathbb{Z}, +, -, 0, \leq)$, the formula will define the graph of the successor function. But in $(\mathbb{Q}, +, -, 0, \leq)$ it will define the empty set, as $\mathbb{Q}$ is a dense ordering.

Definability, interpretability

1.16 Definability of a structure in another one. Let M be an $\mathscr{L}$-structure, and N an $\mathscr{L}'$-structure. We say that N is *definable in* M if there are

– an $\emptyset$-definable set $S \subseteq M^n$ for some n,

– for each m-ary relation R of the language $\mathscr{L}'$, an $\emptyset$-definable subset R^* of S^m,

– for each m-ary function f of $\mathscr{L}'$, an $\emptyset$-definable subset Γ_f of S^{m+1} such that Γ_f is the graph of a function $f^* : S^m \to S$,

– for each constant c, an $\emptyset$-definable tuple $c^* \in S$,

– and a bijection $F : N \to S$, which defines an $\mathscr{L}'$-isomorphism between the structure $N = (N, R, \ldots, f, \ldots, c, \ldots)$ and the structure

$$N^* = (S, R^*, \ldots, f^*, \ldots, c^*, \ldots).$$

1.17 Interpretability of a structure in another one. Let M be an $\mathscr{L}$-structure, N an $\mathscr{L}'$-structure. We say that N is *interpretable in* N if there are

– an $\emptyset$-definable set S of some M^ℓ,

– an $\emptyset$-definable equivalence relation E on S,

– for each m-ary relation R of the language $\mathscr{L}'$, an $\emptyset$-definable subset R' of S^m, which projects to a subset R^* of $(S/E)^m$,

– for each m-ary function f of $\mathscr{L}'$, an $\emptyset$-definable subset Γ_f of S^{m+1} such that Γ_f induces the graph of a function $f^* : (S/E)^m \to S$,

– for each constant c, an $\emptyset$-definable E-equivalence class $c^* \in S/E$,

– and a bijection $F : N \to S/E$, which defines an $\mathscr{L}'$-isomorphism between the structure $N = (N, R, \ldots, f, \ldots, c, \ldots)$ and the structure

$$N^* = (S/E, R^*, \ldots, f^*, \ldots, c^*, \ldots).$$

1.18 Adding parameters. In both definitions, if instead of working in the $\mathscr{L}$-structure M, one works in the $\mathscr{L}(A)$-structure M for some $A \subset M$, one will say that N is A-definable, resp. A-interpretable, in M.

1.19 Bi-interpretability of two structures. Let M be an $\mathscr{L}$-structure, N an $\mathscr{L}'$-structure. We say that M and N are *bi-interpretable* if

(i) N is interpretable in M, and M is interpretable in N,

(ii) the bijections F and F' which give the interpretations of N in M and of M in N respectively, can be chosen so that the maps $F \circ F'$ and $F' \circ F$ are $\emptyset$-definable in M and N respectively.

1.20 Example. Let R be an integral domain, in the usual ring language, and let Q be its field of fractions. Then the ring Q is interpretable in R: indeed, we know that Q is the set of quotients a/b with $a \in R$, $0 \neq b \in R$, so we can identify it with the set of elements of $R \times R^*$ quotiented by the equivalence relation $(a, b) \sim (c, d) \iff ad = bc$. In general, Q is not definable in R, because there is no way of selecting a particular pair in each equivalence class. However, $\mathbb{Q}$ is definable in $\mathbb{Z}$: $\mathbb{Z}$ is a principal ideal domain, in which an ordering is definable, and with only units -1 and 1.

2 Theories, and some important theorems

In this section we will introduce many definitions and important concepts. We will also mention the very important *Compactness theorem*, one of the crucial tools of model theory.

Theories and models

2.1 Theories, models of theories, etc.. Let $\mathscr{L}$ be a language. An $\mathscr{L}$-*theory* (or simply, a theory), is a set of sentences of the language $\mathscr{L}$. A *model of a theory* T is an $\mathscr{L}$-structure M which satisfies all sentences of T, denoted by $M \models T$. The class of all models of T is denoted $\mathrm{Mod}(T)$. If

$\mathscr{K}$ is a class of $\mathscr{L}$-structures, then $\mathrm{Th}(\mathscr{K})$ denotes the set of all sentences true in all elements of $\mathscr{K}$, and $\mathrm{Th}(\{M\})$ is denoted by $\mathrm{Th}(M)$.

A theory T is *consistent* iff it has a model. If φ is a sentence which holds in all models of T, this is denoted by $T \models \varphi$. Two $\mathscr{L}$-structures M and N are *elementarily equivalent*, denoted $M \equiv N$, iff they satisfy the same sentences, iff $\mathrm{Th}(M) = \mathrm{Th}(N)$. A theory is *complete* iff given a sentence φ, either $T \models \varphi$ or $T \models \neg\varphi$. Equivalently, if any two models of T are elementarily equivalent. (Observe that if M is an $\mathscr{L}$-structure, then necessarily $\mathrm{Th}(M)$ is complete.)

Elementary equivalence is an equivalence relation between $\mathscr{L}$-structures. Two isomorphic $\mathscr{L}$-structures are clearly elementarily equivalent, however the converse only holds **for finite $\mathscr{L}$-structures**. A famous theorem (of Keisler-Shelah) states that two structures are elementarily equivalent if and only if they have isomorphic ultrapowers, see definition in Section 2.14.

2.2   Elementary substructures, extensions, embeddings, etc.
Let $M \subseteq N$ be $\mathscr{L}$-structures. We say that M is an *elementary substructure of N*, or that *N is an elementary extension of M*, denoted by $M \prec N$, iff for any formula $\varphi(\bar{x})$ and tuple $\bar{a}$ from M,

$$M \models \varphi(\bar{a}) \iff N \models \varphi(\bar{a}).$$

A map $f : M \to N$ is an *elementary embedding* iff it is an embedding, and if $f(M) \prec N$. In other words, if for any formula $\varphi(\bar{x})$ and tuple $\bar{a}$ from M, $M \models \varphi(\bar{a})$ if and only $N \models \varphi(f(\bar{a}))$.

Using the language of diagrams introduced in 1.11,

$$M \prec N \iff N \models \mathrm{Diag}_{\mathrm{el}}^M(M).$$

Similarly, an *elementary partial map from M to N* is a map f defined on some substructure A of M, with range included in N, and which preserves the formulas in $\mathrm{Diag}_{\mathrm{el}}^M(A)$, i.e., for any formula $\varphi(\bar{x})$ and tuple $\bar{a}$ from A, $M \models \varphi(\bar{a})$ if and only $N \models \varphi(f(\bar{a}))$. A map f which only preserves $\mathrm{Diag}^M(A)$ is called a *partial isomorphism*.

2.3   Comments. (1) Note that one can have $M \subseteq N$ and $M \equiv N$ without having $M \prec N$. Consider the group $\mathbb{Z}$ and its subgroup $2\mathbb{Z}$: they are isomorphic, but the inclusion $2\mathbb{Z} \subset \mathbb{Z}$ is not an elementary map, since 2 is divisible by 2 in $\mathbb{Z}$ but not in $2\mathbb{Z}$.
(2) Similarly, not every partial isomorphism is elementary. Again, the inclusion of $2\mathbb{Z}$ into $\mathbb{Z}$ provides an example.

Some classical results

2.4 **Tarski's test.** *Let M be a substructure of N. Then $M \prec N$ if and only if, for every formula $\varphi(\bar{x}, y)$ and tuple $\bar{a}$ in M, if $N \models \exists y \, \varphi(\bar{a}, y)$, then there exists $b \in M$ such that $N \models \varphi(\bar{a}, b)$.*

Note that while the element b is in M, the satisfaction is taken in N. This theorem is proved using induction on the complexity of formulas.

2.5 **Soundness and completeness theorem.** Given a set of sentences, there is a notion of proof, i.e., which statements are deducible from the given statements using some formal rules of deduction, such as modus ponens (from A and $A \to B$ deduce B), and some substitution rules (from a sentence of the form $\varphi(c)$ where c is a constant, deduce $\exists x \, \varphi(x)$). A proof can be thought of therefore as a finite sequence of sentences, each being obtained from the previous ones by applying some deduction rules. We use the notation

$$T \vdash \varphi$$

to indicate that there is a proof of φ from T. This is not to be confused with the notation

$$T \models \varphi$$

which means that φ is true in all models of T. The first result, the *soundness theorem*, tells us that our notion of satisfaction is well-defined: *If a theory T has a model, then one cannot derive a contradiction from T, i.e., one cannot prove from T the sentence $\forall x (x \neq x)$.*

In other words

$$T \vdash \varphi \Rightarrow T \models \varphi.$$

Gödel's completeness theorem then states the converse:
If from a given theory T, one cannot derive the sentence $\forall x (x \neq x)$, then the theory T has a model.

Another way of stating this result is by saying that
the set of sentences deducible from a given theory T is **exactly** *the set of sentences true in* **all** *models of T, i.e., in the notation introduced above, it coincides with* $\mathrm{Th}(\mathrm{Mod}(T))$.

2.6 **Decidability.** A theory T is *decidable*, if there is an algorithm allowing to decide whether a sentence φ holds in all models of T or not. If one can enumerate a theory T and one knows (somehow) that T is complete, then T is decidable: given a sentence φ, start enumerating the proofs from T; eventually you reach a proof of either φ or $\neg\varphi$.

2.7 Compactness theorem. *Let T be a set of sentences in a language $\mathscr{L}$. If every finite subset of T has a model, then T has a model.*

We will present later a proof of this theorem using ultraproducts. Note that it is a consequence of the completeness theorem, since any proof involves only finitely many elements of T. It also has for consequence the first half of the next theorem.

2.8 Löwenheim-Skolem theorems. *Let $\mathscr{L}$ be a language, T a theory, and let M be an infinite model of T.*

(1) Let κ be an infinite cardinal, $\kappa \geq |M| + |\mathscr{L}|$. Then M has an elementary extension N with $|N| = \kappa$.

(2) Let X be a subset of M. Then M has an elementary substructure N containing X, with $|N| \leq |X| + |\mathscr{L}| + \aleph_0$.

2.9 Comments. These results allow us to use large models with good properties. For instance, assume that we have a set $\Sigma(x_1, \ldots, x_n)$ of formulas in the variables $(x_1, \ldots, x_n)$, and that we know that every finite fragment of $\Sigma(x_1, \ldots, x_n)$ is satisfiable in some model M of T, i.e., there is a tuple $\bar{a}$ of M which satisfies all formulas of that finite fragment. Then there is a model N of T containing a tuple $\bar{b}$ which satisfies simultaneously all formulas of $\Sigma(\bar{x})$. This is connected to saturation, see below for a definition.

Using other techniques, one can show that if $\bar{a}$ and $\bar{b}$ are tuples of an $\mathscr{L}$-structure M, which satisfy the same formulas in M, then M has an elementary extension M^*, in which there is an automorphism which sends $\bar{a}$ to $\bar{b}$.

2.10 Craig's interpolation theorem. *Let $\mathscr{L}_1$ and $\mathscr{L}_2$ be two languages. Let φ be a sentence of $\mathscr{L}_1$ and ψ a sentence of $\mathscr{L}_2$. If $\varphi \models \psi$, then there is a sentence θ of $\mathscr{L}_1 \cap \mathscr{L}_2$ such that $\varphi \models \theta$ and $\theta \models \psi$.*

A somewhat different interpolation theorem is given by Robinson:
Let $\mathscr{L}_1$ and $\mathscr{L}_2$ be two languages, and $\mathscr{L}_0 = \mathscr{L}_1 \cap \mathscr{L}_2$. Assume that T_1 and T_2 are theories in $\mathscr{L}_1$ and $\mathscr{L}_2$ respectively, such that $T_0 = T_1 \cap T_2$ is complete. Then $T_1 \cup T_2$ is consistent.

Types, saturated models

Fix a complete theory T in a language $\mathscr{L}$, a subset A of a model M of T. A *(partial) n-type over A* (in the variables $\bar{x} = (x_1, \ldots, x_n)$) is a collection $p(\bar{x})$ of $\mathscr{L}(A)$-formulas which is finitely consistent in M. A *complete type over A* is an n-type $p(\bar{x})$ which is maximally consistent, i.e., given an $\mathscr{L}(A)$-formula $\varphi(\bar{x})$, one of $\varphi(\bar{x})$, $\neg\varphi(\bar{x})$ belongs to $p(\bar{x})$. The set of complete n-types over A is denoted $S_n(A)$. Here is an example: let $\bar{a}$ be an n-tuple in M. Then

$$tp(\bar{a}/M) := \{\varphi(\bar{x}) \in \mathscr{L}(A) \mid M \models \varphi(\bar{a})\},$$

the *type of $\bar{a}$ over A*, is a complete type.

Warning: depending on the context a *type* can mean either a partial type, or a complete type. There is no set usage.

Given an n-type $p(\bar{x})$ over A, a *realisation of p in M* is an n-tuple a in M which satisfies all formulas of $p(\bar{x})$. In any case, there will be an elementary extension N of M in which $p(\bar{x})$ will be realised.

2.11 Topology on the space of types. Given $A \subset M$ and $n > 0$ as above, one defines a topology on $S_n(A)$, whose basic open sets are

$$\langle \varphi(\bar{x}) \rangle = \{p(\bar{x}) \in S_n(A) \mid \varphi(\bar{x}) \in p(\bar{x})\}.$$

Then $S_n(A)$ is compact, totally disconnected. A type $p(\bar{x}) \in S_n(A)$ is *isolated* if and only if there is an $\mathscr{L}(A)$-formula which implies all formulas in $p(\bar{x})$.

2.12 Saturated models. Let κ be an infinite cardinal, M an $\mathscr{L}$-structure. We say that M is *κ-saturated* if for every subset A of M of cardinality $< \kappa$, every n-type over A is realised in M. We say that M is *saturated* if it is $|M|$-saturated. Observe that an infinite $\mathscr{L}$-structure M can never be $|M|^+$-saturated: consider the set of $\mathscr{L}(M)$-formulas $\{x \neq m \mid m \in M\}$.

Expressed in terms of definable sets: let $\mathscr{S}_n$ be the set of $\mathscr{L}(A)$-definable subsets of M^n. Then the $|A|^+$-saturation[2] of M means that if $(D_i)_{i \in I} \subset \mathscr{S}_n$ is such that the intersection of any finite collection of D_i's is non-empty[3], then there is a tuple $\bar{a}$ in the intersection of all D_i's.

---

[2] If A is finite, one considers instead $\aleph_0$-saturation.
[3] One then says that $\{D_i \mid i \in I\}$ has the *finite intersection property*.

Non-example. Consider the ordered group $(\mathbb{R}, +, -, 0 <)$. It is not $\aleph_0$-saturated: take $A = \{1\}$, and consider

$$\Sigma(x) = \{x > n \mid n \in \mathbb{N}\}.$$

This set of formulas is finitely consistent: for any n, the finite fragment $\{x > m \mid 0 \le m \le n\}$ is satisfied in $\mathbb{R}$ by $n + 1$. However, no element of $\mathbb{R}$ is greater than all elements of $\mathbb{N}$. In fact, a (non-trivial) ordered abelian group which is $\aleph_0$-saturated cannot be archimedean. Note that this argument only works because the elements of $\mathbb{N}$ can be obtained as terms in $\mathscr{L}_{og}(A)$; one can show that the ordered **set** $(\mathbb{R}, <)$ is $\aleph_0$-saturated (but not $\aleph_1$-saturated, since $\mathbb{N}$ is countable and cofinal in $\mathbb{R}$).

2.13   Important results concerning saturated models:

Let κ be an infinite cardinal, M an infinite $\mathscr{L}$-structure. Then M has an elementary extension M^* which is κ-saturated.

In contrast, given an infinite cardinal κ and a theory T, there does not always exist a saturated model of T of cardinality κ. Under GCH[4], a theory T with infinite models has uncountable saturated models of any cardinality.

A saturated model M of T has the following properties:

(i) (Universality) Any model of T of cardinality $< |M|$ embeds elementarily into M.

(ii) (Homogeneity) If $f : A \to B$ is an elementary partial map between subsets A and B of M of cardinality $< |M|$, then f extends to an automorphism of M.

2.14   Definable and algebraic closures. Let T be a complete $\mathscr{L}$-theory, A a subset of a model M of T. We say that an element $a \in M$ is *algebraic over A*, noted $a \in \mathrm{acl}(A)$, if there is an $\mathscr{L}(A)$-formula $\varphi(x)$ which defines a finite subset of M containing a. We say that a is *definable over A*, noted $a \in \mathrm{dcl}(A)$, if there is such a formula $\varphi(x)$ which defines $\{a\}$. An *algebraic, resp. definable, tuple* is one whose elements are algebraic, resp. definable. If $\bar{a} \in \mathrm{acl}(A)$, then $tp(\bar{a}/A)$ is isolated. Clearly one has

$$\mathrm{dcl}(A) \subseteq \mathrm{acl}(A), \ \mathrm{dcl}(\mathrm{dcl}(A)) = \mathrm{dcl}(A), \ \mathrm{acl}(\mathrm{acl}(A)) = \mathrm{acl}(A).$$

---

[4] The General Continuum Hypothesis, which says that given an infinite cardinal κ, a set I of cardinality κ, the successor cardinal of κ is the cardinality (2^κ) of the set of subsets of I. That is: $\kappa^+ = 2^\kappa$ for all $\kappa \ge \aleph_0$.

An alternate way of defining definable and algebraic closures is via automorphism groups: let M be a saturated model of cardinality $> |A|$, and $G = \mathrm{Aut}(M/A)$. Then $a \in \mathrm{dcl}(A)$ if and only if the G-orbit of a has only one element, and $a \in \mathrm{acl}(A)$ if and only if the G-orbit of a is finite.

Ultraproducts, Łos theorem

In this section we introduce an important tool: ultraproducts. They are at the centre of many applications, within and outside model theory.

2.15   Filters and ultrafilters. Let I be a set. A *filter on I* is a subset $\mathscr{F}$ of $\mathscr{P}(I)$ (the set of subsets of I), satisfying the following properties:

(1) $I \in \mathscr{F}$, $\emptyset \notin \mathscr{F}$.
(2) If $U \in \mathscr{F}$ and $V \supseteq U$, then $V \in \mathscr{F}$.
(3) If $U, V \in \mathscr{F}$, then $U \cap V \in \mathscr{F}$.

An *ultrafilter on I* is a filter on I which is maximal for inclusion. Equivalently, it is a filter $\mathscr{F}$ such that for any $U \in \mathscr{P}(I)$, either $U \in \mathscr{F}$ or $I \setminus U \in \mathscr{F}$.

2.16   Remarks. (1) Note that condition (1) above forbids that both U and $I \setminus U$ belong to the same filter on I.

(2) Using Zorn's lemma (and therefore the axiom of choice), every filter on I is contained in an ultrafilter.

(3) If $\mathscr{G} \subset \mathscr{P}(I)$ has the *finite intersection property* (i.e., the intersection of finitely many elements of $\mathscr{G}$ is never empty), then $\mathscr{G}$ is contained in a filter. The *filter generated by $\mathscr{G}$* is then the set of elements of $\mathscr{P}(I)$ containing some finite intersection of elements of $\mathscr{G}$.

2.17   Principal and non-principal ultrafilters, Fréchet filter. Let I be a set. An ultrafilter $\mathscr{F}$ on I is *principal* if there is $i \in I$ such that $\{i\} \in \mathscr{F}$ (and then we will have: $U \in \mathscr{F} \iff i \in U$). An ultrafilter is *non-principal* if it is not principal. Note that if I is finite, then every ultrafilter on I is principal.

Let I be infinite. The *Fréchet filter on I* is the filter $\mathscr{F}_0$ consisting of all cofinite subsets of I. An ultrafilter $\mathscr{F}$ on I is then non-principal if and only if contains the Fréchet filter on I. Note that if $S \subseteq I$ is infinite, then $\mathscr{F}_0 \cup \{S\}$ has the finite intersection property, so that it is contained in an ultrafilter.

2.18 Cartesian products of $\mathscr{L}$-structures. Fix a language $\mathscr{L}$. Let I be an index set, and (M_i), $i \in I$, a family of $\mathscr{L}$-structures. We define the $\mathscr{L}$-structure $M = \prod_{i \in I} M_i$ as follows:

— The universe of M is simply the cartesian product of the M_i's, i.e., the set of sequences $(a_i)_{i \in I}$ such that $a_i \in M_i$ for each $i \in I$. One sometimes uses the functional notation $a(i)$ instead of a_i.

— If c is a constant symbol of $\mathscr{L}$, then $c^M = (c^{M_i})_{i \in I}$.

— If R is an n-ary relation symbol, then $R^M = \prod_{i \in I} R^{M_i}$.

— If f is an n-ary function symbol and $((a_{1,i})_i, \ldots, (a_{n,i})_i) \in M^n$, then

$$f^M((a_{1,i})_i, \ldots, (a_{n,i})_i) = (f^{M_i}(a_{1,i}, \ldots, a_{n,i}))_{i \in I}.$$

2.19 Reduced products of $\mathscr{L}$-structures. Let I be a set, $\mathscr{F}$ a filter on I, and (M_i), $i \in I$, a family of $\mathscr{L}$-structures. The *reduced product of the M_i's over $\mathscr{F}$*, denoted by $\prod_{i \in I} M_i / \mathscr{F}$, is the $\mathscr{L}$-structure defined as follows:

— The universe of $\prod_{i \in I} M_i / \mathscr{F}$ is the quotient of $\prod_{i \in I} M_i$ by the equivalence relation $\equiv_{\mathscr{F}}$ defined by

$$(a_i)_i \equiv_{\mathscr{F}} (b_i)_i \iff \{i \in I \mid a_i = b_i\} \in \mathscr{F}.$$

We denote by $(a_i)_{\mathscr{F}}$ the equivalence class of the element $(a_i)_i$ for this equivalence relation.

The structure on $\prod_{i \in I} M_i / \mathscr{F}$ is then simply the "quotient structure", i.e.,

- The interpretation of c is $(c^{M_i})_{\mathscr{F}}$, for c a constant symbol of $\mathscr{L}$.
- If R is an n-ary relation symbol, and if $a_1, \ldots, a_n \in \prod_{i \in I} M_i / \mathscr{F}$ are represented by $(a_{1,i})_i, \ldots, (a_{n,i})_i \in \prod_{i \in I} M_i$, then we set

$$\prod_{i \in I} M_i / \mathscr{F} \models R(a_1, \ldots, a_n) \iff \{i \in I \mid (a_{1,i}, \ldots, a_{n,i}) \in R^{M_i}\} \in \mathscr{F}.$$

- If f is an n-ary function symbol and if $a_1, \ldots, a_n \in \prod_{i \in I} M_i / \mathscr{F}$ are represented by $(a_{1,i})_i, \ldots, (a_{n,i})_i \in \prod_{i \in I} M_i$, then we set

$$f^M(a_1, \ldots, a_n) = (f^{M_i}(a_{1,i}, \ldots, a_{n,i}))_{\mathscr{F}}.$$

The properties of filters guarantee that the quotient structure is well-defined. Note that the quotient map : $\prod_{i \in I} M_i \to \prod_{i \in I} M_i / \mathscr{F}$, $(a_i)_i \mapsto (a_i)_{\mathscr{F}}$, is a morphism of $\mathscr{L}$-structures.

Definitions. If all structures M_i are equal to the same structure M, then we write $M^I/\mathscr{F}$ instead of $\prod_i M_i/\mathscr{F}$, and the structure is called a *reduced power of M*. If the filter $\mathscr{F}$ is an ultrafilter, then $\prod_i M_i/\mathscr{F}$ is called the *ultraproduct of the M_i's* (with respect to $\mathscr{F}$), and $M^I/\mathscr{F}$ the *ultrapower of M* (with respect to $\mathscr{F}$).

2.20 Łos theorem. *Let I be a set, $\mathscr{F}$ an ultrafilter on I, and (M_i), $i \in I$, a family of $\mathscr{L}$-structures. Let $\varphi(x_1, \ldots, x_n)$ be an $\mathscr{L}$-formula, and let $a_1, \ldots, a_n \in \prod_{i \in I} M_i/\mathscr{F}$ be represented by $(a_{1,i})_i, \ldots, (a_{n,i})_i \in \prod_{i \in I} M_i$. Then*

$$\prod_{i \in I} M_i/\mathscr{F} \models \varphi(a_1, \ldots, a_n) \iff \{i \in I \mid M_i \models \varphi(a_{1,i}, \ldots, a_{n,i})\} \in \mathscr{F}.$$

2.21 Corollary. *Let I be a set, $\mathscr{F}$ an ultrafilter on I, and M an $\mathscr{L}$-structure. Then the natural map $M \to M^I/\mathscr{F}$, $a \mapsto (a)_{\mathscr{F}}$, is an elementary embedding. (Here $(a)_{\mathscr{F}}$ is the equivalence class of the sequence with all terms equal to a.)*

2.22 Remarks and comments. Let I be an infinite index set, and $\mathscr{F}$ an ultrafilter on I.

(1) If $\mathscr{F}$ is principal, say $\{j\} \in \mathscr{F}$, then $\prod_{i \in I} M_i/\mathscr{F} \simeq M_j$ for any family of $\mathscr{L}$-structures M_i, $i \in I$.

(2) Suppose that the M_i's are fields, with maybe additional structure (e.g., an ordering, new functions, etc.). Consider the ideal $\mathscr{M}$ of $\prod_i M_i$ generated by all elements $(a_i)_i$ such that $\{i \in I \mid a_i = 0\} \in \mathscr{F}$. Then $\mathscr{M}$ is a maximal ideal of $\prod_i M_i$, and quotienting by the equivalence relation $\equiv_{\mathscr{F}}$ is equivalent to quotienting by the maximal ideal $\mathscr{M}$. The strength of Łos theorem is to tell you that the elementary properties of the M_i's, including the ones depending on the additional structure, are preserved. E.g., that $\mathbb{R}^I/\mathscr{F}$ is a real closed field.

2.23 Keisler and Shelah's isomorphism theorem. *Let M and N be two $\mathscr{L}$-structures. Then $M \equiv N$ if and only if there is an ultrafilter $\mathscr{F}$ on a set I such that $M^I/\mathscr{F} \simeq N^I/\mathscr{F}$.*

Note the following immediate consequence: if $M \equiv N$, then there is M^* in which both M and N embed elementarily.

2.24 Application 1: another proof of the compactness theorem. *Let T be a theory in a language $\mathscr{L}$, and assume that every finite subset s of T has a model M_s. Then T has a model.*

Proof. If T is finite, there is nothing to prove, so we will assume that T is infinite. Let I be the set of all finite subsets of T. For every $\varphi \in T$, let $S(\varphi) = \{s \in I \mid \varphi \in s\}$. Then the family $\mathscr{G} = \{S(\varphi) \mid \varphi \in T\}$ has the finite intersection property, and therefore is contained in an ultrafilter $\mathscr{F}$. We claim that $\prod_{s \in I} M_s/\mathscr{F}$ is a model of T: let $\varphi \in T$. Then, by assumption, $\{s \in I \mid M_s \models \varphi\}$ contains $S(\varphi)$, and therefore belongs to $\mathscr{F}$. By Łos's theorem, $\prod_{s \in I} M_s/\mathscr{F} \models \varphi$.

2.25 Application 2: $\aleph_1$-saturated models. *If I is an infinite countable set, $\mathscr{U}$ is a non-principal ultrafilter on I, and $(M_i)_{i \in I}$ is a family of $\mathscr{L}$-structures where $\mathscr{L}$ is a countable language, then the ultraproduct $\prod_{i \in I} M_i/\mathscr{U}$ is $\aleph_1$-saturated.*

Proof. If there is a finite bound on the cardinalities of the M_i's, then $M^* = \prod_{i \in I} M_i/\mathscr{U}$ is finite, and there is nothing to prove, so assume this is not the case. Let $A \subset \prod_{i \in I} M_i/\mathscr{U}$ be countable, and $\Sigma(x)$ be a set of $\mathscr{L}(A)$-formulas which is finitely consistent. Then $\Sigma(x)$ is countable, and we choose an enumeration $\varphi_n(x, \bar{a}_n)$, $n \in \mathbb{N}$, of $\Sigma(\bar{x})$ ($\varphi(x, \bar{y}) \in \mathscr{L}$, $\bar{a}_n$ a finite tuple in A, represented by $(\bar{a}_n(i))_i \in \prod_i M_i$). We may assume that $I = \mathbb{N}$. For each n, let

$$S(n) = \{j \in I \mid M_j \models \exists x \bigwedge_{i \leq n} \varphi_i(x, \bar{a}_i(j))\}.$$

By assumption, each $S(n)$ is in $\mathscr{U}$, and $S(n)$ contains $S(n+1)$. For $n \in I = \mathbb{N}$, we choose $b_n \in M_n$ in the following fashion: if $n \in S(n)$, take some $b_n \in M_n$ such that $M_n \models \bigwedge_{i \leq n} \varphi_i(b_n, \bar{a}_i(n))$; if $n \notin S(n)$, take for b_n any element of M_n. Then, for each n,

$$\{j \in I \mid M_j \models \varphi_n(b_j, \bar{a}_n(j))\} \supseteq S(n) \cap [n, +\infty),$$

and is therefore in $\mathscr{U}$. Hence, $M^* \models \varphi_n((b_j)_{\mathscr{U}}, \bar{a}_n)$.

Elimination of quantifiers

2.26 Elimination of quantifiers. Formulas with more than two alternations of quantifiers are fairly awkward, and usually difficult to decide the truth of. One therefore tries to "eliminate quantifiers".

Definition. A theory T *eliminates quantifiers* if for any formula $\varphi(\bar{x})$ there is a quantifier-free formula $\psi(\bar{x})$ which is *equivalent* to $\varphi(\bar{x})$ *modulo* T, i.e., is such that

$$T \models \forall \bar{x}(\varphi(\bar{x}) \leftrightarrow \psi(\bar{x})).$$

Note that the set of free variables in φ and ψ are the same. Thus if φ is a sentence, so is ψ. (If the language has no constant symbol, then one allows ψ to be either $\top$ (true) or $\bot$ (false); if the language contains a constant symbol c, then one can use instead the formulas $c = c$ or $c \neq c$.)

Expressed in terms of definable sets, this means: whenever M is a model of T, $S \subset M^{n+1}$ is *quantifier-free definable* (i.e., definable by a formula without quantifiers), and if $\pi : M^{n+1} \to M^n$ is the projection on the first n coordinates, then $\pi(S)$ is also quantifier-free definable.

Expressed in terms of diagrams, this is equivalent to: whenever M is a model of T and $A \subset M$, then $T \cup \text{Diag}^M(A)$ is complete (in the language $\mathscr{L}(A)$).

2.27 Criterion for quantifier elimination: back and forth arguments. *Let T be a theory in a language $\mathscr{L}$, and Δ a set of $\mathscr{L}$-formulas, closed under finite conjunctions and disjunctions. The following are equivalent:*

(1) Every $\mathscr{L}$-formula is equivalent modulo T to a formula of Δ.

(2) Whenever M and N are $\aleph_1$-saturated models of T, $A \subset M$ and $B \subset N$ are countable (non-empty) substructures and $f : A \to B$ is a morphism which preserves the formulas in Δ (i.e., if $\bar{a}$ is a tuple in M, and $\varphi(\bar{x}) \in \Delta$, then $M \models \varphi(\bar{a}) \Rightarrow N \models \varphi(f(\bar{a})))$, then

- *(forth) for any $a \in M$ there is an extension of f with a in its domain and which preserves the formulas in Δ,*
- *(back) for any $b \in M$, there is an extension of f with b in its range and which preserves the formulas in Δ.*

2.28 Preservation theorems.

Let T be a theory in a language $\mathscr{L}$, and Δ a set of formulas in the (free) variables $(x_1, \ldots, x_n)$, closed under finite disjunctions. Let $\Sigma(x_1, \ldots, x_n)$ be a set of formulas in the free variables $(x_1, \ldots, x_n)$, such that every finite fragment of $\Sigma(x_1, \ldots, x_n)$ is satisfiable in a model of T. The following conditions are equivalent:

(1) There is a subset $\Gamma(\bar{x})$ of Δ such that, if $\bar{c} = (c_1, \ldots, c_n)$ are new constant symbols, then

$$T \cup \Gamma(\bar{c}) \models \Sigma(\bar{c}), \qquad T \cup \Sigma(\bar{c}) \models \Gamma(\bar{c}).$$

(2) For all models M and N of T, and n-tuples $\bar{a}$ in M and $\bar{b}$ in N, if $N \models \Sigma(\bar{b})$ and $\bar{a}$ satisfies (in M) all formulas of Δ that are satisfied by $\bar{b}$ (in N), then $M \models \Sigma(\bar{a})$.

Remark. If the set $\Sigma(\bar{x})$ is finite, then so is $\Gamma(\bar{x})$. Hence, taking $\varphi(\bar{x})$ to be the conjunction of the formulas of $\Sigma(\bar{x})$, one obtains that $\varphi(\bar{x})$ is equivalent, modulo T, to a finite conjunction of formulas of Δ.

2.29 These two results allow to prove classical preservation theorems. Here are a few:

(1) A sentence [formula] is preserved under extensions if and only if it is equivalent to an existential sentence [formula].
(2) A sentence [formula] is preserved under substructures if and only if it is equivalent to a universal sentence [formula].
(3) A sentence [formula] is preserved under union of chains if and only it is equivalent to a $\forall\exists$-sentence [formula].
(4) A sentence [formula] is preserved under homomorphisms if and only if it is equivalent to a positive sentence [formula].

Comments. First a word of explanation of what it means for a formula to be preserved. For instance, the formula $\varphi(\bar{x})$ is preserved under union of chains if whenever $\bar{a} \in M_0$, and $(M_i)_{i\in\mathbb{N}}$ is an increasing chain of $\mathscr{L}$-structures such that for each i, $M_i \models \varphi(\bar{a})$, then $\bigcup_{i\in\mathbb{N}} M_i \models \varphi(\bar{a})$.
If in the above definition, one restrict one's attention to models of T, one will obtain equivalences modulo the theory T.

Examples of complete theories, of quantifier elimination

Here are some complete and incomplete theories, together with an axiomatisation.

2.30 Divisible ordered abelian groups. One has the obvious axioms. It is complete and eliminates quantifiers (in $\mathscr{L}_{og}$). Here is a proof that it eliminates quantifiers, using the criterion 2.27.

Let M and N be two ordered abelian divisible groups, which we assume $\aleph_1$-saturated. In particular, their dimension as $\mathbb{Q}$-vector spaces is $\geq \aleph_1$. We assume that $A \subset M$ and $B \subset N$ are countable substructures, and that $f : A \to B$ is an $\mathscr{L}_{og}$-isomorphism. Let $a \in M$. We want to show that there is $b \in N$ such that by setting $f(a) = b$, we define an isomorphism between the ordered groups $\langle A, a \rangle$ and $\langle B, b \rangle$. This will give us the forth direction, and the back direction is symmetric.

Case 1. There is an integer $n > 0$ such that $na \in A$. Take the smallest such n; because N is divisible, there is some $b \in N$ such that $nb = nf(a)$. One verifies easily that setting $f(a) = b$ gives us the desired extension of f. Indeed the elements of $\langle A, a \rangle$ are of the form $c + ma$, where $c \in A$, $0 \leq m < n$, and if $c' + m'a$ is another such element with $m \leq m'$, and $\square$ is one of $=$, $<$, or $>$, we have

$$c + ma \,\square\, c' + m'a \iff nc + mna \,\square\, nc' + m'na.$$

This remark implies easily that we have an $\mathscr{L}_{og}$ isomorphism.

Case 2. Not case 1. Then, as a group, we have $\langle A, a \rangle = A \oplus \langle a \rangle \simeq A \oplus \mathbb{Z}$. First, using case 1, we may assume that A is divisible. Let $C = \{c \in A \mid c < a\}$, and consider the following set of formulas:

$$\Sigma(x) = \{x > f(c) \mid c \in C\} \cup \{x < f(c) \mid c \in A \setminus C\}.$$

This set is finitely consistent, since the ordering on N is dense. As A is countable and N is $\aleph_1$-saturated, there is some $b \in N$ which satisfies all formulas of Σ. We define $f(a) = b$. Then, as $b \notin f(A)$, $f(A)$ is divisible, and N is torsion free, this f defines a group isomorphism $A \oplus \langle a \rangle \to B \oplus \langle b \rangle$. It remains to show that it preserves the ordering: use the same type of argument as in case 1.

2.31   Ordered Z-groups. An *ordered Z-group* is an $\mathscr{L}_{og}$-structure G which is an ordered abelian group, with a (unique) smallest positive element, which we denote by 1; moreover it satisfies that $[G : nG] = n$ for any integer $n > 1$: we use the axiom

$$\forall x \bigvee_{i=0}^{n-1} \exists y \; x = ny - i.$$

Clearly, $\mathbb{Z}$ is a model of these axioms. This theory does not eliminate quantifiers: note that $2\mathbb{Z}$ is also a model of this theory, and the smallest element of $2\mathbb{Z}$ is $2 \neq 1$.

To eliminate quantifiers, one needs to augment the language, first by adding a constant symbol for 1 (the smallest positive element), and binary relation symbols $\equiv_n$ for congruence modulo n. This language is called the *Presburger* language, $\mathscr{L}_{\text{Pres}}$. The $\mathscr{L}_{\text{Pres}}$-theory of $\mathbb{Z}$ is then obtained by adding to the above axioms the following:

$$\forall x \; x > 0 \to x \geq 1,$$
$$\forall x, y \; x \equiv_n y \leftrightarrow \exists z \; x - y = nz,$$

for all $n > 1$.

2.32   Algebraically closed fields. The theory ACF of algebraically closed fields (in the language $\mathcal{L}_R$ of rings) is axiomatised by saying that the structure is a (commutative) field, and for each $n > 1$, by adding the axiom

$$\forall x_0, x_1, \ldots, x_n \, \exists y \, (x_n = 0 \vee \sum_{i=0}^{n} x_n y^n = 0).$$

(Every polynomial of degree $n > 1$ has a root). Note that this theory is not complete. It becomes complete if one specifies the characteristic: ACF_p says $p = 0$; ACF_0 says that $p \neq 0$ for all primes p. The completeness of ACF_0 is also known as the Lefschetz principle. Note that by compactness, if a sentence φ holds in the field $\mathbb{C}$, it will hold in all algebraically closed fields of characteristic p for p sufficiently large.

The theory ACF eliminates quantifiers, this is a classical result of algebraic geometry: quantifier-free definable sets are called *constructible sets* by geometers; a famous theorem states that the projection of a constructible set is constructible.

It can also be easily proved using a back and forth argument.

2.33   Real closed fields. We will first look at real closed fields in the language of rings. The theory RCF of real closed fields (in $\mathcal{L}_R$) is axiomatised by saying that the structure is a (commutative) field; $\forall x \, \exists y \, y^4 = x^2$; for all $n \geq 1$ the axiom $\forall x_0, \ldots, x_{2n+1}, \, (x_{2n+1} = 0 \vee \exists y \, \sum_{i=0}^{2n+1} x_n y^n = 0)$. (Every polynomial of odd degree has a root.) This is a complete theory. Observe that the ordering is definable: an element is positive if and only if it is a non-zero square. However this definition needs a quantifier (existential; or universal: say that $-x$ is not a square), and the $\mathcal{L}_R$-theory of $\mathbb{R}$ does not eliminate quantifiers. For instance, there are two $\mathcal{L}_R$-embeddings of $\mathbb{Q}(\sqrt{2})$ into $\mathbb{R}$, but inside $\mathbb{R}$, the two square roots of 2 do not satisfy the same formulas (since one of them is a square, while the other is not).

However, if one looks at real closed fields in the language $\mathcal{L}_{or}$ of ordered rings, then their theory eliminates quantifiers. This is a consequence of Sturm's algorithm. The $\mathcal{L}_{or}$-theory of real closed fields is obtained by adding to the above axioms the definition of the ordering: $x < y \leftrightarrow \exists z \, (x - y) = z^2 \wedge x \neq y$.

2.34   Algebraically closed valued fields. Let $\mathcal{L}_{\mathrm{div}} = \{+, -, \cdot, 0, 1, |\}$, and view algebraically closed valued fields as $\mathcal{L}_{\mathrm{div}}$-structures. The axiomatisation is the obvious one: the theory ACVF says that the structure is an algebraically closed field, and that $\mid$ is the divisibility relation coming from a valuation.

Theorem. *The theory ACVF eliminates quantifiers in the language $\mathscr{L}_{\mathrm{div}}$. Its completions are obtained by specifying the characteristics of the valued field and of the residue field.*

Going back to the usual 2-sorted language, this means that every formula (of $\mathscr{L}_{\mathrm{div}}$ or even of the 2-sorted language introduced in Example 4 of 1.6 as long as the free variables are all of the valued field sort) is equivalent to a Boolean combination of formulas of the form

$$v(f(\bar{x})) \leq v(g(\bar{x})), \quad h(\bar{x}) = 0,$$

where f, g and h are polynomials over $\mathbb{Z}$. Note that we can work in either language, as we have a direct translation of atomic formulas in one language by quantifier-free formulas of the other language:

$$v(x) \geq v(y) \Longleftrightarrow y \mid x.$$

The proof of quantifier-elimination can be done using a back-and-forth argument, see 2.27. We are given two $\aleph_1$-saturated algebraically closed valued fields M and N, and a valued field isomorphism f between two countable non-empty subrings A and B of M and N respectively. Note that A and B both contain 1, and therefore: they have the same characteristic, and the same residual characteristic (since in a valued field of characteristic 0 with residual characteristic $p > 0$ we have $p \neq 0 \wedge \neg p \mid 1$). We are also given $c \in M$, and wish to extend f to $A[c]$. First note that an $\mathscr{L}_{\mathrm{div}}$-isomorphism between two domains extends uniquely to an isomorphism of their field of fractions which respects the valuation. Furthermore, elementary properties of valuations on fields imply that f extends to an isomorphism of valued fields between the algebraic closures of A and B (in M and N respectively). We may therefore assume that A and B are algebraically closed, and if $c \in A$, there is nothing to do. Let $C = A(c)$. Then the extension C/A is of one of the following type:

a. C/A pure residual,
b. C/A totally ramified,
c. C/A *immediate* (same value group, same residue field).

Extending f in each case follows from general results on valuation theory (in the immediate case, use Kaplansky's results on pseudo-convergent sequences [56]). $\qquad\square$

The original proof of this result by A. Robinson [75] is slightly different, and uses a 2-sorted language. Even though the theory ACVF is not complete, ACVF is decidable. Indeed, let φ be a sentence, we wish to decide whether φ holds in all algebraically closed valued fields. Let $\text{ACVF}_{(0,0)}$ be the completion of ACVF obtained by saying that the residual characteristic is 0. Either φ is false in all (some) algebraically closed fields of residue characteristic 0, and we find a proof of $\neg\varphi$ from $\text{ACVF}_{(0,0)}$; else, we find a proof of φ from $\text{ACVF}_{(0,0)}$; this proof uses only finitely many axioms expressing that the residual characteristic is 0, i.e., for some integer N, if the residual characteristic is $p > N$, then φ is true in all algebraically closed fields of residual characteristic p. It now remains to check if all of the finitely many theories $\text{ACVF}_{(0,p)}$, $\text{ACVF}_{(p,p)}$, $p < N$, prove φ, and if they do, then we can give a positive answer: φ is true in all algebraically close valued fields. (Here $\text{ACVF}_{(0,p)}$, $\text{ACVF}_{(p,p)}$, denote the theory of algebraically closed fields whose residue field is of characteristic p, and which are of characteristic 0, resp. p. And of course, if one of these theories does not prove φ, then it will prove $\neg\varphi$.) This reasoning is of course absolutely non-effective. S.S. Brown [12] has some effective results on bounds on transfer principles for algebraically closed and complete discretely valued fields.

Imaginary elements

2.35 Definition. Let M be an $\mathscr{L}$-structure, let n be an integer, and E an $\emptyset$-definable equivalence relation on M^n. The E-equivalence class of an n-tuple $\bar{a}$, denoted $\bar{a}/E$, will be called an *imaginary element* of M.

To M we associate a structure M^{eq}, in the multi-sorted language $\mathscr{L}^{eq}$ whose set of sorts is indexed by the $\emptyset$-definable equivalence relations on cartesian powers of M. On the *home sort* M, we have the original $\mathscr{L}$-structure, on the new sorts M^n/E no structure other than the one induced by the natural projections $\pi_E : M^n \to M^n/E$ which are also in the language. So our structure is

$$M^{eq} = \big((M, \mathscr{L}), M^n/E, \ldots, \pi_E, \ldots\big).$$

Clearly, each finite cartesian product of sorts is interpretable in the original structure M, and if $T = \text{Th}(M)$, then we obtain a theory T^{eq} in the language $\mathscr{L}^{eq}$. One shows that $(M^{eq})^{eq}$ is definable in M^{eq}, and that if $M \prec N$ then $M^{eq} \prec N^{eq}$.

2.36 Examples

1. This first example is fundamental. Let $\varphi(\bar{x}, \bar{y})$ be an $\mathscr{L}$-formula, $\bar{x}$ an m-tuple of variables, $\bar{y}$ an n-tuple of variables, M an $\mathscr{L}$-structure. Define the equivalence relation E_φ on M^n by

$$E_\varphi(\bar{y}_1, \bar{y}_2) := \forall \bar{x}(\varphi(\bar{x}, \bar{y}_1) \leftrightarrow \varphi(\bar{x}, \bar{y}_2)).$$

This is clearly an equivalence relation, and it associates to the subset of M^m defined by the formula $\varphi(\bar{x}, \bar{a})$ the class $\bar{a}/E$, i.e., a *canonical parameter*, or *code*, for the set $\varphi(M, \bar{a})$. It is sometimes denoted by $\ulcorner \varphi(\bar{x}, \bar{a}) \urcorner$

2. Let M be a structure. Then the n-tuples are imaginary elements: M^n quotiented by the trivial equivalence relation. But also, any n-element subset of M is an imaginary element: consider the subset S of M^n consisting of n-tuples of distinct elements, and quotient by the (action of the) symmetric group on n elements.

3. In general, anything interpretable in a structure will be imaginary. For instance, let G be a group, H a definable subgroup (in any language containing the language of groups). Then any left-coset of H in G will be an imaginary element. I.e., the quotient G/H with an action of G by left translation, lives in G^{eq}.

4. In the particular case of valued fields, we already saw two examples of imaginary elements: note that the two-sorted language we introduced in Example 4 of 1.6 can be obtained from one of the basic languages by adding sorts of M^{eq}; in certain cases, the same will be true of the language of Pas that we will introduce later. There are other imaginaries we didn't add, e.g., the elements $K/\mathscr{O}_\alpha$, where $\mathscr{O}_\alpha$ is the set $\{x \in K \mid v(x) \geq \alpha\}$ (the *closed ball of radius α centered at 0*; also noted $B(0; \geq \alpha)$). There are many other imaginaries, for a description of imaginaries of algebraically closed fields, see below 2.39.5.

2.37 Elimination of imaginaries. Let T be a complete theory in a language $\mathscr{L}$. We say that T *eliminates imaginaries* if whenever M is a model of T, E is a $\emptyset$-definable equivalence relation on M^n, then there is a $\emptyset$-definable function $f : M^n \to M^\ell$ for some $\ell > 0$, such that the fibers of f are exactly the E-equivalence classes.

An equivalent statement is as follows: a theory T eliminates imaginaries if whenever M is a saturated model of T (hence, having many automorphisms), and $D \subseteq M^r$ an M-definable set, there is a finite tuple $\bar{c}$ in M such that for any $\sigma \in \text{Aut}(M)$, $\sigma(D) = D$ if and only if σ fixes the elements of the tuple $\bar{c}$. In other words: if D is defined over $\bar{a}$ and over $\bar{b}$, then it is defined over $\text{dcl}(\bar{a}) \cap \text{dcl}(\bar{b})$. Working in M^{eq} this becomes: $\text{dcl}^{eq}(\ulcorner D \urcorner) = \text{dcl}^{eq}(\bar{c})$.

The theory T *weakly eliminates imaginaries* if given any model M of T and M-definable set D, there is a smallest algebraically closed set $A \subset M$ over which D is defined. In other words: if D is defined over $\bar{a}$ and over $\bar{b}$, then it is defined over $\text{acl}(\bar{a}) \cap \text{acl}(\bar{b})$. Working in M^{eq}, this becomes: $\text{acl}^{eq}(\ulcorner D \urcorner) = \text{acl}^{eq}(\bar{c})$.

Elimination of imaginaries implies weak elimination of imaginaries. This is enough for many applications. The property of (weakly) eliminating imaginaries is preserved under adjunction of constants to the language: if the $\mathscr{L}$-theory T (weakly) eliminates imaginaries, and A is a subset of a model M of T, then so does the $\mathscr{L}(A)$-theory $\text{Diag}^M_{el}(A)$. If one knows that a theory T weakly eliminates imaginaries, then to show that it eliminates imaginaries, it suffices to show that, for all $n, m > 0$, one can code m-element subsets of M^n.

2.38 Galois theory. If a theory T eliminates imaginaries, then, given $A \subset M \models T$, if G is the profinite group consisting of all $\mathscr{L}(A)$-automorphisms of $\text{acl}(A)$ which are elementary in M, then there is a Galois correspondence between closed subgroups of G and definably closed subsets of $\text{acl}(A)$. See e.g. B. Poizat [72].

2.39 Examples

1. Clearly the theory T^{eq} eliminates imaginaries in the language $\mathscr{L}^{eq}$.

2. Consider the theory T of an infinite set, in the empty language $\mathscr{L}$. This theory eliminates quantifiers: any definable set will be defined by a Boolean combination of formulas of the form $x = y$, or $x = a$. In this language, T does not eliminate imaginaries: let M be an infinite set, $a \neq b$ two elements of M, and consider the definable set $\{a, b\}$; consider any permutation σ of M which sends a to b, b to a, and has no fixed point. One can however show that T weakly eliminates imaginaries.

3. If K is a field (maybe with extra structure), then any finite subset of a cartesian power of K has a code. Indeed, let $\bar{a}_i = (a_{i,1}, \ldots, a_{i,n})$, $1 \leq i \leq m$, n-tuples in K. Consider the polynomial $g(\bar{X}) = \prod_{i=1}^{m}(X_0 + \sum_{j=1}^{n} a_{i,j} X_j)$. Then the tuple of coefficients of $g(\bar{X})$ is a code for the finite set $\{\bar{a}_1, \ldots, \bar{a}_m\}$.

4. Many theories of fields eliminate imaginaries:
– the theory of algebraically closed fields of a given characteristic,
– the theory of real closed fields,
– the theory of differentially closed fields of characteristic 0,
– any complete theory of pseudo-finite field, in the language of fields to which one adds enough constant symbols to be able to describe for each $n > 1$ the unique algebraic extension of degree n.
– the theory of separably closed fields of characteristic $p > 0$ and finite degree of imperfection e, in the language of fields to which one adds e new constant symbols, which will be interpreted by the elements of a p-basis.

5. Let T be a complete theory of algebraically closed valued fields, in one of the languages $\mathscr{L}$ introduced before (in 1.6). We already saw examples of imaginaries which did not have real representatives in that language. D. Haskell, E. Hrushovski and H.D. Macpherson describe in [50] a language $\mathscr{L}_{\mathscr{G}}$ in which the natural expansion of T eliminates imaginaries. Let K be a model of T, $\mathcal{O}$ its valuation ring, $\mathcal{M}$ its maximal ideal, and $k = \mathcal{O}/\mathcal{M}$. $\mathscr{L}_{\mathscr{G}}$ is obtained by adding to $\mathscr{L}$ two sets of sorts: for each $n > 0$,
(i) S_n is the set of $\mathcal{O}$-submodules of K^n which are free of rank n. Thus an element of S_n corresponds to the $\mathrm{GL}_n(\mathcal{O})$-orbit of a basis of the K-vector space K^n.
(ii) If $N \in S_n$, define $\mathrm{red}(N) = N/\mathcal{M}N$. Thus $\mathrm{red}(N)$ is isomorphic to k^n. Then T_n is the disjoint union of all $\mathrm{red}(N)$, $N \in S_n$. We add to the language the natural projection $T_n \to S_n$, $(a + \mathcal{M}N) \mapsto N$.

T. Mellor shows in [68] that the theory of real closed valued fields eliminates imaginaries in the language $\mathscr{L}_{\mathscr{G}}$. E. Hrushovski and B. Martin ([55]) show that the field of p-adic numbers eliminates imaginaries in a sublanguage of $\mathscr{L}_{\mathscr{G}}$, and they use this result to show that certain p-adic integrals are rational functions. One should also be able to use the language $\mathscr{L}_{\mathscr{G}}$ to eliminate imaginaries in other valued fields.

3 The results of Ax, Kochen, and Eršov

In this section we will briefly state some early results by Ax and Kochen, and independently by Eršov. These results are the inspiration for the later study of the model theory of valued fields. Recall that a valued field is *Henselian* if it satisfies Hensel's lemma (or equivalently, the valuation has a unique extension to the algebraic closure of the field). The references are [2], [3], [4] and [40], [34]–[39].

3.1 Theorem. *Let $\mathscr{U}$ be any non-principal ultrafilter on the set P of prime numbers. The valued fields $\prod_{p\in P}\mathbb{Q}_p/\mathscr{U}$ and $\prod_{p\in P}\mathbb{F}_p((t))/\mathscr{U}$ are elementarily equivalent.*

In fact, the proof of Ax and Kochen gives more: assuming CH (the continuum hypothesis, which states that the smallest uncountable cardinal $\aleph_1$ is $2^{\aleph_0}$), they prove that these two valued fields are isomorphic. Note that these two fields already have isomorphic residue field ($\prod_{p\in P}\mathbb{F}_p/\mathscr{U}$) and value group ($\mathbb{Z}^P/\mathscr{U}$). Under CH, these fields are furthermore saturated, and the proof uses this fact.

3.2 Consequences of Ax and Kochen. One of the motivations for their study was Artin's conjecture, that the fields $\mathbb{Q}_p$ are C_2, i.e., for every d, a form of degree d in $> d^2$ variables has a non-trivial zero. While the conjecture was later proved to be false (see [78]), their result shows that for every d, there is a number N such that whenever $p > N$, the statement holds for all forms of degree d. Furthermore, they obtain that the theory of all $\mathbb{Q}_p$ is decidable, using results of Ax on the theory of all finite fields [1].

3.3 Other results: the AKE-principle. The AKE-principle is fairly easy to state:
Two Henselian valued fields K and L are elementarily equivalent if and only if their residue fields are elementarily equivalent and their value groups are elementarily equivalent. The henselianity condition (or some additional condition) is necessary: $\mathbb{Q}$ with the p-adic valuation is not elementarily equivalent to $\mathbb{Q}_p$, even though they have isomorphic residue field and valuation group. However, the AKE principle does not always work. Here is a more precise statement of the results of Ax and Kochen [2]–[4], which were also obtained independently by Eršov:

Theorem. *Let K and L be valued fields, with residue fields k_K and k_L respectively, and value groups Γ_K, Γ_L respectively. Assume they satisfy one of the following set of conditions:*

(a) The residue fields of K and L are of characteristic 0.

(b) K and L are of characteristic 0, the residue fields are of characteristic $p > 0$, the value groups have a smallest positive element, and in both fields the value of p is a finite multiple e of this smallest positive element.

Then

(1)

$$K \equiv L \iff k_K \equiv k_L \text{ and } \Gamma_K \equiv \Gamma_L.$$

(2) *If K is a valued subfield of L, then*

$$K \prec L \iff k_K \prec k_L \text{ and } \Gamma_K \prec \Gamma_L.$$

Here K and L are equipped with any of the languages we discussed before, the residue fields are equipped with the ring structure, and the value group with the ordered group structure (in the languages $\mathscr{L}_R$ and $\mathscr{L}_{og}$ respectively).

3.4 Valued fields of positive characteristic. Note that all fields in the above result are of characteristic 0. Results in characteristic $p > 0$ are few, except for the algebraically closed case. An early result was obtained by Y. Eršov and states that the AKE-principle holds for valued fields of positive characteristic which satisfy *Kaplansky's condition A* (see [56] for a definition) and are *defectless* (i.e., if L is a finite extension of K, then L has no proper algebraic immediate extension). There are a few other positive results, see the work of F. Delon [20] and of F.V. Kuhlmann [57]. And undecidability results if one adds for instance a section of the valuation to the language.

When the residue characteristic is positive, but the field is of characteristic 0, the nicest results are for bounded ramification. For unbounded ramification, there are still some elementary equivalence results, see section 4 of [80], and structure results for definable sets in [17].

4 More results on valued fields

In this section, we will introduce the languages of Macintyre and of Denef–Pas. The Macintyre language is a language in which the theory of the field of p-adic numbers $\mathbb{Q}_p$ eliminates quantifiers[5]. This result is instrumental in subsequent proofs of rationality of Poincaré series (see [22]). The Denef–Pas language is a language which is 3-sorted, and in which one obtains relative quantifier-elimination, from which an AKE-principle can be reobtained.

---

[5] Other people gave languages in which $\mathbb{Q}_p$ eliminates quantifiers, e.g. Ax and Kochen [4] and Cohen [19].

Results on the p-adics, the language of Macintyre

4.1   The language of Macintyre. One of the languages in which the field of p-adic numbers eliminates quantifiers is the language of Macintyre, $\mathscr{L}_{\text{Mac}}$, which is obtained by adding to $\mathscr{L}_{\text{div}}$ predicates P_n, $n > 1$, which are interpreted by

$$P_n(x) \leftrightarrow \exists y \; y^n = x \wedge x \neq 0.$$

In fact, the relation $|$ is unnecessary, as it is quantifier-free definable in $\mathbb{Q}_p$: for instance, if $p \neq 2$, we have:

$$v(x) \leq v(y) \iff y = 0 \vee P_2(x^2 + py^2).$$

The definition however depends on p, and for uniformity questions it is better to include $|$ in the language.

4.2   Axioms for the p-adics. The $\mathscr{L}_{\text{div}}$ theory of the valued field $\mathbb{Q}_p$ is axiomatised by expressing the following properties:

K is a Henselian valued field of characteristic 0, with residue field $\mathbb{F}_p$. Its value group is an ordered Z-group, with $v(p)$ the smallest positive element.

4.3   Comments.

- Let K be a subfield of $\mathbb{Q}_p$, relatively algebraically closed in $\mathbb{Q}_p$. Then $K \prec \mathbb{Q}_p$. This follows from quantifier elimination in $\mathscr{L}_{\text{Mac}}$. Thus, the relative algebraic closure of $\mathbb{Q}$ inside $\mathbb{Q}_p$ is an elementary substructure.
- By adding constant symbols, one may obtain a quantifier-elimination result for the theory of a finite algebraic extension of $\mathbb{Q}_p$.
- The elimination is uniform in p, see [64].
- A valued field satisfying the axioms given above is said to be *p-adically closed*.

A detailed study of formally p-adic fields and p-adically closed fields appears in [73].

The language of Denef–Pas

**4.4   The splitting of the proof of the back-and-forth argument into three cases, residual, ramified and immediate, is also apparent in the proofs of the results of Ax and Kochen, and of Eršov. This suggests passing to three sorts: the valued field, the value group, and the residue field, with additional maps the valuation and the residue map. It turns out that for quantifier-elimination results this is not quite enough. One language, which is quite convenient, is the language $\mathscr{L}_{\text{Pas}}$:

- It has three sorts: the valued field, the value group and the residue field.
- The language of the field sort is the language of rings.
- The language of the value group is any language containing the language of ordered abelian groups (and ∞).
- The language of the residue field is any language containing the language of rings.
- In addition, we have a map v from the field sort to the value group (the valuation), and a map $\overline{ac}$ from the field sort to the residue field (*angular component map*).

4.5 Definition. The *angular component map* is a map $\overline{ac} : K \to k_K$ (where k_K is the residue field of K), which is multiplicative, sends 0 to 0, and on the valuation ring $\mathcal{O}_K^\times$ coincides with the residue map. It therefore suffices to know this map on a set of representatives of the value group.

If the valuation map has a *cross-section*, i.e., a map $s : \Gamma_K \to K^\times$ satisfying $s(\gamma + \delta) = s(\gamma)s(\delta)$ and $vs = id_{\Gamma_K}$, then the natural way of defining an angular component map is by setting

$$\overline{ac}(x) = xsv(x^{-1}).$$

In all natural examples, there is a natural coefficient map:

- On the valued field $k((t))$, ($v(t) = 1$, v trivial on k), define $\overline{ac}(0) = 0$, and $\overline{ac}(t) = 1$. Thus, if $a_j \neq 0$ then

$$\overline{ac}\left(\sum_{i \geq j} a_i t^i\right) = a_j.$$

- On $\mathbb{Q}_p$, define $\overline{ac}$ by $\overline{ac}(p) = 1$.

In most cases we do strengthen the language by adding the angular component map. In the case of $\mathbb{Q}_p$ however, $\overline{ac}$ is definable in the valued field $\mathbb{Q}_p$: indeed, $\overline{ac}$ equals 1 on the $(p-1)$-th powers. It therefore suffices to specify the values of $\overline{ac}$ on a system of generators of the finite group $\mathbb{Q}_p^\times / \mathbb{Q}_p^{\times\, p-1}$. A similar result holds for finite algebraic extensions of $\mathbb{Q}_p$.

It is not true that every valued field has an angular component map. However, every valued field K has an *elementary extension* K^* which has an angular component map. This is because in every $\aleph_1$-saturated valued field L, the group morphism $v : L^\times \to \Gamma_L$ has a cross-section, and we know that every valued field has an elementary extension which is $\aleph_1$-saturated. See also [70] for a discussion on the non-definability of angular component maps.

4.6   Theorem. *Let (K, Γ_K, k_K) be an $\mathscr{L}_{\mathrm{Pas}}$-structure, where K is a Henselian valued field, and k_K has characteristic 0. Then every formula $\varphi(x, \xi, \bar{x})$ (x, ξ, $\bar{x}$, tuples of variables of the valued field, valued group, residue field sort) of the language is equivalent to a Boolean combination of formulas*

$$\varphi_1(x) \wedge \varphi_2(v(f(x)), \xi) \wedge \varphi_3(\overline{\mathrm{ac}}(f(x)), \bar{x}),$$

where $f(x)$ is a tuple of elements of $\mathbb{Z}[x]$, φ_1 is a **quantifier-free** *formula of the language of rings, φ_2 is a formula of the language of the group sort, and φ_3 is a formula of the language of the residue field sort.*

4.7   Other results of Pas include a cell decomposition of definable sets. See [69]. Even though the theorem speaks about valued fields of residual characteristic 0, by compactness, it also applies to valued fields $\mathbb{Q}_p$ for p sufficiently large. See also [71].

4.8   Adding angular component maps to $\mathbb{Q}_p$. If one wishes to study the p-adics in a three-sorted language with angular component maps, one is obliged to add angular component maps of higher order, namely, for each n, a multiplicative map $\overline{\mathrm{ac}}_n : K \to \mathbb{Z}/p^n\mathbb{Z}$, which on $\mathbb{Z}_p$ coincides with the usual   mod p^n reduction. We also require that $\overline{\mathrm{ac}}_n = \overline{\mathrm{ac}}_{n+1}$ mod p^n. Note that this requires introducing many countably many new sorts, but that, as we saw above, these maps are interpretable in the valued field $\mathbb{Q}_p$. See [8], [9] for details.

4.9   Application to power series. Consider the natural $\overline{\mathrm{ac}}$ map on $k((t))$, where k is a field of characteristic 0, and look at the $\mathscr{L}_{\mathrm{Pas}}$-structure

$$(K, \mathbb{Z} \cup \{\infty\}, k)$$

where the language of the group sort is the Presburger language $\mathscr{L}_{\mathrm{Pres}}$, see 2.31. Then in the above the formula φ_2 will be a formula without quantifiers.

A definable (with parameters) function $K \to \Gamma_K$ will therefore be locally defined by expressions of the form

$$(\sum_i m_i v(f_i(x)) + \alpha)/N$$

where the f_i's are polynomials over K, $\alpha \in \Gamma_K$, and N is some integer. By locally, I mean that there is a partition of K into definable sets, and on each definable set of the partition, the function is given by an expression as above.

Assume now in addition that the theory of the residue field k eliminates quantifiers in the language of the residue field. Then, in the theorem, the formula φ_3 can also be taken to be quantifier-free. Thus we then obtain full elimination of quantifiers. Useful examples are: $\mathbb{C}$ in the language of rings; $\mathbb{R}$ in the language of ordered rings; and also ... $\mathbb{Q}_p$ in the language $\mathscr{L}_{\mathrm{Mac}}$ of Macintyre. Thus we know a language in which the valued field $\mathbb{Q}_p((t))$ eliminates quantifiers.

Further reading

The (model) theory of valued fields is extremely rich, and has grown in several directions. We will here indicate some of the existing literature. Omissions are most of the time due to the writer's ignorance.

Quantifier-elimination. Many people worked on quantifier elimination for valued fields, sometimes with a cross-section for the valuation map, to cite a few: Ax and Kochen [4], Ziegler [81], Basarab [7], Delon [20], Weispfenning [80]. The paper of Weispfenning contains a very good bibliography.

Analytic structures. Complete valued fields can be endowed with an analytic structure, and several model theorists studied these enriched valued fields. On the field of p-adic numbers this was done first by Denef and Van den Dries [24]. On other fields, one of the earliest papers is by L. Lipshitz, and to-date, the most complete treatment is probably the one by R. Cluckers and L. Lipshitz [15], which in particular encompasses earlier results by Lipshitz, Robinson, Schoutens, ...; the paper contains an excellent bibliography. J. Denef gives in [23] an excellent survey of results obtained using quantifier-elimination. (One should be aware that later on, a mistake was discovered in the quantifier-elimination result of Gardener and Schoutens; see [63]).

Valued differential fields. Valued differential fields occur naturally in analysis. Work on differential valued fields was done by N. Guzy, F. Point and C. Rivière, see [42] – [47], and also by Bélair [10], Scanlon [76].

Valued difference fields. Classical examples of difference valued fields are the maximal unramified extension $\mathbb{Q}_p^{unr}$ of $\mathbb{Q}_p$ or its completion $W(\mathbb{F}_p^{alg})$, with a lifting of the Frobenius automorphism on the residue field. The theory of these difference fields was studied by Bélair, Macintyre and Scanlon [11], who prove a relative quantifier-elimination result for $W(\mathbb{F}_p^{alg})$, as well as an AKE-principle. Azgin and Van den Dries [5] improve slightly their result. In that connection we should also mention earlier work by Van den Dries [29] on $W(\mathbb{F}_p^{alg})$ with a predicate for the set of Teichmüller representatives. Also, Scanlon studies the model theory of D-valued fields (here D is an operator, which on the valued field originates from an automorphism via $\sigma(x) = eDx + x$ for some e in the valuation ring, and which on the residue field defines either a derivation or again yields an automorphism), and obtains AKE-type results, see [76]. Let me also mention an unpublished result of Hrushovski [52] on (algebraically closed) valued fields with an automorphism which is ω-increasing, such as a "non-standard Frobenius".

Various notions of minimality. Searching to generalize the properties of strong minimality (of algebraically closed fields) and of o-minimality (of real closed fields), several notions of minimality were studied. First by Macpherson and Steinhorn [67], then by other people, see e.g. [49], [17], [53]. See also the paper by Delon in this volume for more details.

Main omissions. I have not at all spoken about some of the main developments of the model theory of valued fields, which are taking place at this very moment and are in constant progress. Some of these developments started with the work of Denef and Loeser on motivic integration [25], or, should I say already with the work of Denef on the rationality of the Poincaré series [22]? This initial work was followed by many others, by Denef and Loeser, then joined by Cluckers, Hrushovski, Kazhdan, … . I should also mention on-going work around NIP, metastable theories, etc., which has already given important results (e.g., the space of types of Hrushovski and Loeser [54]). Other people are better qualified to talk about them.

References

[1] J. Ax, The elementary theory of finite fields, *Annals of Math.* **88** (1968), 239–271.

[2] J. Ax, S. Kochen, Diophantine problems over local fields. I, *Amer. J. Math.* **87** 1965 605–630.

[3] J. Ax, S. Kochen, Diophantine problems over local fields II. A complete set of axioms for p-adic number theory, *Amer. J. Math.* **87** 1965 631–648.

[4] J. Ax, S. Kochen, Diophantine problems over local fields III. Decidable fields, *Ann. of Math. (2)* **83** 1966 437–456.

[5] S. Azgin, L. van den Dries, Equivalence of valued fields with value preserving automorphism, to appear in *J. of Institute of Mathematics of Jussieu.*

[6] S. Basarab, Some model theory for Henselian valued fields, *J. of Algebra* **55**, 191–212.

[7] S. Basarab, A model theoretic tranfer theorem for Henselian valued fields, *J. reine angew. Math.* **311/312**, 1–30.

[8] S.A. Basarab, Relative elimination of quantifiers for Henselian valued fields, *Ann. of Pure and Appl. Logic* **53** (1991), 51–74.

[9] L. Bélair, Types dans les corps valués munis d'applications coefficients (French), *Illinois J. Math.* **43** (1999), no. 2, 410–425.

[10] L. Bélair, Approximation for Frobenius algebraic equations in Witt vectors, *J. Algebra* **321** (2009), no. 9, 2353–2364.

[11] L. Bélair, A.J. Macintyre, T. Scanlon, Model theory of the Frobenius on the Witt vectors, *Amer. J. of Math.* **129** (2007), 665–721.

[12] S. S. Brown, *Bounds on transfer principles for algebraically and complete discretely valued fields*, Memoirs of the AMS, Vol **15** Nr 104, July 1978.

[13] R. Cluckers, Analytic p-adic cell decomposition and integrals, *Trans. Amer. Math. Soc.* **356** (2004), No. 4, 1489–1499.

[14] R. Cluckers, G. Comte, F. Loeser, Lipschitz continuity properties for p-adic semi-algebraic and subanalytic functions, *Geom. Funct. Anal.* **20**, 68–87 (2010).

[15] R. Cluckers, L. Lipshitz, Fields with analytic structure, to appear in *Journal of the European Mathematical Society.* Available at arXiv:0908.2376.

[16] R. Cluckers, F. Loeser, Constructible exponential functions, motivic Fourier transform and transfer principle, *Annals of Mathematics* **171**, 1011–1065 (2010).

[17] R. Cluckers, F. Loeser, b-minimality, *J. Math. Log.* **7** (2007), no. 2, 195–227.

[18] R. Cluckers, L. Lipshitz, Z. Robinson, Analytic cell decomposition and analytic motivic integration, *Ann. Sci. École Norm. Sup. (4)* **39** (2006), no. 4, 535–568.

[19] P.J. Cohen, Decision procedures for real and p-adic fields, *Comm. on Pure and Applied Math.* **XXII** (1969), 131–151.

[20] F. Delon, *Quelques propriétés des corps valués en théorie des modèles*, Thèse de Doctorat d'Etat, Université Paris VII, Paris, 1982.

[21] F. Delon, Hensel fields in equal characteristic $p > 0$, in: *Model Theory of Algebra and Arithmetic, Proc. Karpacz 1979*, Lecture Notes in Mathematics Nr **834**, Springer-Verlag Berlin Heidelberg 1980.

[22] J. Denef, The rationality of the Poincaré series associated to the p-adic points on a variety, *Inventiones Math.* **77** (1984), 1–23.

[23] J. Denef, Arithmetic and geometric applications of quantifier elimination for valued fields, in: *Model theory, algebra, and geometry*, 173–198, Math. Sci. Res. Inst. Publ., **39**, Cambridge Univ. Press, Cambridge, 2000.

[24] J. Denef, L. van den Dries, p-adic and real subanalytic sets, *Annals of Math.*, **128** (1988), 79–138.

[25] J. Denef, F. Loeser, Geometry on arc spaces of algebraic varieties, in: *Proceedings of 3rd European Congress of Mathematics, Barcelona*, Progress in Mathematics **201**, 327–348 (2001), Birkhäuser.

[26] J. Denef, F. Loeser, Definable sets, motives and p-adic integrals, *Journal of the Amer. Math. Soc.* **14**, 429–469 (2001).

[27] J. Denef, F. Loeser, Motivic integration and the Grothendieck group of pseudo-finite fields, in: *Proceedings of the International Congress of Mathematicians, Vol. II (Beijing, 2002)*, 13–23, Higher Ed. Press, Beijing, 2002.

[28] Jan Denef, Hans Schoutens, On the decidability of the existential theory of $\mathbb{F}_p[\![t]\!]$, in: *Valuation theory and its applications, Vol. II (Saskatoon, SK, 1999)*, 43–60, Fields Inst. Commun., **33**, Amer. Math. Soc., Providence, RI, 2003.

[29] L. van den Dries, On the elementary theory of rings of Witt vectors with a multiplicative set of representatives for the residue field, *Manuscripta Math.* **98**, 133–137 (1999).

[30] L. van den Dries, D. Haskell, H.D. Macpherson, One-dimensional p-adic subanalytic sets, *J. London Math. Soc.* (2) **59** (1999), 1–20.

[31] Lou van den Dries, Franz-Viktor Kuhlmann, Images of additive polynomials in $\mathbb{F}_q((t))$ have the optimal approximation property, *Canad. Math. Bull.* **45** (2002), no. 1, 71–79.

[32] Lou van den Dries, Philip Scowcroft, On the structure of semialgebraic sets over p-adic fields, *J. Symbolic Logic* **53** (1988), no. 4, 1138–1164.

[33] Marcus P.F. du Sautoy, Finitely generated groups, p-adic analytic groups and Poincaré series, *Ann. of Math.* (2) **137** (1993), no. 3, 639–670.

[34] Ju. L. Eršov, On elementary theories of local fields, (Russian) *Algebra i Logika Sem.* **4** 1965 no. 2, 5–30.

[35] Ju. L. Eršov, On elementary theory of maximal normalized fields, (Russian) *Algebra i Logika Sem.* **4** 1965 no. 3, 31–70.

[36] Ju. L. Eršov, On the elementary theory of maximal normed fields, II, (Russian) *Algebra i Logika Sem.* **4** 1965 no. 6, 47–48.

[37] Ju. L. Eršov, On the elementary theory of maximal normed fields, II, (Russian) *Algebra i Logika Sem.* **5** 1966 no. 1, 5–40.

[38] Ju. L. Eršov, On the elementary theory of maximal normed fields, III, (Russian. English summary) *Algebra i Logika Sem.* **6** 1967 no. 3, 31–38.

[39] Ju. L. Eršov, Rational points over Hensel fields, (Russian. English summary) *Algebra i Logika Sem.* **6** 1967 no. 3, 39–49.

[40] Ju. L. Eršov, On the elementary theory of maximal normed fields, (Russian) *Dokl. Akad. Nauk SSSR* **165** 1965 21–23.

[41] F.J. Grunewald, D. Segal, G.C. Smith, Subgroups of finite index in nilpotent groups, *Invent. Math.* **93** (1988), no. 1, 185–223.

[42] N. Guzy, Note sur les corps différentiellement clos valués, *C. R. Math. Acad. Sci. Paris* **341** (2005), no. 10, 593–596.

[43] N. Guzy, 0-*D*-valued fields, *J. Symbolic Logic* **71** (2006), no. 2, 639–660.

[44] N. Guzy, Quelques remarques sur les corps *D*-valués, *C. R. Math. Acad. Sci. Paris* **343** (2006), no. 11-12, 689–694

[45] N. Guzy, Valued fields with *K* commuting derivations, *Comm. Algebra* **34** (2006), no. 12, 4269–4289.

[46] N. Guzy, F. Point, Topological differential fields, *Ann. of Pure and Appl. Logic* **161** (2010) Nr 4, 570–598.

[47] N. Guzy, C. Rivière, Geometrical axiomatization for model complete theories of differential topological fields, *Notre Dame J. Formal Logic* **47** (2006), no. 3, 331–341.

[48] D. Haskell, H.D. Macpherson, Cell decompositions of C-minimal structures, *Ann. Pure and Appl. Logic* **66** (1994), 113–162.

[49] D. Haskell, H.D. Macpherson, A version of o-minimality for the *p*-adics, *J. of Symb. Logic* **62**, No. 4 (1997), 1075–1092.

[50] D. Haskell, E. Hrushovski, H.D. Macpherson, Definable sets in algebraically closed valued fields: elimination of imaginaries, *J. reine angew. Math.* **597** (2006), 175–236.

[51] D. Haskell, E. Hrushovski, H.D. Macpherson, *Stable domination and independence in algebraically closed valued fields*, Lecture notes in Logic, Cambridge U. Press, Cambridge 2008.

[52] E. Hrushovski, private communication, 2001.

[53] E. Hrushovski, D. Kazhdan, Integration in valued fields, in *Algebraic geometry and number theory*, 261–405, Progr. Math., **253**, Birkhäuser Boston, Boston, MA, 2006.

[54] E. Hrushovski, F. Loeser, Non archimedean tame topology and stably dominated types, preprint 2010, arXiv:1009.0252.

[55] E. Hrushovski, B. Martin, Zeta functions from definable equivalence relations, available at `arxiv:0701011`.

[56] I. Kaplansky, Maximal fields with valuations, *Duke Math. J.* **9**, (1942), 303–321.

[57] F.-V. Kuhlmann, Quantifier elimination for henselian fields relative to additive and multiplicative congruences, *Israel Journal of Mathematics* **85** (1994), 277–306.

[58] Franz-Viktor Kuhlmann, Elementary properties of power series fields over finite fields, *J. Symbolic Logic* **66** (2001), no. 2, 771–791.

[59] Franz-Viktor Kuhlmann, Salma Kuhlmann, Saharon Shelah, Exponentiation in power series fields, *Proc. Amer. Math. Soc.* **125** (1997), no. 11, 3177–3183.

[60] L. Lipshitz, Rigid subanalytic sets, *Amer. J. Math.* **115** (1993), 77–108.

[61] L. Lipshitz, Z. Robinson, One-dimensional fibers of rigid subanalytic sets, *J. Symb. Logic* **63** (1998), 83–88.

[62] L. Lipshitz, Z. Robinson, *Rings of separated power series and quasi-affinoid geometry*, Astérisque No. **264** (2000).

[63] L. Lipshitz, Z. Robinson, Flattening and analytic continuation of affinoid morphisms. Remarks on a paper of T. S. Gardener and H. Schoutens: "Flattening and subanalytic sets in rigid analytic geometry", *Proc. London Math. Soc. (3)* **83** (2001), no. 3, 681–707.

[64] A. Macintyre, On definable subsets of p-adic fields, *J. Symbolic Logic* **41** (1976), no. 3, 605–610.

[65] A. Macintyre, Twenty years of p-adic model theory, in: *Logic colloquium '84 (Manchester, 1984)*, 121–153, Stud. Logic Found. Math., **120**, North-Holland, Amsterdam, 1986.

[66] A. Macintyre, Rationality of p-adic Poincaré series: uniformity in p, *Ann. of Pure and Applied Logic* **49** (1990), 31–74.

[67] H.D. Macpherson, C. Steinhorn, On variants of o-minimality, *Ann. Pure and Appl. Logic* **79** (1996), 165–209.

[68] T. Mellor, Imaginaries in real closed valued fields, *Ann. Pure Appl. Logic* **139** (2006), no. 1–3, 230–279.

[69] J. Pas, Uniform p-adic cell decomposition and local zeta functions, *J. reine angew. Math.* **399** (1989), 137–172.

[70] J. Pas, On the angular component modulo p, *J. of Symb. Logic* **55** (1990), 1125–1129.

[71] Johan Pas, Cell decomposition and local zeta functions in a tower of unramified extensions of a p-adic field, Proc. London Math. Soc. (3) 60 (1990), no. 1, 37–67.

[72] B. Poizat, Une théorie de Galois imaginaire, *J. of Symbolic Logic* **48** (1983), no. 4, 1151–1170 (1984).

[73] A. Prestel and P. Roquette, *Formally p-adic Fields*, Lecture Notes in Mathematics 1050, Springer-Verlag Berlin Heidelberg, 1984.

[74] A. Prestel, M. Ziegler, Model-theoretic methods in the theory of topological fields, *J. reine angew. Math.* **299/300** (1978), 318–341.

[75] A. Robinson, *Complete theories*, North-Holland, Amsterdam 1956.

[76] T. Scanlon, A model complete theory of valued D-fields, *Journal of Symbolic Logic*, **65** , no. 4, December 2000, 1758–1784.

[77] P. Simonetta, On non-abelian C-minimal groups, *Ann. Pure and Appl. Logic* **122** (2003), 263–287.

[78] G. Terjanian, Un contre-exemple à une conjecture d'Artin, *C.R. Acad. Sci. Paris, Série A*, **262** (Mars 1966), p.612.

[79] V. Weispfenning, On the elementary theory of Hensel fields, *Annals of Mathematical Logic* **10**, Issue 1, (1976), 59–93.

[80] V. Weispfenning, Quantifier elimination and decision procedures for valued fields, in: *Logic Colloquium Aachen 1983*, Lecture Notes in Mathematics, **1103**, Springer (1984) pp. 419–472.

[81] M. Ziegler, *Die elementare Theorie henselscher Körper*, Dissertation, Köln, 1972.

3

On the definition of rigid analytic spaces

Siegfried Bosch

In his seminar notes [5] on *Rigid Analytic Spaces*, Tate developed a powerful analogue of classical complex analysis over complete non-Archimedean fields. The toplogy of such a field K is quite weak, as it is totally disconnected. For example, any disk in K may be viewed as a disjoint union of smaller disks. Therefore a local definition of analyticity cannot yield useful global results of the type we know them from complex analysis. For example, this concerns global identity theorems and also the fact that analytic functions on classical projective spaces are constant.

Following the method from complex analysis and looking at functions admitting local power series expansions, we end up with so-called *locally analytic spaces*, or *wobbly analytic spaces* in the terminology of [5]. Since these are lacking connectivity in the non-Archimedean case, Tate equipped them with an additional structure, a so-called *h-structure*, which provides a certain substitute for connectivity. All this is guided by the idea that polydisks in affine n-spaces should be viewed as being "connected" and that their analytic functions should admit globally convergent power series expansions.

Having such an idea in mind, there is another theory, which provides useful guidance, namely, the theory of Grothendieck's schemes (of locally finite type over K). Indeed, replace the polynomial ring $K[\zeta]$, where $\zeta = (\zeta_1, \ldots, \zeta_n)$ is a system of n variables, by the *Tate algebra*

$$T_n = K\langle \zeta_1, \ldots, \zeta_n \rangle = \left\{ \sum_{\nu \in \mathbb{N}^n} a_\nu \zeta^\nu \; ; \; \lim_{|\nu| \to \infty} |a_\nu| = 0 \right\}$$

Motivic Integration and its Interactions with Model Theory and Non-Archimedean Geometry (Volume I), ed. Raf Cluckers, Johannes Nicaise, and Julien Sebag. Published by Cambridge University Press. © Cambridge University Press 2011.

of *restricted power series* in $\zeta_1, \ldots, \zeta_n$ over K; the latter consists of all formal power series in $\zeta_1, \ldots, \zeta_n$ over K that converge on the "closed" unit polydisk around 0 in K^n. Taking quotients of Tate algebras by ideals, we arrive at the category of so-called *affinoid algebras* (i. e., algebras of topologically finite type over K in the terminology of [5]); its counterpart in algebraic geometry is the category of algebras of finite type over K. Passing to opposite categories, the category of *affinoid spaces* is defined as the opposite of the category of affinoid algebras and thereby may be seen as an analogue of the category of affine schemes of finite type over K. However, in order that such a concept gets a geometric meaning, a few things have to be settled. For any affinoid algebra A, its corresponding affinoid space, denoted by $\operatorname{Sp} A$, has to be interpreted in terms of an underlying point set. The spectrum of maximal ideals in A is a good choice for this, while the whole prime spectrum would be too big for certain reasons. A crucial step is then to approach the topology of $\operatorname{Sp} A$ by specifying certain basic open subsets $U \subset \operatorname{Sp} A$ that can be viewed as affinoid spaces themselves, say $U = \operatorname{Sp} A_U$. This is done in such a way that the inclusion $\operatorname{Sp} A_U \hookrightarrow \operatorname{Sp} A$ is given by a homomorphism of affinoid algebras $A \longrightarrow A_U$, which can be viewed as a completed localization morphism, at least in the case of so-called *Weierstraß, Laurent,* or *rational* domains, as we will define them in 1.4 and 1.5. The topology generated by such basic open subsets $U \subset \operatorname{Sp} A$ is the *canonical topology,* the same as the one used in the classical complex case, and the functor $U \longmapsto A_U$ from basic open subsets to associated affinoid algebras is a presheaf $\mathcal{O}_{\operatorname{Sp} A}$ inducing as associated sheaf the sheaf $\mathcal{O}_{\operatorname{Sp} A}^w$ of *wobbly analytic functions* on $\operatorname{Sp} A$.

As we have explained above, the latter sheaf is much too big for our purposes, since the canonical morphism $A \longrightarrow \Gamma(\operatorname{Sp} A, \mathcal{O}_{\operatorname{Sp} A}^w)$, although injective, will be far from being surjective, except for very special cases. The key to resolving this problem consists in Tate's Acyclicity Theorem [5], 8.2. In simplified terms, it says that the presheaf $\mathcal{O}_{\operatorname{Sp} A}$ behaves like a *sheaf* if we restrict ourselves to testing the sheaf conditions for *finite* coverings only, namely for coverings of type $U = \bigcup_{i=1}^{n} U_i$, where U and the U_i are basic open subsets in $\operatorname{Sp} A$. In algebraic geometry the analogous theorem is needed to construct the structure sheaf on the affine scheme $\operatorname{Spec} A$ associated to a ring A. Here basic open subsets $U \subset \operatorname{Spec} A$ are of type $D(f) = \{x \in \operatorname{Spec} A \,; f(x) \neq 0\}$ for some $f \in A$, and $\mathcal{O}_{\operatorname{Spec} A}(D(f))$ is defined as the localization of A by the multiplicative system generated by f in A.

It is such a restriction to *finite* coverings, reflected in the add-on *rigid*, which puts Tate's Acyclicity Theorem at the center of the theory of rigid analytic spaces. This way the presheaf $\mathcal{O}_{\mathrm{Sp}\,A}$ is ennobled to become the *structure sheaf* of the affinoid space $\mathrm{Sp}\,A$ and, moreover, a meaningful context of connectedness becomes available for rigid analytic spaces. But, most important, instead of being merely objects in some abstract category, affinoid spaces are equipped with a geometric structure such that global objects can be constructed by glueing local affinoid ones. As a technical tool for handling the finiteness condition on coverings, it is common practice today to use Grothendieck topologies [1]. But the same effect can be obtained by resorting to *h*-structures, as introduced by Tate. We will explain both approaches in this paper and show in Theorem 4.2 that they lead to the same result. In particular, this answers a question put forward by Tate himself, when he was looking back at the definition of rigid analytic spaces given by him in [5].

Since we cannot develop all details of the theory at this place, we will use the lectures [2] as a convenient reference source for some standard facts. Of course, we could have referred just as well to Tate's notes [5] or to the monography [3].

1 Affinoid spaces

Let K be a field with a complete non-Archimedean absolute value $|\cdot|$, which is supposed to be non-trivial. We will keep K fixed throughout this paper.

Definition 1.1 *Any K-algebra of type $K\langle \zeta_1, \ldots, \zeta_n \rangle/\mathfrak{a}$, where $\mathfrak{a}$ is an ideal of the Tate algebra in some variables $\zeta_1, \ldots, \zeta_n$ over K, is called an* affinoid algebra.

Affinoid algebras have many properties in common with K-algebras of finite type. For example, they are Noetherian, and for any maximal ideal $\mathfrak{m}$ in an affinoid algebra A the quotient $A/\mathfrak{m}$ is finite algebraic over K. This follows from the fact that Tate algebras satisfy Noether Normalization and are Noetherian themselves; see [2], 1.2/10 and 1.2/13.

On first sight it is not so important to know that affinoid algebras are, in fact, topological algebras. Namely, on a Tate algebra $K\langle \zeta_1, \ldots, \zeta_n \rangle$ we have the so-called *Gauß norm* given by

$$|f| = \max_\nu |a_\nu| \qquad \text{for} \qquad f = \sum_\nu a_\nu \zeta^\nu \in K\langle \zeta_1, \ldots, \zeta_n \rangle,$$

which makes $K\langle\zeta_1,\ldots,\zeta_n\rangle$ a Banach K-algebra. Furthermore, there is a *residue norm* on any quotient $A = K\langle\zeta_1,\ldots,\zeta_n\rangle/\mathfrak{a}$ by some ideal $\mathfrak{a} \subset K\langle\zeta_1,\ldots,\zeta_n\rangle$. For $g \in A$ its residue norm $|g|$ is given by the infimum of all values $|f|$, where $f \in K\langle\zeta_1,\ldots,\zeta_n\rangle$ varies over all representatives of g. The residue norm yields a structure of Banach K-algebra on A and it turns out that all K-algebra morphisms $A \longrightarrow B$ between affinoid algebras are automatically continuous, regardless of the chosen residue norms; see [2], 1.4/19. In particular, affinoid algebras are canonically equipped with a structure of a topological K-algebra, and there is no difference in defining a morphism of affinoid algebras as just a morphism of K-algebras or as a continuous one. Keeping this in mind, let us give a very provisional definition of affinoid spaces.

Definition 1.2   *The category of* affinoid spaces *is the opposite of the category of affinoid algebras.*

We will write $\operatorname{Sp} A$ for the affinoid space corresponding to an affinoid algebra A and even, in a redundant way, $\operatorname{Sp} A = (\operatorname{Max} A, A)$, where $\operatorname{Max} A$ is the spectrum of maximal ideals of A. Likewise, we denote by $\operatorname{Sp}\varphi \colon \operatorname{Sp} A \longrightarrow \operatorname{Sp} B$ the morphism corresponding to a morphism of affinoid algebras $\varphi \colon B \longrightarrow A$. Also here we can interpret $\operatorname{Sp}\varphi$ as a pair $({}^a\varphi, \varphi)$, where ${}^a\varphi \colon \operatorname{Max} A \longrightarrow \operatorname{Max} B$ is given by $\mathfrak{m} \longmapsto \varphi^{-1}(\mathfrak{m})$. That $\varphi^{-1}(\mathfrak{m})$ is maximal in B, when $\mathfrak{m}$ is maximal in A is easily seen. Indeed, if $\mathfrak{m} \subset A$ is maximal, the field $A/\mathfrak{m}$ is finite over K, see [2], 1.4/3, and the inclusions $K \hookrightarrow B/\varphi^{-1}(\mathfrak{m}) \hookrightarrow A/\mathfrak{m}$ show that $B/\varphi^{-1}(\mathfrak{m})$ is a field as well so that $\varphi^{-1}(\mathfrak{m})$ must be maximal in B. In the following we will simplify our terminology by using $\operatorname{Sp} A$ also in the sense of $\operatorname{Max} A$, for any affinoid algebra A. In particular, we are able then to talk about "subsets" of an affinoid space $X = \operatorname{Sp} A$. Of particular interest are certain subsets, which will play the role of basic open subsets in X.

Definition 1.3 (Tate [5], 7.1)   *Let $X = \operatorname{Sp} A$ be an affinoid space. A subset $U \subset X$ is called an* affinoid subdomain *of X if there exists a morphism of affinoid spaces $\iota \colon X' \longrightarrow X$ with $\iota(X') \subset U$ such that the following universal property holds:*
   Every morphism of affinoid spaces $Y \longrightarrow X$ with image contained in U admits a unique factorization through $\iota \colon X' \longrightarrow X$ via a morphism of affinoid spaces $Y \longrightarrow X'$.

Of course, if $U \subset X$ is an affinoid subdomain of X, then the corresponding morphism $\iota \colon X' \longrightarrow X$, as required in the definition, is uniquely determined by U. Furthermore, it is not too hard to show that, pointwise, ι induces a bijection from X' onto U; see [2], 1.6/10. Thus, any affinoid subdomain $U \subset X$ is automatically equipped with a unique structure of an affinoid space. In order to exhibit explicit examples of affinoid subdomains, let us specify some classes of subsets of affinoid spaces. To do this, note that for any point x of an affinoid space $X = \mathrm{Sp}\, A$ and any element $f \in A$ the residue class $f(x) \in A/\mathfrak{m}$, where $\mathfrak{m} \subset A$ is the maximal ideal designated by x, has a well-defined value $|f(x)|$. Namely, the field $A/\mathfrak{m}$ is finite algebraic over K by [2], 1.4/3, and, thus, admits a unique extension of the absolute value from K to $A/\mathfrak{m}$.

Definition 1.4 *Let $X = \mathrm{Sp}\, A$ be an affinoid space.*
(i) *A subset in X of type*

$$X(f_1, \ldots, f_r) = \{x \in X \,;\, |f_1(x)|, \ldots, |f_r(x)| \leq 1\}$$

for functions $f_1, \ldots, f_r \in A$ is called a Weierstraß domain *in X.*
(ii) *A subset in X of type*

$$X(f_1, \ldots, f_r, g_1^{-1}, \ldots, g_s^{-1})$$
$$= \{x \in X \,;\, |f_1(x)|, \ldots, |f_r(x)| \leq 1, |g_1(x)|, \ldots, |g_s(x)| \geq 1\}$$

for functions $f_1, \ldots, f_r, g_1, \ldots, g_s \in A$ is called a Laurent domain *in X.*
(iii) *A subset in X of type*

$$X\left(\frac{f_1}{f_0}, \ldots, \frac{f_r}{f_0}\right) = \{x \in X \,;\, |f_1(x)|, \ldots, |f_r(x)| \leq |f_0(x)|\}$$

for functions $f_0, \ldots, f_r \in A$ without common zeros is called a rational domain *in X.*

Proposition 1.5 *Weierstraß, Laurent, and rational domains in an affinoid space X are examples of affinoid subdomains.*

For a proof, see [2], 1.6/11. Let us just mention that the affinoid algebra corresponding to a Weierstraß domain $X(f_1, \ldots, f_r) \subset X$ is given by

$$A\langle f_1, \ldots, f_r \rangle = A\langle \zeta_1, \ldots, \zeta_r \rangle / (\zeta_i - f_i \,;\, i = 1, \ldots, r),$$

for a Laurent domain $X(f_1, \ldots, f_r, g_1^{-1}, \ldots, g_s^{-1}) \subset X$ by

$$A\langle f_1, \ldots, f_r, g_1^{-1}, \ldots, g_s^{-1} \rangle$$
$$= A\langle \zeta_1, \ldots, \zeta_r, \xi_1, \ldots, \xi_s \rangle / (\zeta_i - f_i, 1 - g_j \xi_j \,;\, i = 1, \ldots, r; j = 1, \ldots, s),$$

and for a rational domain $X(\frac{f_1}{f_0}, \ldots, \frac{f_r}{f_0}) \subset X$ by

$$A\left\langle \frac{f_1}{f_0}, \ldots, \frac{f_r}{f_0} \right\rangle = A\langle \zeta_1, \ldots, \zeta_r \rangle / (f_i - f_0 \zeta_i \, ; \, i = 1, \ldots, r).$$

Here the ζ_i and ξ_j are variables, and

$$A\langle \zeta_1, \ldots, \zeta_r \rangle, \qquad A\langle \zeta_1, \ldots, \zeta_r, \xi_1, \ldots, \xi_s \rangle$$

are meant as algebras of restricted power series with coefficients in A, where A is equipped with any possible residue norm.

It is clear that any Weierstraß domain in X is also Laurent. Furthermore, one can show that Laurent domains in X are rational. Namely they can be viewed as finite intersections of rational domains, and any such intersection is rational again; see [2], 1.6/14. Rational domains are not yet mentioned in Tate's notes [5]. However, they appear quite naturally. If we consider a Laurent domain $X' \subset X$ and a Weierstraß domain $X'' \subset X'$, then, in general, X'' will be neither Weierstraß nor Laurent in X. However, we can see that X'' is rational in X. Furthermore, one knows that any rational domain in X occurs in this way; see [2], 1.6/16 and 1.6/17.

Now let us consider on affinoid spaces $X = \operatorname{Sp} A$ their so-called *canonical topology*, which is obtained by taking the Weierstraß domains in X as a basis. One can show that this topology is the literal counterpart of the one considered on classical analytic spaces. Indeed, let $\overline{K}$ be an algebraic closure of K and provide it with the (unique) extension of the absolute value on K. Then the unit polydisk in $\overline{K}^n$ becomes a topological space and the same applies to its Zariski closed subspaces $V(\mathfrak{a})$, for ideals $\mathfrak{a} \subset K\langle \zeta_1, \ldots, \zeta_r \rangle$. Identifying the spectrum of maximal ideals $\operatorname{Max} A$, where $A = K\langle \zeta_1, \ldots, \zeta_r \rangle / \mathfrak{a}$, with the quotient $V(\mathfrak{a})/\Gamma$ by the automorphism group $\Gamma = \operatorname{Aut}(\overline{K}/K)$, see [3], 7.1.1/1, we may consider on the point set of $\operatorname{Sp} A$ the quotient topology of the one on $V(\mathfrak{a})$. The resulting topology coincides with the *canonical topology* on $\operatorname{Sp} A$; see [2], 1.6/2. Furthermore, one knows ([2], 1.6/19):

Proposition 1.6 *Let U be an affinoid subdomain of an affinoid space X. Then U is open in X and the canonical topology of X restricts to the one of U.*

More precise information on the structure of general affinoid subdomains is provided by the following result ([2], 1.8/12):

Theorem 1.7 (Gerritzen - Grauert [4])   *Let X be an affinoid space and $U \subset X$ an affinoid subdomain. Then U is a finite union of rational subdomains of X.*

The proof of this theorem in [4] or [2] provides quite precise information on the functions needed to describe the rational subdomains covering a given affinoid subdomain $U \subset X$. Surprisingly, a more rapid, but less specific proof was recently given by Temkin [6] from the viewpoint of Berkovich theory.

Using the fact that affinoid subdomains carry a well-defined structure of affinoid space, we are able now to introduce the sheaf of wobbly analytic functions on affinoid spaces $X = \operatorname{Sp} A$. Indeed, associating to an affinoid subdomain $U \subset X$ its corresponding affinoid algebra constitutes a presheaf $\mathscr{O}_X$ on a basis of the canonical topology on X, the so-called *presheaf of affinoid functions* on X. The associated sheaf $\mathscr{O}_X^w$ is called the *sheaf of wobbly analytic functions* on X. However, since X is totally disconnected with respect to the canonical topology, the algebra of global sections $\mathscr{O}_X^w(X)$ will be substantially larger than the affinoid algebra A giving rise to the definition of X, except for very special situations.

One may ask if there exists a reasonable class of coverings $\mathfrak{U} = (U_i)_{i \in I}$ of X by affinoid subdomains $U_i \subset X$ such that the associated diagram

$$\mathscr{O}_X(U) \to \prod_{i \in I} \mathscr{O}_X(U_i) \rightrightarrows \prod_{i,j \in I} \mathscr{O}_X(U_i \cap U_j), \qquad (*)$$

$$f \longmapsto (f|_{U_i})_{i \in I}, \qquad (f_i)_{i \in I} \longmapsto \begin{cases} (f_i|_{U_i \cap U_j})_{i,j \in I} \\ (f_j|_{U_i \cap U_j})_{i,j \in I} \end{cases}$$

is always exact (note that the intersection of two affinoid subdomains in X is an affinoid subdomain in X again by [2], 1.6/14). The answer is given by the following result, which is part of Tate's Acyclicity Theorem.

Theorem 1.8 (Tate)   *Let X be an affinoid space and $\mathfrak{U} = (U_i)_{i \in I}$ a finite covering of X by affinoid subdomains $U_i \subset X$. Then the above diagram $(*)$ is exact.*

To state the full version of the theorem, consider the augmented Čech complex

$$0 \longrightarrow \mathscr{F}(X) \xrightarrow{\varepsilon} C^0(\mathfrak{U}, \mathscr{F}) \xrightarrow{d^0} C^1(\mathfrak{U}, \mathscr{F}) \xrightarrow{d^1} \dots$$

for any covering $\mathfrak{U} = (U_i)_{i \in I}$ of X by affinoid subdomains $U_i \subset X$ and any presheaf $\mathscr{F}$ on the category of affinoid subdomains of X. If the sequence is exact, $\mathfrak{U}$ is called *acyclic* for $\mathscr{F}$. Using this terminology, Tate's Acyclicity Theorem reads as follows:

Theorem 1.9 (Tate [5], 8.2)   *Let X be an affinoid space and $\mathfrak{U}$ a finite covering of X by affinoid subdomains. Then $\mathfrak{U}$ is acyclic for the presheaf $\mathscr{O}_X$ of affinoid functions on X.*

For the *proof* of both theorems we refer to [2], Sect. 1.9. The strategy consists in simplifying the affinoid covering $\mathfrak{U}$ as much as possible, with the help of some general facts about Čech cohomology. Moreover, it is enough to consider the Čech complex of alternating cochains. Then, for $X = \mathrm{Sp}\, A$, it remains to actually do the proof in the case where $\mathfrak{U}$ is a Laurent covering generated by a single function $f \in A$; i. e., where

$$\mathfrak{U} = (X(f), X(f^{-1})).$$

This case can easily be settled by explicit computation.

More generally, for an affinoid space $X = \mathrm{Sp}\, A$ and an A-module M, one can consider the presheaf $M \otimes_A \mathscr{O}_X$ on the category of affinoid subdomains of X, which is given by

$$U \longmapsto M \otimes_A \mathscr{O}_X(U).$$

A simple argument, see [2], 1.9/11, shows that the assertion of Tate's Acyclicity Theorem can be generalized to such a presheaf in place of $\mathscr{O}_X$:

Corollary 1.10   *Let $X = \mathrm{Sp}\, A$ be an affinoid space, M an A-module, and $\mathfrak{U}$ a finite covering of X by affinoid subdomains. Then $\mathfrak{U}$ is acyclic for the presheaf $M \otimes_A \mathscr{O}_X$.*

2 Rigid analytic spaces via Grothendieck topologies

Let X be an affinoid space and consider on it the canonical topology as well as on a basis of the latter the presheaf of affinoid functions $\mathscr{O}_X$, which associates to each affinoid subdomain $U \subset X$ its corresponding affinoid algebra A_U. Then the situation allows the following interpretation: there are certain *distinguished* or *admissible* open subsets in X, on which the presheaf $\mathscr{O}_X$ is defined. Moreover, by Tate's Acyclicity Theorem 1.8, this presheaf is already a *sheaf* if we restrict ourselves to *finite* coverings of type $U = \bigcup_{i \in I} U_i$, where U and all U_i are admissible open in X; such open coverings might be called *admissible* as well. Thus, enriching the structure of X as a topological space by means of the data furnished by the admissible open sets and coverings, we may interpret $\mathcal{O}_X$ as the *structure sheaf* of the affinoid space X. Extending this concept to more general topological spaces, we will see that one can construct global rigid analytic spaces by glueing local affinoid ones. As a technical tool it is convenient to use the formalism of *Grothendieck topologies* [1], refining the notion of a topology.

Definition 2.1 *A* Grothendieck topology $\mathfrak{T}$ *on a set X consists of a category* $\operatorname{Cat}\mathfrak{T}$ *of subsets in X, with inclusions as morphisms, and a set* $\operatorname{Cov}\mathfrak{T}$ *of families* $(U_i \longrightarrow U)_{i \in I}$ *of morphisms in* $\operatorname{Cat}\mathfrak{T}$ *satisfying* $U = \bigcup_{i \in I} U_i$, *called* coverings, *such that the following hold:*

 (i) *If* $\Phi \colon U \longrightarrow V$ *is an isomorphism in* $\operatorname{Cat}\mathfrak{T}$, *then* $(\Phi) \in \operatorname{Cov}\mathfrak{T}$.

 (ii) *If* $(U_i \longrightarrow U)_{i \in I}$ *and* $(V_{ij} \longrightarrow U_i)_{j \in J_i}$ *for* $i \in I$ *belong to* $\operatorname{Cov}\mathfrak{T}$, *then the same is true for the composition* $(V_{ij} \longrightarrow U_i \longrightarrow U)_{i \in I, j \in J_i}$.

 (iii) *If* $(U_i \longrightarrow U)_{i \in I}$ *is in* $\operatorname{Cov}\mathfrak{T}$ *and if* $V \longrightarrow U$ *is a morphism in* $\operatorname{Cat}\mathfrak{T}$, *then the fibred products* $U_i \times_U V = U_i \cap V$ *exist in* $\operatorname{Cat}\mathfrak{T}$, *and* $(U_i \times_U V \longrightarrow V)_{i \in I}$ *belongs to* $\operatorname{Cov}\mathfrak{T}$.

The objects of $\operatorname{Cat}\mathfrak{T}$ are usually referred to as the *admissible open sets* of X, assuming tacitly that we provide X with the topology generated by all these sets. Likewise, the elements of $\operatorname{Cov}\mathfrak{T}$ are the *admissible open coverings* of admissible open subsets in X. As indicated above, we are particularly interested in the so-called *weak Grothendieck topology* $\mathfrak{T}_X$ on affinoid spaces X, where $\operatorname{Cat}\mathfrak{T}_X$ is the category of affinoid subdomains in X, and $\operatorname{Cov}\mathfrak{T}_X$ the set of all finite coverings of affinoid subdomains in X by sets of the same type. One knows that every morphism of affinoid spaces $f \colon Z \longrightarrow X$ is continuous with respect to the weak Grothendieck topologies on X and Z, in the sense that f-inverses of admissible open sets and coverings of $\mathfrak{T}_X$ are admissible open with respect to $\mathfrak{T}_Z$. This follows from the fact that $f^{-1}(U)$ is an affinoid subdomain in Z, for any affinoid subdomain $U \subset X$; see [2], 1.6/13.

A *presheaf* with respect to a Grothendieck topology $\mathfrak{T}$ is a contravariant functor $\mathscr{F}$ on $\operatorname{Cat}\mathfrak{T}$ with values in some category $\mathfrak{C}$. Such a functor is called a *sheaf* if the diagram

$$\mathscr{F}(U) \to \prod_{i \in I} \mathscr{F}(U_i) \rightrightarrows \prod_{i,j \in I} \mathscr{F}(U_i \times_U U_j)$$

is exact for every covering $(U_i \longrightarrow U)_{i \in I}$ in $\operatorname{Cov}\mathfrak{T}$ (assuming that $\mathfrak{C}$ admits cartesian products). Thus, Tate's Acyclicity Theorem 1.8 just says that, for an affinoid space X, the functor $\mathcal{O}_X$, which associates to an affinoid subdomain $U \subset X$ its corresponding affinoid algebra, is a sheaf with respect to the weak Grothendieck topology on X.

There is a canonical best possible way to refine the weak Grothendieck topology on affinoid spaces by adding more admissible open sets and more admissible coverings such that morphisms of affinoid spaces remain continuous and sheaves extend uniquely to sheaves with respect to this new topology. The resulting Grothendieck topology is the *strong Grothendieck topology* on affionid spaces, which we define now.

Definition 2.2 *Let X be an affinoid space. The* strong Grothendieck *topology on X is given as follows.*

(i) *A subset $U \subset X$ is called* admissible open *if there is a (not necessarily finite) covering $U = \bigcup_{i \in I} U_i$ of U by affinoid subdomains $U_i \subset X$ such that for all morphisms of affinoid spaces $f \colon Z \longrightarrow X$ satisfying $f(Z) \subset U$ the covering $(f^{-1}(U_i))_{i \in I}$ of Z admits a refinement, which is a finite covering of Z by affinoid subdomains.*

(ii) *A covering $V = \bigcup_{j \in J} V_j$ of some admissible open subset $V \subset X$ by means of admissible open sets V_j is called* admissible *if for each morphism of affinoid spaces $f \colon Z \longrightarrow X$ satisfying $f(Z) \subset V$ the covering $(f^{-1}(V_j))_{j \in J}$ of Z admits a refinement, which is a finite covering of Z by affinoid subdomains.*

Note that any covering $(U_i)_{i \in I}$ as in (i) is admissible by (ii). It is easily checked that the strong Grothendieck topology on X really is a Grothendieck topology, and that any *finite* union of affinoid subdomains of X is admissible open. Furthermore, one knows that Zariski open subsets of X are admissible open, and that each Zariski open covering of such a subset is admissible; see [2], 1.10/9. Thus, we can say that the strong Grothendieck topology on X is finer than the Zariski topology.

If $\mathscr{F}$ is a presheaf with respect to the weak or strong Grothendieck topology on an affinoid space X, we write as usual

$$\mathscr{F}_x = \varinjlim \mathscr{F}(U),$$

for the stalk of $\mathscr{F}$ at a point $x \in X$, where the limit extends over all admissible open subsets $U \subset X$ containing x. One knows that the stalks $\mathscr{O}_{X,x}$ of the sheaf of affinoid functions on X are local K-algebras ([2], 1.7/1).

For any set X with a Grothendieck topology $\mathfrak{T}$ on it and a sheaf $\mathscr{O}_X$ of K-algebras with respect to $\mathfrak{T}$, we call the pair $(X, \mathscr{O}_X)$ a *G-ringed space*. Furthermore, we talk about a *locally G-ringed space* if all stalks of $\mathscr{O}_X$ are local. In particular, for any affinoid space, we can consider its associated locally G-ringed space $(X, \mathscr{O}_X)$, assuming tacidly that affinoid spaces are always equipped with their strong Grothendieck topology. Using the appropriate notion of morphisms between such spaces (see [2], 1.12/1), it is clear that any morphism of affinoid spaces $f \colon X \longrightarrow Y$ induces in a natural way a morphism $(f, f^{\#}) \colon (X, \mathscr{O}_X) \longrightarrow (Y, \mathscr{O}_Y)$ of locally G-ringed spaces. One can show by the usual argument ([2], 1.12/2):

Proposition 2.3 *The functor $X \longmapsto (X, \mathscr{O}_X)$ from the category of affinoid spaces to the category of locally G-ringed spaces is fully faithful. Thus, the former category can be viewed as a full subcategory of the latter.*

Now it is more or less straightforward how to define the category of rigid analytic spaces. We could just say that such a space is a locally G-ringed space $(X, \mathscr{O}_X)$ with a Grothendieck topology on X such that X admits an admissible open covering $(X_i)_{i \in I}$ where $(X_i, \mathscr{O}_X|_{X_i})$ is affinoid for all $i \in I$. However, if we proceed like this, a similar effect will occur, as the one we have encountered when passing from the weak to the strong Grothendieck topology on affinoid spaces: in general, it will be possible to introduce additional admissible open sets and coverings, without changing sheaves on X and without changing the topology on the defining affinoid pieces X_i. To remedy this, we observe that the strong Grothendieck topology on affinoid spaces satisfies certain completeness conditions, namely:

Proposition 2.4 *Let X be an affinoid space. Then:*

(G_0) $\emptyset$ *and X are admissible open.*

(G_1) *Let $(U_i)_{i \in I}$ be an admissible covering of an admissible open subset $U \subset X$. Furthermore, let $V \subset U$ be a subset such that $V \cap U_i$ is admissible open for all $i \in I$. Then V is admissible open in X.*

(G_2) *Let $(U_i)_{i \in I}$ be a covering of an admissible open set $U \subset X$ by admissible open subsets $U_i \subset X$ which admits an admissible covering of U as refinement. Then $(U_i)_{i \in I}$ itself is admissible.*

If X is a set with a Grothendieck topology $\mathfrak{T}$ on it satisfying the above conditions (G_0), (G_1), and (G_2), and if $(X_i)_{i \in I}$ is an admissible open covering of X, then the Grothendieck topology on X can be recovered from the ones induced on the spaces X_i ([2], 1.10/10). Even better, if X is a set admitting a covering $X = \bigcup_{i \in I} X_i$ and if each X_i is equipped with a Grothendieck topology satisfying (G_0), (G_1), and (G_2), compatible in the sense that all intersections $X_i \cap X_j$ are admissible open in X_i and X_j and that their topologies restrict to the same Grothendieck topology on

$X_i \cap X_j$, then there is a unique Grothendieck topology on X satisfying (G_0), (G_1), and (G_2), and containing all X_i as admissible open subspaces ([2], 1.10/11). Therefore it makes sense to put the definition of rigid analytic spaces as follows:

Definition 2.5 *A rigid analytic space is a locally G-ringed space $(X, \mathscr{O}_X)$ with respect to a Grothendieck topology on X such that*
  (i) *the Grothendieck topology of X satisfies the completeness conditions (G_0), (G_1), and (G_2) of Proposition 2.4, and*
  (ii) *X admits an admissible open covering $(X_i)_{i \in I}$ where $(X_i, \mathscr{O}_X|_{X_i})$ is an affinoid space for all $i \in I$.*
  A morphism of rigid analytic spaces $(X, \mathscr{O}_X) \longrightarrow (Y, \mathscr{O}_Y)$ is a morphism in the sense of locally G-ringed spaces.

Note that the category of *locally* G-ringed spaces can just as well be replaced by the one of G-ringed spaces, thus, defining a morphism of rigid analytic spaces as a morphism in the sense of G-ringed spaces; this follows from the explanations given after 3.1. Furthermore, due to the fact that we require the completeness conditions (G_0), (G_1), and (G_2), we can conclude from the properties mentioned above that global rigid spaces can be constructed in the usual way by glueing local affinoid parts. In particular, we thereby see that the category of rigid analytic spaces admits fibred products. Namely, the completed tensor product of affinoid algebras, as dealt with in [3], 2.1.7, 3.1.1, and 6.1.1/10, provides a fibred product

$$\operatorname{Sp} A \times_{\operatorname{Sp} B} \operatorname{Sp} C = \operatorname{Sp}(A \hat{\otimes}_B C)$$

in the category of affinoid spaces, and one can construct fibred products of more general type via the usual glueing process, see [3], 9.3.5/2.

3 Rigid analytic spaces via Tate's h-structures

The original construction of rigid analytic spaces given by Tate in [5] starts out from the category of *wobbly analytic spaces*. Recall that in Sect. 1 we have constructed for any affinoid algebra A the sheaf $\mathscr{O}^w_{\operatorname{Max} A}$ of wobbly analytic functions on the maximal spectrum of A. It is a sheaf with respect to the canonical topology on $\operatorname{Max} A$, and is induced from the presheaf associating to any affinoid subdomain $U \subset \operatorname{Max} A$ its corresponding affinoid algebra A_U. This characterization coincides with the one given in [5], § 9, due to the openness of affinoid subdomains

1.6, a result not yet available, when the notes [5] were set up. Just as in the classical case, the stalks of the sheaf $\mathscr{O}^w_{\mathrm{Max}\,A}$ are local K-algebras, see [2], 1.7/1, and the locally ringed space $\mathrm{Sp}^w A = (\mathrm{Max}\,A, \mathscr{O}^w_{\mathrm{Max}\,A})$ is referred to as the *wobbly analytic space* associated to A. As a convention, (locally) ringed spaces and their morphisms will always be meant over the field K.

Definition 3.1 (Tate [5], 9.4) *A wobbly analytic space is a locally ringed space, which is locally isomorphic to wobbly analytic spaces of type $\mathrm{Sp}^w A$ for affinoid algebras A. A morphism of wobbly analytic spaces is a morphism in the context of locally ringed spaces.*

It is clear that any open subspace of a wobbly analytic space is wobbly analytic again. Furthermore, we like to point out that the category of wobbly analytic spaces can just as well be seen within the context of ringed spaces. This is due to the fact that any morphism of ringed spaces between wobbly analytic spaces is automatically a morphism of locally ringed spaces. To explain this fact, observe that for any affinoid algebra A and a point $x \in \mathrm{Max}\,A$ given by a maximal ideal $\mathfrak{m} \subset A$, the canonical morphism $A \longrightarrow \mathscr{O}^w_{\mathrm{Max}\,A,x}$ gives rise to an isomorphism $A/\mathfrak{m} \overset{\sim}{\longrightarrow} \mathscr{O}^w_{\mathrm{Max}\,A,x}/\mathfrak{m}\mathscr{O}^w_{\mathrm{Max}\,A,x}$; see [2], 1.7/2. In particular, $\mathscr{O}^w_{\mathrm{Max}\,A,x}$ is a local K-algebra whose residue field is *finite* over K, since the same is true for $A/\mathfrak{m}$; see [2], 1.2/11. Now any K-homomorphism $\varphi\colon R' \longrightarrow R$ between local K-algebras, where the residue field $R/\mathfrak{m}$ of the maximal ideal $\mathfrak{m} \subset R$ is finite over K, will automatically be local. Indeed, the induced map $\overline{\varphi}\colon R'/\varphi^{-1}(\mathfrak{m}) \hookrightarrow R/\mathfrak{m}$ shows that $R'/\varphi^{-1}(\mathfrak{m})$ is finite over K and, hence, a field, so that $\varphi^{-1}(\mathfrak{m})$ must coincide with the maximal ideal of R'. Thus, indeed, any morphism of ringed spaces between wobbly analytic spaces will be a morphism of locally ringed spaces.

Next consider a homomorphism of affinoid algebras $\varphi\colon A' \longrightarrow A$. Similarly as we have explained in Sect. 2 for associated affinoid spaces, there is a canonical morphism in the sense of wobbly analytic spaces $\mathrm{Sp}^w \varphi\colon \mathrm{Sp}^w A \longrightarrow \mathrm{Sp}^w A'$ associated to φ. This is proved by adapting the arguments given in [2], Sect. 1.12, to the present situation where the canonical topology on affinoid spaces is used instead of the weak or the strong Grothendieck topology. Furthermore, the resulting functor from affinoid algebras to ringed spaces, or wobbly analytic spaces, is faithful, as we will see in Proposition 3.2 below. Morphisms of type $\mathrm{Sp}^w \varphi$ for a homomorphism of affinoid algebras φ will later be referred to as *structural*.

More generally, it should be noted that there is a forgetful functor, which associates to every rigid analytic space X (in the sense of 2.5) its underlying wobbly analytic space X^w. Just consider on X the topology generated by all affinoid open subspaces and take as structure sheaf the one induced from the presheaf associating to any affinoid open subspace $U \subset X$ its corresponding affinoid algebra A_U.

Proposition 3.2 *The forgetful functor $X \longmapsto X^w$ from rigid to wobbly analytic spaces is faithful, but not fully faithful.*

Proof. For any admissible open subset $U \subset X$ look at the canonical diagram

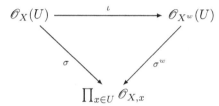

where ι is given by the canonical map from $\mathscr{O}_X$ into its sheafification $\mathscr{O}_{X^w}$ with respect to the canonical topology and σ as well as σ^w are obtained from canonical maps into stalks. Then σ^w is injective, since $\mathscr{O}_{X^w}$ is a sheaf with respect to a true topology. But the same holds for σ by [2], 1.7/3. Therefore ι will be injective as well. Keeping this in mind, we see immediately that two morphisms of rigid analytic spaces $f, g\colon X \longrightarrow Y$ coincide as soon as the corresponding morphisms of underlying wobbly analytic space $f^w, g^w\colon X^w \longrightarrow Y^w$ coincide.

To show that the functor $X \longmapsto X^w$ is not fully faithful, look at the unit disk $\mathbb{B}^1_K = \operatorname{Sp} K\langle \zeta \rangle$ and its admissible open subsets

$$X_1 = \{x \in \mathbb{B}^1_K \,;\, |\zeta(x)| < 1\}, \qquad X_2 = \{x \in \mathbb{B}^1_K \,;\, |\zeta(x)| = 1\}.$$

We know that the covering $\mathbb{B}^1_K = X_1 \cup X_2$ is *not* admissible. However, we can equip the point set of $\mathbb{B}^1_K$ with the structure of a new rigid analytic space X by taking $X = X_1 \cup X_2$ as a defining admissible covering. Then $\mathbb{B}^1_K$ and X have the same underlying wobbly analytic space, although they cannot be isomorphic. □

If $\varphi\colon A \longrightarrow A'$ is a morphism of affinoid algebras, we may look at the corresponding morphism of affinoid spaces $\operatorname{Sp}\varphi\colon \operatorname{Sp} A' \longrightarrow \operatorname{Sp} A$, as well as at the morphism of wobbly analytic spaces $\operatorname{Sp}^w \varphi\colon \operatorname{Sp}^w A' \longrightarrow \operatorname{Sp}^w A$ associated to φ. Then it is easily seen that $\operatorname{Sp}^w \varphi$ is obtained from $\operatorname{Sp}\varphi$

by applying the above mentioned forgetful functor. Furthermore, if $\operatorname{Sp} \varphi$ defines $\operatorname{Sp} A'$ as an affinoid subdomain in $\operatorname{Sp} A$, the openness of $\operatorname{Sp} A'$ in $\operatorname{Sp} A$ from 1.6 shows that $\operatorname{Sp}^w \varphi$ is an open immersion of ringed spaces; see also [5], 9.2. The converse of this is true as well by the Theorem of Gerritzen-Grauert (in the version of [2], 1.8/11) in conjunction with Tate's Acyclicty Theorem 1.8; see [3], 8.2.1/4. Thus, $\operatorname{Sp}^w \varphi$ is an open immersion of ringed spaces if and only if $\operatorname{Sp} \varphi$ defines $\operatorname{Sp} A'$ as an affinoid subdomain in $\operatorname{Sp} A$. This result was not yet available in [5], where it occurs as an open question after the proof of Prop. 7.3.

In 2.5 we have defined rigid analytic spaces in terms of Grothendieck topologies. Thus, a rigid analytic space X differs from its underlying wobbly analytic space X^w by an additional topological structure, namely the one given by the admissible open subsets and coverings in the sense of the Grothendieck topology. In contrast to this, Tate defines a rigid analytic space as a wobbly analytic space X^w, which is equipped with a certain selection of so-called *structural* morphisms $\operatorname{Sp}^w B \longrightarrow X^w$, for B varying over all affinoid algebras. Such an additional structure is called an *h-structure*.

Definition 3.3 (Tate [5], 10.1) *Let X be a wobbly analytic space. An h-structure on X is a selection $X(B)$, for each affinoid algebra B, of a set of certain morphisms of wobbly analytic spaces $\operatorname{Sp}^w B \longrightarrow X$; these are referred to as the* structural morphisms *from $\operatorname{Sp}^w B$ to X. The selection is subject to the following conditions:*

(i) For each $x \in X$ there exists a certain affinoid algebra B and a structural morphism $\operatorname{Sp}^w B \longrightarrow X$ that is an open immersion of ringed spaces with image containing x.

(ii) The composition of a morphism $\operatorname{Sp}^w B' \longrightarrow \operatorname{Sp}^w B$, derived from a morphism of affinoid algebras $\varphi : B \to B'$, with a structural morphism $\operatorname{Sp}^w B \longrightarrow X$ yields a structural morphism $\operatorname{Sp}^w B' \longrightarrow X$.

Definition 3.4 (Tate [5], 10.1) *An h-space is a wobbly analytic space X together with an h-structure; it will be denoted by X again. An h-morphism, or morphism of h-spaces, $X \longrightarrow Y$ is a morphism of the underlying wobbly analytic spaces such that composition with it maps structural morphisms $\operatorname{Sp}^w B \longrightarrow X$ to structural morphisms $\operatorname{Sp}^w B \longrightarrow Y$, for any affinoid algebra B.*

It is easily seen that on any open subset X' of an h-space X there is an induced h-structure making X' itself an h-space. We call X' an *open h-subspace* of X. Furthermore, having developed rigid analytic spaces in Sect. 2, we can derive h-spaces from them.

Example 3.5 *Let X be a rigid analytic space in the sense of 2.5 and write X^w for its underlying wobbly analytic space. For any affinoid algebra B, call a morphism of wobbly analytic spaces $\operatorname{Sp}^w B \longrightarrow X^w$ structural if it is induced from a morphism of rigid analytic spaces $\operatorname{Sp} B \longrightarrow X$. Then these structural morphisms define an h-structure on X^w making it an h-space X^h.*

Furthermore, any morphism of rigid spaces $X \longrightarrow Y$ induces a morphism between associated h-spaces $X^h \longrightarrow Y^h$. Thereby we get a faithful functor

$$\mathfrak{H}\colon (\mathrm{Rig}) \longrightarrow (h\text{-Spaces})$$

from the category of rigid spaces to the one of h-spaces; the functor is even fully faithful, as will be seen in 4.4.

The *proof* of the first assertion is quite easy. Choosing a point $x \in X$, there is an admissible open neighborhood $U \subset X$ of x that is affinoid, say $U = \operatorname{Sp} B_U$. Then the canonical injection $\operatorname{Sp} B_U \hookrightarrow X$ gives rise to an open immersion of ringed spaces $\operatorname{Sp}^w B_U \hookrightarrow X^w$. This settles condition (i) in Definition 3.3, whereas (ii) is trivial. The remaining assertion on morphisms is trivial as well, while the faithfulness of $\mathfrak{H}$ follows from Proposition 3.2. □

From now on, we will denote by $\operatorname{Sp}^h A$, for any affinoid algebra A, the wobbly analytic space $\operatorname{Sp}^w A$ equipped with the canonical h-structure induced from the affinoid space $\operatorname{Sp} A$, as specified in Example 3.5. Then, by 2.3, the structural morphisms $\operatorname{Sp}^w B \longrightarrow \operatorname{Sp}^w A$ of $\operatorname{Sp}^h A$ are just those morphisms that are induced from homomorphisms of affinoid algebras $A \longrightarrow B$. Furthermore, there is no difference in saying that a morphism of wobbly analytic spaces $\operatorname{Sp}^h B \longrightarrow \operatorname{Sp}^h A$ is a structural morphism of $\operatorname{Sp}^h A$ or that it is an h-morphism. We will call $\operatorname{Sp}^h A$ together with its canonical h-structure in the sense of Example 3.5 an *affinoid h-space*.

Next, affinoid coverings come into play, in a manner not too far from the one used in the context of Grothendieck topologies. Let X^h be an h-space; i. e., a wobbly analytic space equipped with an h-structure. By an open covering of X^h we understand a covering by open h-subspaces X_i^h, $i \in I$. Such a covering $(X_i^h)_{i \in I}$ is called *affinoid* if all X_i^h are affinoid, say of type $X_i^h = \operatorname{Sp}^h A_i$ for affinoid algebras A_i. Of particular interest is the case, where X^h is affinoid itself, say $X^h = \operatorname{Sp}^h A$. Then the covering $(X_i^h)_{i \in I}$ of $\mathrm{Sp}^h A$ is called *special* if the index set I is finite and all X_i^h are affinoid open h-subspaces in $\mathrm{Sp}^h A$. The definition of special coverings in [5], 10.10, is much more technical, since the notion of rational subdomains and the Gerritzen-Grauert Theorem 1.7 on the structure of affinoid subdomains were not yet available at that time. Our notion of special coverings is slightly more general than the one in [5], but does the same job. To do exactly the same thing as in [5], we would have to require from a special covering $(X_i^h)_{i \in I}$ of $\mathrm{Sp}^h A$ that, in addition, each X_i^h corresponds to a *rational subdomain* in $\mathrm{Sp}\, A$.

Definition 3.6 (Tate [5], 10.13)   *An open covering $\mathfrak{U}$ of an h-space X^h is called* admissible *if for every structural morphism $\mathrm{Sp}^h B \longrightarrow X^h$ the inverse image covering $f^{-1}(\mathfrak{U})$ of $\mathrm{Sp}^h B$ admits a refinement that is special.*

Definition 3.7 (Tate [5], 10.13)   *An h-space X^h is called* special *if its h-structure satisfies the following condition:*
    Given an affinoid algebra B and a special covering $(\mathrm{Sp}^h B_i)_{i \in I}$ of $\mathrm{Sp}^h B$, a morphism of wobbly analytic spaces $f : \mathrm{Sp}^h B \longrightarrow X^h$ is structural if and only if all restrictions $f|_{\mathrm{Sp}^h B_i} : \mathrm{Sp}^h B_i \longrightarrow X^h$ are structural.

Now we are ready to give the definition of rigid analytic spaces in terms of h-spaces:

Definition 3.8 (Tate [5], 10.17)   *A semi rigid h-space is a special h-space X^h admitting an admissible open covering by affinoid open h-subspaces $X_i^h \subset X^h$, $i \in I$. It is called a rigid h-space if, in addition, all intersections $X_i^h \cap X_j^h$ for $i, j \in I$ are semi rigid. Morphisms of such spaces are meant as morphisms of h-spaces.*

4 Grothendieck topologies versus h-structures

The purpose of this section is to show that both approaches to rigid analytic spaces, as we have explained them in Sects. 2 and 3, lead to the same result. To do this, let us look at the functor

$$\mathfrak{H} : (\mathrm{Rig}) \longrightarrow (h\text{-Spaces}), \qquad X \longmapsto X^h,$$

of 3.5, which associates to a rigid analytic space its underlying wobbly analytic space, equipped with the canonical h-structure. As a basic technical tool, let us show that the functor $\mathfrak{H}$ is compatible with open immersions.

Lemma 4.1 *Let $f: X \longrightarrow Y$ be a morphism of rigid analytic spaces and denote by $f^h: X^h \longrightarrow Y^h$ the corresponding morphism of h-spaces, as obtained by applying the functor $\mathfrak{H}$. Then f is an open immersion of rigid analytic spaces if and only if f^h is an open immersion of associated h-spaces.*

Proof. First, observe that f and f^h coincide on the level of maps between sets. Thus, if one of them is injective, the same will be true for the other. Now, starting with the only-if part, let $f: X \longrightarrow Y$ be an open immersion of rigid analytic spaces, i. e., a morphism defining X as an admissible open subspace of the rigid analytic space Y. Then, clearly, $f^h: X^h \longrightarrow Y^h$ is an open immersion, at least on the level of wobbly analytic spaces. To see that f^h identifies X^h with an open h-subspace of Y^h, observe that any morphism of rigid analytic spaces $g: Z \longrightarrow Y$ satisfying $g(Z) \subset f(X)$ admits a unique factorization through X. Therefore, by the definition of the functor $\mathfrak{H}$ in 3.5, each structural morphism $g^h: \operatorname{Sp}^h B \longrightarrow Y^h$ satisfying $g^h(\operatorname{Sp}^h B) \subset f^h(X^h)$, where B is an affinoid algebra, admits a factorization through X^h. This shows that X^h is an open h-subspace of Y^h via f^h, since the corresponding fact is already known on the level of wobbly analytic spaces.

To verify the if part, assume that $f^h: X^h \longrightarrow Y^h$ is an open immersion of h-spaces. In a first step, let us consider the special case, where X and Y are affinoid. Using the fact that $\mathfrak{H}$ is faithful, we see that any morphism of affinoid spaces $g: Z \longrightarrow Y$ satisfying $g(Z) \subset f(X)$ admits a unique factorization through X. But then, comparing with 1.3, it follows that f defines X as an affinoid subdomain in Y. In particular, f is an open immersion.

Next, let X be general, but assume that Y is still affinoid. Then we can choose an admissible affinoid covering $(X_i)_{i \in I}$ of X. As each X_i gives rise to an open h-subspace of X^h, by the only-if part of our assertion, we see that the restriction $f_i: X_i \longrightarrow Y$ of f to X_i gives rise to an open immersion $f_i^h: X_i^h \longrightarrow Y^h$. The special case just dealt with shows that also f_i is an open immersion. We claim that $f(X) = \bigcup_{i \in I} f(X_i)$ is admissible open in Y. To check this, consider a morphism of rigid analytic spaces $g: Z \longrightarrow Y$, where Z is affinoid and where $g(Z) \subset f(X)$. Then the associated morphism of h-spaces $g^h: Z^h \longrightarrow Y^h$ factors through X^h and, by the definition of the h-structure of X^h in conjunction with 3.2, we see that g factors through X via a unique morphism of rigid analytic spaces $g': Z \longrightarrow X$. Since $(X_i)_{i \in I}$ is an admissible covering of X, its pull-back under g' is admissible as well and, thus, admits a finite affinoid covering as a refinement. But then, by the definition in 2.2, $f(X)$ is admissible open in Y and admits $(f(X_i))_{i \in I}$ as an admissible affinoid covering. Equipping $f(X)$ with its canonical structure as an admissible open subspace of Y, we get a canonical morphism of rigid analytic spaces $f' : X \longrightarrow f(X) \subset Y$. Furthermore, we can conclude from the glueing of morphisms in [2], 1.12/6, that f' is an isomorphism. Thus, $f : X \longrightarrow Y$ is an open immersion of rigid analytic spaces.

Finally, assume that $f^h : X^h \longrightarrow Y^h$ is an open immersion of h-spaces, where X and Y are of general type. Choosing an admissible affinoid covering $(Y_i)_{i \in I}$ of Y, we can look at the induced morphisms of rigid analytic spaces $f_i : f^{-1}(Y_i) \longrightarrow Y_i$ for $i \in I$. Then, using the only-if part of our assertion, the associated morphisms of h-spaces $f_i^h : (f^h)^{-1}(Y_i^h) \longrightarrow Y_i^h$ are induced from f^h and, hence, are open immersions. Since Y_i is affinoid, we know already that all f_i are open immersions. Furthermore $f(X)$ is admissible open in Y and admits $(Y_i \cap f(X))_{i \in I}$ as an admissible covering, since the Grothendieck topology of Y satisfies the completeness condition (G_1) of 2.4, respectively 2.5. Then, just as before, we can equip $f(X)$ with its canonical structure as an admissible open subspace of Y, getting from f a canonical morphism of rigid analytic spaces $f' : X \longrightarrow f(X) \subset Y$. Using the glueing of morphisms in [2], 1.12/6, we see that f' is an isomorphism. Consequently, $f : X \longrightarrow Y$ is an open immersion of rigid analytic spaces. $\qquad \square$

Now we are ready to access our main result on the equivalence between rigid analytic spaces and rigid h-spaces.

Theorem 4.2 *The functor* $\mathfrak{H} : (\text{Rig}) \longrightarrow (h\text{-Spaces})$ *factors through the full subcategory* $(h\text{-Rig}) \subset (h\text{-Spaces})$ *consisting of rigid h-spaces. It restricts to an equivalence*

$$(\text{Rig}) \overset{\sim}{\longrightarrow} (h\text{-Rig})$$

between the category of rigid analytic spaces and the one of rigid h-spaces.

We divide the proof of the theorem into several steps. Let us first show that the functor $\mathfrak{H}$ factors, indeed, through the category $(h\text{-Rig})$.

Lemma 4.3 *For any rigid analytic space X, the associated h-space $X^h = \mathfrak{H}(X)$ is rigid.*

Proof. Fix a rigid analytic space X and look at its associated h-space X^h. To see that X^h is special in the sense of 3.7, choose an affinoid algebra B with a special (i. e., finite affinoid) covering $(\operatorname{Sp}^h B_i)_{i \in I}$ of $\operatorname{Sp}^h B$ and consider a morphism of wobbly analytic spaces $f^h \colon \operatorname{Sp}^h B \longrightarrow X^h$. Clearly, if f^h is structural, i.e., induced from a morphism of rigid analytic spaces $\operatorname{Sp} B \longrightarrow X$, we can restrict the latter to $\operatorname{Sp} B_i$ for any $i \in I$ and thereby arrive at a morphism $\operatorname{Sp} B_i \longrightarrow X$ that is mapped via $\mathfrak{H}$ to the restriction of f^h to $\operatorname{Sp}^h B_i$. Hence, $f^h|_{\operatorname{Sp}^h B_i}$ is structural for all $i \in I$.

Conversely, assume that $f^h|_{\operatorname{Sp}^h B_i}$ is structural for all $i \in I$. Then each of these morphisms is induced via $\mathfrak{H}$ from a morphism of rigid analytic spaces $f_i \colon \operatorname{Sp} B_i \longrightarrow X$. Furthermore, we know that the underlying morphisms of wobbly analytic spaces of f_i and f_j coincide on $\operatorname{Sp} B_i^h \cap \operatorname{Sp} B_j^h$ for all $i, j \in I$, since $\mathfrak{H}(f_i) = f^h|_{\operatorname{Sp}^h B_i}$ as well as $\mathfrak{H}(f_j) = f^h|_{\operatorname{Sp}^h B_j}$ are restrictions of the same morphism f^h. Now using the fact that the forgetful functor from rigid to wobbly analytic spaces is faithful, 3.2, we see that $f_i|_{\operatorname{Sp} B_i \cap \operatorname{Sp} B_j} = f_j|_{\operatorname{Sp} B_i \cap \operatorname{Sp} B_j}$ for all $i, j \in I$. Then, taking into account that $(\operatorname{Sp} B_i)_{i \in I}$ is an admissible covering of the affinoid space $\operatorname{Sp} B$, we can apply [2], 1.12/6, and thereby glue the f_i to obtain a morphism of rigid analytic spaces $f \colon \operatorname{Sp} B \longrightarrow X$ such that $f|_{\operatorname{Sp} B_i} = f_i$ for all $i \in I$. Clearly, $\mathfrak{H}(f) = f^h$, and we see that $f^h \colon \operatorname{Sp}^h B \longrightarrow X^h$ is structural. This shows that X^h is a special h-space.

Now, using Lemma 4.1, it is easy to see that X^h is a rigid h-space. Any admissible affinoid covering $(X_i)_{i \in I}$ of X gives rise to an admissible covering $(X_i^h)_{i \in I}$ of X^h, consisting of affinoid open h-subspaces $X_i^h \subset X^h$. Of course, the same applies to the intersections $X_i \cap X_j$, $i, j \in I$, in place of X and we are done. $\qquad\square$

Thus, we have seen that the functor $\mathfrak{H} \colon (\text{Rig}) \longrightarrow (h\text{-Spaces})$ restricts to a functor $(\text{Rig}) \longrightarrow (h\text{-Rig})$; the latter will be denoted by $\mathfrak{H}$ again.

Lemma 4.4 *The functor $\mathfrak{H} \colon (\text{Rig}) \longrightarrow (h\text{-Rig})$ is fully faithful.*

Proof. It follows from 3.2 that $\mathfrak{H}$ is faithful. To show that it is even fully faithful, consider an h-morphism $f^h \colon X^h \longrightarrow Y^h$ between rigid h-spaces X^h and Y^h that are induced from given rigid analytic spaces X and Y. Furthermore, choose an admissible affinoid covering $(\operatorname{Sp} A_i)_{i \in I}$ of X. Then $(\operatorname{Sp}^h A_i)_{i \in I}$ is an admissible covering of X^h consisting of affinoid open h-subspaces $\operatorname{Sp}^h A_i \subset X^h$. It follows that the restrictions

$$f_i^h = f^h|_{\operatorname{Sp}_i^h A_i} \colon \operatorname{Sp}^h A_i \longrightarrow Y^h, \qquad i \in I,$$

are structural for Y^h and, thus, are induced from morphisms of rigid analytic spaces

$$f_i \colon \operatorname{Sp} A_i \longrightarrow Y.$$

Now we conclude similarly as in the proof of Lemma 4.3. Since the morphisms f_i^h and f_j^h coincide on $\operatorname{Sp}^h A_i \cap \operatorname{Sp}^h A_j$ for indices $i, j \in I$, we see from 3.2 that f_i and f_j coincide on $\operatorname{Sp} A_i \cap \operatorname{Sp} A_j$. Thus, applying [2], 1.12/6, the f_i, $i \in I$, can be glued to define a morphism of rigid analytic spaces $f \colon X \longrightarrow Y$. Clearly f^h equals the h-morphism associated to f so that the functor $\mathfrak{H}$ is seen to be fully faithful. $\qquad\square$

To end the proof of Theorem 4.2, it remains to show:

Lemma 4.5 *The functor* $\mathfrak{H}\colon$ (Rig) $\longrightarrow$ (h-Rig) *is essentially surjective.*

Proof. Fix a rigid h-space X^h and, using the fact that it is semi rigid as well, consider an admissible covering $(\operatorname{Sp}^h A_i)_{i \in I}$ of X^h consisting of affinoid open h-subspaces $\operatorname{Sp}^h A_i \subset X^h$. Let us first look at the case, where $\operatorname{Sp}^h A_i \cap \operatorname{Sp}^h A_j$ is affinoid, say $\operatorname{Sp}^h A_i \cap \operatorname{Sp}^h A_j = \operatorname{Sp}^h A_{ij}$, for all indices $i, j \in I$. Then, applying Lemma 4.1, we can view $\operatorname{Sp} A_{ij}$ as an admissible open subspace and, hence, as an affinoid subdomain in each of the affinoid spaces $\operatorname{Sp} A_i$ and $\operatorname{Sp} A_j$. Furthermore, we can try to apply the glueing theorem 1.12/5 in [2] in order to glue the $\operatorname{Sp} A_i$ along the "intersections" $\operatorname{Sp} A_{ij}$, thereby obtaining a rigid analytic space X that admits $(\operatorname{Sp} A_i)_{i \in I}$ as an admissible affinoid covering. For this to work well, we need to know that the cocycle condition is satisfied by our glueing data. However, the corresponding glueing problem on the level of h-spaces has a solution, namely X^h. In particular, the cocycle condition is satisfied on the level of h-spaces. Since the functor $\mathfrak{H}$ is faithful by Lemma 4.4, the cocyle condition is satisfied on the level of rigid analytic spaces just as well and, indeed, we can glue the $\operatorname{Sp} A_i$, $i \in I$, along the "intersections" $\operatorname{Sp} A_{ij}$ to construct a rigid analytic space X admitting $(\operatorname{Sp} A_i)_{i \in I}$ as an admissible affinoid covering.

It remains to show that the image $\mathfrak{H}(X)$ coincides canonically with X^h. By our construction, there is a canonical identification $\mathfrak{H}(X) \simeq X^h$ on the level of wobbly analytic spaces. Thus, we have to show that this identification respects structural morphisms $\operatorname{Sp}^h B \longrightarrow \mathfrak{H}(X)$, respectively $\operatorname{Sp}^h B \longrightarrow X^h$, for any affinoid algebra B. Now observe that, by our choice, $(\operatorname{Sp}^h A_i)_{i \in I}$ is an admissible covering of X^h consisting of affinoid open h-subspaces and that the same is true for $\mathfrak{H}(X)$ by its construction. In particular, structural morphisms of type $\mathrm{Sp}^h B \longrightarrow X^h$ or $\mathrm{Sp}^h \longrightarrow \mathfrak{H}(X)$ are the same, provided they factor through a certain member of the admissible covering $(\mathrm{Sp}^h A_i)_{i \in I}$. However, the condition of a morphism of wobbly analytic spaces $\mathrm{Sp}^h B \longrightarrow X^h$ to be structural for X^h can be tested locally with respect to admissible open coverings of X^h and their pull-backs to $\mathrm{Sp}^h B$; this follows from the fact that X^h, as a rigid h-space is special in the sense of 3.7. The corresponding fact remains true replacing X^h by $\mathfrak{H}(X)$, since a morphism of rigid spaces $\mathrm{Sp}\, B \longrightarrow X$ can be defined locally with respect to admissible open coverings of X and their pull-backs to $\mathrm{Sp}\, B$, using the glueing of morphisms in [2], 1.12/6. Thus, the canonical identification $\mathfrak{H}(X) \simeq X^h$ is an isomorphism of h-spaces, and it follows that the functor $\mathfrak{H}$ is essentially surjective, provided the covering $(\mathrm{Sp}^h A_i)_{i \in I}$ of X^h has the property that the intersection of any two of its members is affinoid again.

For example, the latter is the case if X^h can be viewed as an open h-subspace of some ambient affinoid h-space Y^h. This follows from two facts, namely, first, that a morphism $\mathrm{Sp}^h A' \longrightarrow \mathrm{Sp}^h A$ between affinoid h-spaces is an open immersion if an only if the corresponding morphism of affinoid spaces defines $\mathrm{Sp}\, A'$ as an affinoid subdomain in $\mathrm{Sp}\, A$, see Lemma 4.1, and, second, that the intersection of two affinoid subdomains in $\mathrm{Sp}\, A$ yields an affinoid subdomain again; see [2], 1.6/14.

Now, turning to the general case, consider an admissible covering $(\mathrm{Sp}^h A_i)_{i \in I}$ of the rigid h-space X^h by affinoid open h-subspaces such that all intersections $X_{ij}^h = \mathrm{Sp}^h A_i \cap \mathrm{Sp}^h A_j$, $i, j \in I$, are semi rigid. Then X_{ij}^h admits an ambient affinoid h-space, and it follows, as we just have seen, that there exists a rigid analytic space X_{ij} together with a canonical isomorphism of h-spaces $\mathfrak{H}(X_{ij}) \simeq X_{ij}^h$. Furthermore, using Lemma 4.4, the canonical inclusions $X_{ij}^h \hookrightarrow \mathrm{Sp}^h A_i$ and $X_{ij}^h \hookrightarrow \mathrm{Sp}^h A_j$ are induced from morphisms of rigid analytic spaces $X_{ij} \longrightarrow \mathrm{Sp}\, A_i$ and $X_{ij} \longrightarrow \mathrm{Sp}\, A_j$. The latter are open immersions of rigid analytic spaces by Lemma 4.1. Now, having settled this point, we can proceed in exactly the same way, as we did in the special case above. One constructs a rigid analytic space X by glueing the $\mathrm{Sp}\, A_i$, $i \in I$, via the "intersections" X_{ij} and shows that $\mathfrak{H}(X)$ is canonically isomorphic to X^h. □

The proof of Theorem 4.2 demonstrates quite nicely the role played by the different conditions, which are used for the definition of rigid h-spaces à la Tate. Constructing rigid spaces within the context of wobbly analytic spaces, it is absolutely necessary to restrict morphisms. This is done by using *h*-structures and by taking affinoid spaces as prototypes, the latter being viewed as objects in the category opposite to the one of affinoid algebras. This way an *h*-structure becomes a bifunctor from the product of the category of affinoid algebras and the one of wobbly analytic spaces to the category of sets. It describes so to speak the *allowed* morphisms from affinoid spaces to the wobbly analytic spaces under consideration. Such morphisms have been called *structural*. Then the condition "special" from 3.7 is used in order to enable the local construction of structural morphisms relative to *admissible* open coverings of affinoid spaces. The latter are those coverings which have been recognized as *good* from the viewpoint of Tate's Acyclicity Theorem; they are at the core of the whole theory. Finally, some construction principle for global rigid spaces is needed: the fact that there is an admissible covering consisting of open affinoid subspaces and that the same is true for the intersections of the members of such a covering.

References

[1] M. Artin: Grothendieck topologies. Notes on a seminar by M. Artin, Harvard University (1962)

[2] S. Bosch: Lectures on formal and rigid geometry. Preprint 378 of the SFB Geometrische Strukturen in der Mathematik, Münster (revised version 2008), http://wwwmath.uni-muenster.de/sfb/about/publ/heft378.pdf

[3] S. Bosch, U. Güntzer, R. Remmert: Non-Archimedean Analysis. Grundlehren der Mathematischen Wissenschaften Vol. 261, Springer (1984)

[4] L. Gerritzen, H. Grauert: Die Azyklizität der affinoiden Überdeckungen. Global Analysis, Papers in Honor of K. Kodaira, 159–184. University of Tokyo Press, Princeton University Press (1969)

[5] J. Tate: Rigid analytic spaces. Private notes, reproduced with(out) his permission by I. H. E. S. (1962). Reprinted in Invent. Math. 12, 257–289 (1971)

[6] M. Temkin: A new proof of the Gerritzen-Grauert theorem. Math. Ann. 333, 261–269 (2005)

4
Topological rings in rigid geometry

Fumiharu Kato

1 Introduction

This paper gives a partial survey of the joint project [7] with K. Fujiwara (Nagoya Univ.), dealing with the part consisting of topological-ring theoretical aspects in rigid geometry, which has not been presented in our previous survey [6].

In classical algebraic geometry, finite type algebras over a field play a cornerstone role as the so-called 'coordinate rings', that is, the rings of regular functions on affine varieties. Scheme theory replaces affine varieties by affine schemes, and thus deals with arbitrary rings as basic building blocks. Still in scheme theory, however, fields and finite type algebras over a field keep their privileged position; fields are 'point objects', and finite type algebras over a field are 'fiber objects' over a point for locally of finite type morphisms between schemes.

In rigid geometry, on the other hand, we usually start with the so-called *affinoids*, that is, certain 'affine-like' objects, which come from topologically of finite type algebras over a complete non-archimedean valued field. This situation can be seen as an analogue of classical algebraic geometry, and thus one wants to ask for a scheme-theory-like generalization of rigid geometry. There are already several attempts to this goal; one of such attempts is via the *relative rigid spaces* by Bosch and Lütkebohmert [1]. The most important question in these attempts is: *what kind of topological rings should one start with?*

Motivic Integration and its Interactions with Model Theory and Non-Archimedean Geometry (Volume I), ed. Raf Cluckers, Johannes Nicaise, and Julien Sebag. Published by Cambridge University Press. © Cambridge University Press 2011.

In order to discuss this point, it is natural to investigate 'good' adic rings, which are supposed to be 'formal models' of generalized affinoid rings. This viewpoint is, of course, based on the famous theorem by Raynaud, which bridges rigid geometry with formal geometry. The main problem here is that the rings that are expected to be 'good' in our rigid analytic intuition are almost always *not* Noetherian, and thus may be quite difficult to deal with.

This paper proposes a certain kind of adic rings, the so-called *topologically universally adhesive rings*, which are quite well-behaved and general enough for most of the working situations. This new class of adic rings allows us to give a useful and sound foundations of formal geometry, and therefore of rigid geometry. The goal of this paper is to give the precise definition of this class of adic rings, and to explain the application to rigid geometry. Combined with some other results, which are to be presented in our forthcoming book [7], we will also give a brief survey of new results in our approach to rigid geometry.

The class of topologically universally adhesive rings is, in fact, still not satisfactory; there are some situations in which one wants to use a much more general class of topological rings. For example, topologically universally adhesive rings are not enough to recover the important core of the theory of adic spaces [9] that treats the so-called *strongly Noetherian Tate rings*. Corresponding to this, we will introduce in our book [7] (and in the paper [5] in preparation) the notion of *topologically universally rigid-Noetherian rings*. Many of the results in this paper (especially those in §4.3 and §5) are still valid, possibly with some minor modifications, in this more generalized sense. In this survey, however, we limit ourselves to treat topologically universally adhesive rings, for they comprise the particularly good foundations of topological rings both in topological and homology algebraic aspects, and moreover, it is quite non-trivial to develop the general theory of them.

This survey is based on the author's talk at the ICMS workshop "Motivic Integration and its Interactions with Model Theory and Non-Archimedean Geometry" (May 12–17, 2008). The author is grateful to the organizers. The author thanks the referee for the careful reading and useful comments.

Notation

- All rings considered in this paper are assumed to be commutative with unit, and all ring homomorphisms are assumed to map the unit element into the unit element.
- For a local ring R, the unique maximal ideal is denoted by $\mathfrak{m}_R$.

- For a ring A, we denote by $\mathrm{Frac}(A)$ the total ring of fractions, that is, the localization of A with respect to the multiplicative subset of all non-zero-factors.

2 Birational approach to rigid geometry: survey

2.1 Raynaud's theorem and formal birational patching

The construction of rigid spaces in the classical rigid geometry by Tate [15] starts with the notion of affinoids. An *affinoid* is a Grothendieck-topologized locally ringed spaces of the form $\mathrm{Spm}\,\mathscr{A}$, where $\mathscr{A}$ is an *affinoid algebra*, that is, a topologically of finite type algebra over a complete non-archimedean valued field K (with a non-trivial valuation). The next step of the construction is to 'patch' affinoids by the Grothendieck topology to get more general rigid analytic spaces.

Raynaud's approach, on the other hand, starts with *coherent* (= quasi-compact and quasi-separated) (admissible) formal schemes, based on his famous theorem (2.1), which says that any coherent rigid space is of the form 'X^{rig}' by a coherent formal scheme X. Moreover, the formal scheme X, the so-called *formal model*, is determined up to *admissible blow-ups*. In this approach, there is no particular reason to start with affinoids, because 'affineness' is not preserved by admissible blow-ups. But still in this situation, one has a well-behaved notion of 'patching' coming from 'birational[1] patching' of formal schemes, and thus one gets the notion of more general (not necessarily coherent) rigid spaces. It is here the first trait of 'birational aspects' of rigid geometry comes to play.

Raynaud's vewpoint

Let us recall a slightly more precise picture of Raynaud's viewpoint of rigid geometry:

- from this viewpoint, rigid geometry in totality is induced from a geometry of "models" (Figure 4.1);

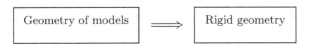

Figure 4.1 Raynaud's viewpoint

---

[1] Here 'birational' means 'up to admissible blow-ups'.

- as the geometry of models, Raynaud suggests the geometry of formal schemes over complete valuation rings.

For instance, if K is the fractional field of a complete discrete valuation ring V, then theorems in rigid geometry over K should follow from theorems in formal geometry over V, some of which are already worked out in [EGA, **III**, §4, §5].

Slightly more generally, we consider an a-adically complete valuation ring V of height 1 (not necessarily discretely valued), where $a \in \mathfrak{m}_V \setminus \{0\}$, and the category of all coherent formal schemes of finite type over V.[2] There is the so-called *Raynaud's functor*

$$\cdot^{\mathrm{rig}} : \begin{Bmatrix} \text{coherent } V\text{-formal} \\ \text{schemes of finite type} \end{Bmatrix} \longrightarrow \begin{Bmatrix} \text{coherent rigid} \\ \text{spaces over } K \end{Bmatrix}, \qquad X \longmapsto X^{\mathrm{rig}},$$

where K is the fractional field of V. The topology, points, and structure sheaf of $\mathscr{X} = X^{\mathrm{rig}}$ are characterized as follows:

- topology: a quasi-compact admissible open subset of $\mathscr{X}$ is of the form $\mathscr{U} = U^{\mathrm{rig}}$, where U is a quasi-compact open subset of an admissible blow-up X' of X;
- points: with the notation as above,

$$\mathscr{X}(K) = \{\text{sections } \operatorname{Spf} V \to X\},$$
$$\mathscr{U}(K) = \{\text{sections that factors through } U\};$$

- structure sheaf: when $U = \operatorname{Spf} A$, then $\Gamma(\mathscr{U}, \mathscr{O}_{\mathscr{X}}) = A_K \; (= A \otimes_V K)$.

This viewpoint culminates in the following theorem:

Theorem 2.1 (Raynaud 1972 [14])   *The Raynaud functor $X \mapsto X^{\mathrm{rig}}$ gives rise to the categorical equivalence*

$$\begin{Bmatrix} \textit{coherent V-formal} \\ \textit{schemes of finite type} \end{Bmatrix} \Big/ \begin{pmatrix} \textit{admissible} \\ \textit{blow-ups} \end{pmatrix} \overset{\sim}{\longrightarrow} \begin{Bmatrix} \textit{coherent rigid} \\ \textit{spaces over } K \end{Bmatrix}.$$

Comparison of topology and formal birational patching
From the theorem, one can read off the following 'comparison of patchings': the objects in the right-hand category are defined a priori by 'patching affinoids'; the equivalence shows that this patching turns out

---

[2] One can always kill a-torsions in the structure sheaf $\mathscr{O}_X$, since they form a finite type ideal of $\mathscr{O}_X$ that comes locally on affine neighborhoods from a finitely generated module. The resulting formal schemes are then flat and of finite type over V, which are the ones usually referred to as *admissible formal schemes*.

to be equivalent to the 'birational patching' (that is, 'birational' up to admissible blow-ups). A more precise explanation of this point already involves some of the important ingredients for the proof of the theorem:

- existence of formal birational patching,
- comparison of topologies.

The last one comes from nothing but the famous Gerritzen-Grauert Theorem. For the first, the necessary thing to do is roughly explained as follows: let X and Y be coherent V-formal schemes, and assume that there exist quasi-compact admissible open subsets $\mathscr{U}$ and $\mathscr{V}$ of $\mathscr{X} = X^{\text{rig}}$ and $\mathscr{Y} = Y^{\text{rig}}$, respectively, such that $\mathscr{U} \cong \mathscr{V}$. We want to 'patch' the coherent rigid spaces $\mathscr{X}$ and $\mathscr{Y}$ along $\mathscr{U} \cong \mathscr{V}$. By definition of the topology, we have admissible blow-ups $X' \to X$ and $Y' \to Y$ and quasi-compact open subsets $U \subseteq X'$ and $V \subseteq Y'$ such that $\mathscr{U} = U^{\text{rig}}$ and $\mathscr{V} = V^{\text{rig}}$. Since $\mathscr{U} \cong \mathscr{V}$, there exists a diagram

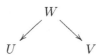

consisting of admissible blow-ups. Now the admissible blow-ups $W \to U$ and $W \to V$ have *extensions* $X'' \to X'$ and $Y'' \to Y'$ that are still admissible blow-ups:

Lemma 2.2 (Extension of admissible blow-ups)   *Let X be a coherent formal scheme of finite type over* $\operatorname{Spf} V$, *and* $U \subseteq X$ *a quasi-compact open subset. Suppose we are given an admissible blow-up* $U' \to U$ *of* U. *Then there exists an admissible blow-up* $X' \to X$ *such that* $U' \cong X' \times_X U$.

The proof refers to the well-known extension of quasi-coherent ideals; cf. [EGA, I_{new}, §6.9]; notice that an admissible ideal (= open ideal of finite type) of a topologically of finite type algebra A over V comes from an ideal of some $A/a^n A$ for $n \geq 1$. We leave the detail of the proof to the reader as an exercise.

This 'extension of admissible blow-ups' is the most crucial point in our formal birational patching. Indeed, after performing these extensions, the rest of the task is just patching X'' and Y'' (with respect to Zariski topology) along the common quasi-compact open subset W to get a coherent V-formal scheme Z; the associated rigid space $\mathscr{Z} = Z^{\text{rig}}$ is the desired patching of $\mathscr{X}$ and $\mathscr{Y}$.

This way of patching rigid spaces allows one to define more general (not necessarily quasi-coherent nor quasi-separated) rigid spaces; for more detail, see [6, §6.5].

2.2 Zariski's technique in classical birational geometry

Raynaud's viewpoint of rigid geometry is one of the two fundamental sources of our approach to rigid geometry. The other one is the so-called Zariski's technique in classical birational geometry, which we are going to explain.

U-admissible birational geometry

In his classical works in birational geometry (especially, resolution of singularities of algebraic surfaces [17][18][19][20]), Zariski entirely works with algebraic varieties. Here, following Raynaud's more modern framework [13, §5.7], let us work in the following set-up: let X be a coherent (= quasi-compact and quasi-separated) scheme, and $U \subseteq X$ a quasi-compact open subscheme. In order to consider the so-called 'U-admissible birational geometry' in a systematic way, we consider the category $\mathbf{MD}_{(X,U)}$ defined as follows: it is the category whose objects are proper morphisms $\pi \colon X' \to X$ that is U-*admissible*, that is, $\pi^{-1}(U) \cong U$, and, for two such maps $X' \to X$ and $X'' \to X$, the arrows between them are X-morphisms.

A typical example of U-admissible maps is provided by U-*admissible blow-ups*: by definition, a U-*admissible ideal* is a quasi-coherent ideal sheaf $\mathscr{J}$ on X of finite type such that the closed subscheme $V(\mathscr{J})$ is set-theoretically disjoint from U; a U-admissible blow-up is a blow-up centered at a U-admissible ideal.

The following facts are due to [13, §5.7]:

Proposition 2.3 (1) *The category* $\mathbf{MD}_{(X,U)}$ *is cofiltered, and* id_X *gives the final object.*

(2) *The set of all* U-*admissible blow-ups is cofinal in* $\mathbf{MD}_{(X,U)}$.

The classical Zariski-Riemann space

The central object in Zariski's technique is the so-called '(classical) Zariski-Riemann space', which is, by definition, the topological space obtained as the projective limit of all U-admissible modifications of X; due to Proposition 2.3 (2), one can equivalently define it as the projective limit of all U-admissible blow-ups:

Definition 2.4 The (classical) *Zariski-Riemann space* associated to the pair (X, U), denoted by $\langle X \rangle_U$, is the underlying topological space of the limit

$$\varprojlim_{\mathscr{J}} X_{\mathscr{J}},$$

taken along the set of all U-admissible ideals, where $X_{\mathscr{J}}$ is the U-admissible blow-up of X centered at $\mathscr{J}$. For any $X' \to X$ in $\mathbf{MD}_{(X,U)}$, the canonical projection (called *specialization map*) $\langle X \rangle_U \to X'$ is denoted by $\mathrm{sp}_{X'}$. The inductive limit sheaf

$$\mathcal{O}_{\langle X \rangle_U} = \varinjlim_{\mathscr{J}} \mathrm{sp}_{X_{\mathscr{J}}}^{-1} \mathcal{O}_{X_{\mathscr{J}}}$$

is a sheaf of local rings, and called the *structure sheaf* of $\langle X \rangle_U$.

Quasi-compactness

The most crucial point in this technique is the fact that the Zariski-Riemann space $\langle X \rangle_U$ is *quasi-compact*. In case X is an algebraic variety, this is the famous theorem by Zariski published in his 1944 paper [20]. Actually, this fact can be proven purely in general topology. Let us see briefly how it is done.

A topological space X is said to be *coherent* if it satisfies the following conditions:

- X has an open basis consisting of quasi-compact open subsets;
- X is quasi-compact, and moreover, quasi-separated, that is, for any quasi-compact open subsets $U, V \subseteq X$, the intersection $U \cap V$ is again quasi-compact.

(Notice that this definition mimics the definition of coherent topoi in [SGA4-2, Exposé VI].) Then one has the following theorem:

Theorem 2.5 *Let $(I, \leq)$ be a directed set, $\{X_i, p_{ij}\}$ a projective system of sober[3] coherent spaces indexed by I such that each transition map p_{ij} is quasi-compact for any $i \leq j$, $i, j \in I$. Set $X = \varprojlim_{i \in I} X_i$.*

(1) The topological space X is sober and coherent, and, for each $i \in I$, the canonical projection map $p_i \colon X \to X_i$ is quasi-compact.

(2) If, moreover, each X_i for $i \in I$ is non-empty, then X is non-empty.

---

[3] A topological space X is said to be *sober* if every irreducible closed subset has a unique generic point.

A short proof of this theorem is given by lattice theory, especially, by Stone's representation theorem (e.g. [10, Chap. II, §3.4]), which shows a beautiful dualism between sober coherent topological spaces and distributive lattices.

The following proposition is useful in describing the topology of the limit space:

Proposition 2.6 *Let $(I, \leq)$ be a directed set, and $\{X_i, p_{ij}\}$ a projective system of topological spaces such that each transition map p_{ij} is quasi-compact for any $i \leq j$, $i, j \in I$. Set $X = \varprojlim_{i \in I} X_i$. Let $U \subseteq X$ be a quasi-compact open subset. Then there exists $i \in I$ and a quasi-compact open subset $U_i \subseteq X_i$ such that $p_i^{-1}(U_i) = U$, where $p_i \colon X \to X_i$ is the canonical projection map.*

Proof By definition of the topology of X and the quasi-compactness of U there exist finitely many indices $i_1, \ldots, i_n$ of I and a quasi-compact open subset U_{i_k} of X_{i_k} such that $U = \bigcup_{k=1}^n p_{i_k}^{-1}(U_{i_k})$ (cf. [2, Chap. I, §4.4, Prop. 9]). Take $i \in I$ such that $i \geq i_1, \ldots, i_n$, and set $U_i = \bigcup_{k=1}^n p_{ii_k}^{-1}(U_{i_k})$. Then U_i is quasi-compact, and $U = p_i^{-1}(U_i)$. □

Points and the structure sheaf

Although the classical Zariski-Riemann space is almost always *not* a scheme, it is an interesting locally ringed space with rich geometric structure.

Let X be a coherent scheme, and U a non-empty quasi-compact open subset of X. Let $\mathscr{I} \subset \mathscr{O}_X$ be a quasi-coherent ideal of finite type such that $X \setminus U = V(\mathscr{I})$. Consider the set

$$\mathrm{Val}(X, U) = \left\{ \begin{array}{l} \text{X-isomorphism classes of morphisms of the} \\ \text{form } \alpha \colon \mathrm{Spec}\, V \to X, \text{ where } V \text{ is a } \mathscr{I}V\text{-} \\ \text{adically separated valuation ring, which maps} \\ \text{the generic point to a point of } U \end{array} \right\}.$$

There exists a map from this set to the Zariski-Riemann space $\langle X \rangle_U$ constructed as follows. Let $\alpha \colon \mathrm{Spec}\, V \to X$ be given, and set $K = \mathrm{Frac}(V)$. For any U-admissible proper map $X' \to X$, we have a morphism $\mathrm{Spec}\, K \to X'$ such that the diagram

$$\begin{array}{ccc} \mathrm{Spec}\, K & \longrightarrow & X' \\ \downarrow & & \downarrow \\ \mathrm{Spec}\, V & \longrightarrow & X \end{array}$$

commutes, since $X' \to X$ is isomorphic on U. Since $X' \to X$ is proper, we have the unique lift $\alpha_{X'} \colon \operatorname{Spec} V \to X'$ by the valuative criterion. Varying X', one has a projective system of morphisms

$$\{\alpha_{X'} \colon \operatorname{Spec} V \to X'\}_{X'},$$

whence a morphism of locally ringed spaces $\operatorname{Spec} V \to \langle X \rangle_U$. The desired point of $\langle X \rangle_U$ is the image of the closed point of $\operatorname{Spec} V$.

Consider the relation $\sim$ on the set $\operatorname{Val}(X, U)$ defined as follows: for two elements $\alpha \colon \operatorname{Spec} V \to X$ and $\beta \colon \operatorname{Spec} W \to X$, $\alpha \sim \beta$ if there exists a local map $f \colon V \hookrightarrow W$ such that $\alpha \circ \operatorname{Spec} f = \beta$. Then, clearly, the above mentioned mapping induces a map

$$(*) \qquad\qquad \operatorname{Val}(X, U)/{\approx} \longrightarrow \langle X \rangle_U,$$

where $\approx$ is the equivalence relation generated by $\sim$.

On the left-hand set, one can consider the weakest topology such that the map $(*)$ is continuous. This topology is actually the same as the one generated by subsets of the following form:

$$\left\{ \begin{array}{l} \approx\text{-equivalence class of } \alpha \colon \operatorname{Spec} V \to X \\ \text{that extends to } \operatorname{Spec} V \to X' \text{ such that} \\ \text{the image is contained in } Y' \end{array} \right\},$$

where $X' \to X$ is a U-admissible proper map, and Y' is an affine open subset of X'.

Theorem 2.7 *The map $(*)$ is a homeomorphism.*

The theorem can be proven with the aid of the following lemma.

Lemma 2.8 *For any point $\xi \in \langle X \rangle_U$, the ring $\mathscr{O}_{\langle X \rangle_U, \xi}$ is $\mathscr{I} \mathscr{O}_{\langle X \rangle_U, \xi}$-valuative.*

Here, in general, for a ring A and a finitely generated ideal I, A is said to be I-*valuative* if any I-admissible ideal (that is, a finitely generated ideal that contains some power of I) is an invertible ideal. The above lemma is not difficult to see, for the blow-up centered at a U-admissible ideal makes the ideal invertible; here, to show the lemma, one has to use the following U-admissible version of 'extension of admissible blow-ups'.

Lemma 2.9 (Extension of U-admissible blow-ups; cf. [EGA, $\mathrm{I}_{\mathrm{new}}$, §6.9]) *Let S be a coherent scheme, and $U \subset S$ a quasi-compact open subset. Let $T \subset S$ be a quasi-compact open subset of S, and $\mathscr{J}$ a $U \cap T$-admissible ideal on T. Then there exists a U-admissible ideal $\widetilde{\mathscr{J}}$ on S such that $\widetilde{\mathscr{J}}|_T = \mathscr{J}$.*

In order to deduce Theorem 2.7 from Lemma 2.8, one has to know that the notion of I-valuative rings is closely related to the notion of valuation rings, as the following (straightforward) proposition shows:

Proposition 2.10   (1) *Let A be a local ring that is I-valuative, where I is a non-zero finitely generated ideal of A. Set $J = \bigcap_{n \geq 1} I^n$. Then:*

(a) *$B = \varinjlim_{n \geq 1} \mathrm{Hom}(I^n, A)$ is a local ring, and $V = A/J$ is a valuation ring, $\bar{a}$-adically separated, for the residue field F of B, where $IV = (\bar{a})$;*

(b) *$A = \{f \in B \mid (f \bmod \mathfrak{m}_B) \in V\}$;*

(c) *$J = \mathfrak{m}_B$.*

(2) *Conversely, for B a local ring, and V an $\bar{a}$-adically separated valuation ring for the residue field F of B with $\bar{a} \neq 0$, the subring A of B defined as in* (b) *above is an I-valuative local ring for any finitely generated ideal I such that $IV = (\bar{a})$, and $B = \varinjlim_{n \geq 1} \mathrm{Hom}(I^n, A)$.*

The proposition shows that I-valuative local rings are 'composites' of local rings with valuation rings. Therefore, Lemma 2.8 implies that, at any point ξ of $\langle X \rangle_U$, there exists the canonically built-in valuation ring, and by this, one sees that the map $(*)$ is bijective; by definition of the topology on the left-hand set, one sees that this map is a homeomorphism.

Zariski's argument

Zariski introduced his Zariski-Riemann space (or, according to his original terminology, the 'generalized Riemann manifold') in order to apply it to his proof of resolution of singularities of algebraic surfaces [19]. Although we have to omit the detail of his argument, we would like to give a brief sketch of the basic idea.

Let X be an algebraic variety. We want to find a blow-up $X' \to X$ such that X' is non-singular. To this end, one considers the Zariski-Riemann space $\langle X \rangle_U$, where U is a non-singular open subset (possibly empty) of X. Then the proof goes along as follows:

(1) first, find at any point of the Zariski-Riemann space a 'good' neighborhood, that is, an open subset of $\langle X \rangle_U$ that is the pull-back of a non-singular open subset of a blow-up X' of X (*local uniformization theorem*);

(2) secondly, still in the Zariski-Riemann space level, ascertain that the 'good' neighborhoods can be patched to a larger 'good' neighborhood;

(3) finally, use the quasi-compactness to deduce that already finitely many 'good' local objects obtained in (1) can be patched together to exhaust the Zariski-Riemann space; if this is done, we have a finite open covering of the Zariski-Riemann space consisting of 'good' neighborhoods and, since each local neighborhood comes from an open subset of a U-admissible blow-up of X, by an argument similar to that in the proof of 2.6, one has a desired blow-up X' by biration patching.

M. Nagata [12] used this technique in his proof of the compactification theorem, nowadays called *Nagata's compactification theorem*. Aside from these famous and important applications, one can also apply the argument to give a proof of *algebraic flattening theorem* by Raynaud and Gruson [13]; cf. [6]. Having this application in mind, K. Fujiwara gave a proof of *formal flattening theorem* using the formal version of this technique.[4]

2.3 Fresh start

Now, having the above discussions in mind, we can make a fresh start of the introduction of rigid spaces.

(1) First, we consider a category of good 'models' with the notion of 'admissible blow-ups': for this, Raynaud suggested the category of coherent V-formal schemes (where V is an a-adically complete valuation ring of height 1) of finite type; but there can be other natural candidates for this category of models, for example:

- we may drop the condition 'height 1' for the base valuation ring V; this leads actually to numerous technical difficulties (including Artin-Rees-like properties, etc), which can be, however, cleared as we will see below in the second half of this paper (especially in §4.2);
- we can also start from the category of Noetherian formal schemes (possibly without reference to any base formal schemes); this is nothing but the situation of *relative rigid spaces* introduced by Bosch and Lütkebohmert [1]; for example, one wants to consider rigid spaces over Spf $\mathbb{Z}[[t]]$ (with t-adic topology), which turns out to be important for many applications, e.g. compactification of arithmetic moduli spaces (cf. [6, §7]);

---

[4] Formal flattening theorem has been also obtained by Bosch-Raynaud, but their method of proof is different from this.

- we may moreover start from a more general category of formal schemes, perhaps most generally, from the category of all coherent adic formal schemes that allow (at least locally) an ideal of definition of finite type; this generalization will be discussed later in this paper;
- of course, we do not have to stick to formal schemes; we can, for example, consider a certain category of *henselian schemes* as the category of 'models';
- perhaps, an example more familiar to birational geometers is the following situation: we consider as the category of 'models' the category of pairs (X, U) consisting of coherent schemes X and a quasi-compact open subset $U \subseteq X$, where the arrows $(X, U) \rightarrow (Y, V)$ between such pairs are defined to be morphisms $f \colon X \rightarrow Y$ such that $f^{-1}(V) = U$; it has the U-*admissible blow-ups* (cf. §2.2) for the natural candidate for admissible blow-ups; In this U-*admissible rigid geometry* framework, one can say, the 'rigid geometry' is nothing but Zariski's birational geometry.[5]

(2) Secondly, define 'coherent rigid spaces' to be objects of the quotient category of the category of 'models' by the collection of all 'admissible blow-ups'. One may denote the related quotient functor by

$$X \longmapsto X^{\mathrm{rig}}$$

for an object X of the category of 'models'. When the category of 'models' is as above, then the category of coherent rigid spaces is naturally topologized by means of Zariski topology on the category of 'models'. For example, if a coherent rigid space $\mathscr{X}$ is the one associated to a coherent V-formal scheme X of finite type (that is, $\mathscr{X} = X^{\mathrm{rig}}$), then:

- a quasi-compact compact open subspace of $\mathscr{X}$ is, by definition, a rigid subspace of $\mathscr{X}$ of the form $\mathscr{U} = U^{\mathrm{rig}}$, where $U \subseteq X'$ is a quasi-compact open subset of an admissible blow-up X' of X:

$$
\begin{array}{c}
U \overset{\text{q-cpt}}{\hookrightarrow} X' \\
\Big\downarrow {\scriptstyle \text{adm. blow-up}} \\
X
\end{array}
\qquad \Longleftrightarrow \qquad \mathscr{U} = U^{\mathrm{rig}} \overset{\text{q-cpt}}{\hookrightarrow} \mathscr{X}.
$$

---

[5] This would imply that rigid geometry is a kind of generalization of birational geometry. In this sense, the 'usual' rigid geometry (coming from formal geometry) can be seen as a certain incarnation of birational geometry, or, in other words, a hybrid of formal geometry and birational geometry.

- a finite covering by quasi-compact open subspace of $\mathscr{X}$ is, by definition, the one induced by the functor $\cdot^{\mathrm{rig}}$ from a finite Zariski covering by quasi-compact open subsets of an admissible blow-up X' of X.

(3) Having thus defined topology (the so-called *admissible topology*) on the quotient category, one can then patch coherent rigid spaces to construct more general rigid spaces, not necessarily quasi-compact, nor quasi-separated.

2.4 Visualization

Another thing that Raynaud's theorem teaches us is that, in the above definition of rigid spaces, one does not need to have the notion of points of rigid spaces; in Tate's rigid geometry, Tate constructed rigid spaces as patching of affinoids, while affinoids are defined to be supported on the set of maximal ideals of affinoid algebras. But, now, having Raynaud's approach to rigid geometry in our disposal, we find that one can really do 'patching' (due to formal birational patching) without notion of points.

To explain this, let us suppose that our rigid spaces are, as in the classical setting, those coming from V-formal schemes, where V is an a-adically complete valuation ring of height 1; we choose this setting, only because it is the most familiar one. So, all rigid spaces in the sequel are Tate's rigid spaces over K, where $K = \mathrm{Frac}(V) = V[\frac{1}{a}]$ is the field of fractions of V.

Let $\mathscr{X}$ be a coherent rigid space. We have the *admissible topology* on $\mathscr{X}$ (described in §2.1), thereby obtaining the small topos $\mathscr{X}_{\mathrm{ad}}$, called the *small admissible topos*. This is a coherent topos in the sense of [SGA4-2, Exposé VI]. Hence, if there exists a sober topological space Z for which the associated topos $\mathbf{top}(Z)$ ($=$ the category of set-valued sheaves on Z) is isomorphic to $\mathscr{X}_{\mathrm{ad}}$, then Z must be a coherent topological space. Since such a topological space Z is, if it exists, unique up to homeomorphism, the underlying set of points is the correct 'point set' for the coherent rigid space $\mathscr{X}$.

The important point is: *one can actually find such a coherent topological space Z for any rigid space $\mathscr{X}$*, and thus can 'visualize' the rigid spaces.[6] Looking at our description of admissible topology as above, it is more or less clear how to construct such Z at least for coherent rigid spaces. Since the topology is defined as 'Zariski topology up to admissible blow-ups', the desired space should be simply the *projective limit of*

---

[6] This fact has been recognized, already in 1990's, by many experts, such as van der Put-Schneider and Huber, and nowadays, seems to be a common knowledge.

all admissible blow-ups:

$$\langle \mathscr{X} \rangle = \varprojlim_{X' \to X} \text{ (the underlying topological space of } X').$$

Here, X' runs over all admissible blow-ups of a fixed formal model X, and the projective limit is taken in the category **Top** of topological spaces. One can show, moreover, that the above limit is a filtered one. By Theorem 2.5, the topological space $\langle \mathscr{X} \rangle$ is sober and coherent. We call this topological space the *Zariski-Riemann space* associated to the rigid space $\mathscr{X}$. By definition, for any admissible blow-up X' of X, there exists the canonical projection map $\langle \mathscr{X} \rangle \to X'$, which we denote by $\mathrm{sp}_{X'}$, and which we call the *specialization map*.

The Zariski-Riemann space $\langle \mathscr{X} \rangle$ comes with a sheaf of local rings $\mathscr{O}_{\mathscr{X}}^{\mathrm{int}}$, that is, simply the structure sheaf of $\langle \mathscr{X} \rangle$ that comes when one take the above projective limit in the category of locally ringed spaces. We call this sheaf the *integral structure sheaf.* Together with this sheaf, the Zariski-Riemann space $\langle \mathscr{X} \rangle$ provides, so to speak, the *canonical formal model* of $\mathscr{X}$.

But, because it is still a model, the sheaf $\mathscr{O}_{\mathscr{X}}^{\mathrm{int}}$ is not the one with which one is able to develop rigid geometry. To obtain the correct one, we need to invert the topological generator a:

$$\mathscr{O}_{\mathscr{X}} = \mathscr{O}_{\mathscr{X}}^{\mathrm{int}}[\tfrac{1}{a}].$$

The sheaf thus obtained is still a sheaf of *local* rings, since one can show that each stalk of $\mathscr{O}_{\mathscr{X}}^{\mathrm{int}}$ is a-valuative:

Proposition 2.11 *Let $\mathscr{X}$ be a coherent rigid space, $\langle \mathscr{X} \rangle$ the associated Zariski-Riemann space, and $x \in \langle \mathscr{X} \rangle$ a point. Consider the ring $A_x = \mathscr{O}_{\mathscr{X},x}^{\mathrm{int}}$, the stalk of the integral structure sheaf at x. Then A_x is a-valuative and a-adically henselian. Therefore, we have (by Proposition 2.10):*

(a) *$B_x = A_x[\tfrac{1}{a}] = \mathscr{O}_{\mathscr{X},x}$ is a local ring, and $V_x = A_x/J_x$ is a valuation ring, a-adically separated, for the residue field F_x of B_x, where $J_x = \bigcap_{n \geq 1} a^n A_x$;*
(b) *$A_x = \{f \in B_x \mid (f \bmod \mathfrak{m}_{B_x}) \in V_x\}$;*
(c) *$J_x = \mathfrak{m}_{B_x}$.*

Furthermore, B_x is a henselian local ring, and V_x is a-adically henselian.

Proof Let $\mathscr{X} = X^{\mathrm{rig}}$. Then A_x is the filtered inductive limit of a-adically complete local rings, that is, the local rings $\mathscr{O}_{X',\mathrm{sp}_{X'}(x)}$ for admissible blow-ups X' of X, and hence is a-adically henselian.

Let J be an a-admissible ideal of A_x, that is, a finitely generated ideal that contains some power of a. We need to show that J is an invertible ideal of A_x. There exists a quasi-compact open neighborhood $\mathfrak{U}$ of x and a finitely generated a-adically open ideal sheaf $\mathscr{J}$ on $\mathfrak{U}$ whose stalk at x coincides with J. By Proposition 2.6, replacing X by an admissible blow-up if necessary, we may assume that such $\mathfrak{U}$ comes from a quasi-compact open subset of X, and furthermore, since we work only locally, we may assume that $\mathfrak{U} = \langle \mathscr{X} \rangle$. Moreover, since J is finitely generated, we may assume that there exists an admissible ideal $\mathscr{J}_X$ such that $\mathscr{J} = \mathscr{J}_X \mathscr{O}_{\mathscr{X}}^{\mathrm{int}}$. Replacing X by $X_{\mathscr{J}}$ (the admissible blow-up along $\mathscr{J}_X$), we may assume $\mathscr{J}_X$ is invertible. Then $(\mathrm{sp}_X^{-1} \mathscr{J}_X) \mathscr{O}_{\mathscr{X}}^{\mathrm{int}} = (\mathrm{sp}_X^{-1} \mathscr{J}_X) \otimes_{\mathrm{sp}^{-1} \mathscr{O}_X} \mathscr{O}_{\mathscr{X}}^{\mathrm{int}}$ is an invertible ideal, and thus $J = \mathscr{J}_x$ is an invertible ideal of A_x, as desired. $\qquad\square$

We call the sheaf $\mathscr{O}_{\mathscr{X}}$ the *rigid structure sheaf*, or simply *structure sheaf* of the coherent rigid space $\mathscr{X}$. Thus we get the triple

$$\mathbf{ZR}(\mathscr{X}) = (\langle \mathscr{X} \rangle, \mathscr{O}_{\mathscr{X}}^{\mathrm{int}}, \mathscr{O}_{\mathscr{X}}),$$

consisting of a coherent topological space and two sheaves of local rings, called the *Zariski-Riemann triple* of $\mathscr{X}$; this is the object that we suggest to be the 'visualization' of the rigid space $\mathscr{X}$.[7]

Points

Having thus the notion of visualization of rigid spaces, we are now interested in how points are described. In Zariski's classical birational geometry, points of Zariski-Riemann spaces are described in terms of valuation rings. Similarly, points of $\langle \mathscr{X} \rangle$ can be described in terms of valuation rings as follows.

By 2.11, we have, for every point $x \in \langle \mathscr{X} \rangle$ of the associated Zariski-Riemann space, a canonical valuation ring V_x that is a-adically separated and a-adically henselian. By the construction of V_x, we moreover have a natural adic map

$$\alpha_x \colon \mathrm{Sph}\, V_x \longrightarrow (\langle \mathscr{X} \rangle, \mathscr{O}_{\mathscr{X}}^{\mathrm{int}})$$

of locally ringed spaces, where Sph denotes the henselian spectrum, or, if we prefer formal schemes, we have an adic map

$$\alpha_x \colon \mathrm{Spf}\, \widehat{V}_x \longrightarrow (\langle \mathscr{X} \rangle, \mathscr{O}_{\mathscr{X}}^{\mathrm{int}})$$

of locally ringed spaces; the both maps send the closed point to x.

---

[7] For more general rigid spaces (not necessarily coherent), the visualization is constructed simply by patching of topological spaces.

An adic map from the formal spectrum of an a-adically complete valuation ring as above is said to be a *rigid point*. If $\alpha\colon \operatorname{Spf} W \to \langle \mathscr{X} \rangle$ is a rigid point, then, composing with the specialization maps, we obtain the compatible system of maps of formal schemes

$$\alpha_{X'}\colon \operatorname{Spf} W \longrightarrow X'$$

for any admissible blow-up X' of X; conversely, if $\alpha_X\colon \operatorname{Spf} W \to X$ is a map of formal schemes from the formal spectrum of an a-adically complete valuation ring W, then by the valuative criterion, we get a compatible system of such maps into the admissible blow-ups X' of X, and thus get a rigid point, which gives rise to a point in $\langle \mathscr{X} \rangle$ as the image of the closed point.

Hence, similarly to what we have seen in §2.2 in the framework of Zariski's classical birational geometry, we get the following valuative description of points of the Zariski-Riemann space $\langle \mathscr{X} \rangle$:

Proposition 2.12 *Define an equivalence relation $\approx$ on the set of all rigid points generated by the relation $\sim$ given as follows: for rigid points $\alpha\colon \operatorname{Spf} W \to \langle \mathscr{X} \rangle$ and $\alpha'\colon \operatorname{Spf} W' \to \langle \mathscr{X} \rangle$, $\alpha \sim \alpha'$ if there exists an injective map $f\colon W \hookrightarrow W'$ such that $\alpha \circ \operatorname{Spf} f = \alpha'$ and W' dominates W (that is, $\mathfrak{m}_W \subseteq \mathfrak{m}_{W'}$). Then we have a canonical bijection*

$$\left\{ \begin{matrix} \text{isomorphism classes} \\ \text{of rigid points of } \mathscr{X} \end{matrix} \right\} \Big/ {\approx} \;\xrightarrow{\sim}\; \langle \mathscr{X} \rangle. \qquad \square$$

Remark 2.13 Even when we are working over the valuation ring V of height 1, the valuation rings V_x for $x \in \langle \mathscr{X} \rangle$ that appear in the description of points of the Zariski-Riemann space are not necessarily of height 1; indeed, in general, it may have higher height, even if our base V is a complete DVR. This means that, if we pick up a point x that is of height > 1, then we cannot take fibers over this point, as long as we only work over height 1 base valuation rings. Hence, in order to get a more coherent theory of rigid spaces, one cannot stay in the classical setting as before, in which everything is defined over a height 1 valuation ring; one has to enlarge the category of 'models' (as we will do later) so that it contains formal schemes of finite type over an a-adically complete valuation ring of *arbitrary* height.

Affinoids

A coherent rigid space $\mathscr{X}$ is called an *affinoid* if it is of the form X^{rig}, where $X = \operatorname{Spf} A$ is an affine formal scheme (A is a topologically of finite type V-algebra). One can always find, simply by dividing out a-torsion, such an A that is V-flat. Every rigid space is covered by affinoid open subspaces.

In our situation (however, not in general), affinoids are *Stein* for coherent $\mathscr{O}_{\mathscr{X}}$-modules[8]; that is, for any coherent $\mathscr{O}_{\mathscr{X}}$-module $\mathscr{F}$ on an affinoid $\mathscr{X}$ (that is, a coherent sheaf on the ringed space $(\langle\mathscr{X}\rangle, \mathscr{O}_{\mathscr{X}})$), we have $H^q(\mathscr{X}, \mathscr{F}) = 0$ for $q \geq 1$.

One can, in fact, prove Theorem A and Theorem B for coherent sheaves on affinoids, by which one can calculate cohomologies of coherent sheaves by affinoid covering. Theorem A says, for example, that, for an affinoid $\mathscr{X} = X^{\mathrm{rig}}$, where $X = \mathrm{Spf}\, A$, we have $H^0(\mathscr{X}, \mathscr{O}_X) = A[\frac{1}{a}]$.

Here, let us see an example of Theorem A (in case V is a complete DVR): Consider $V\langle\langle X_1, \ldots, X_n\rangle\rangle$, the restricted power series ring, that is, the ring of a-adically convergent formal power series with coefficients in V. The rigid space $\mathbb{D}^n := (\mathrm{Spf}\, V\langle\langle X_1, \ldots, X_n\rangle\rangle)^{\mathrm{rig}}$ is the 'closed unit polydisk', which is the most basic building block in Tate's rigid geometry. We want to show the equality for $\mathscr{X} = \mathbb{D}^n$:

$$\Gamma(\mathscr{X}, \mathscr{O}_{\mathscr{X}}) = V\langle\langle X_1, \ldots, X_n\rangle\rangle[\tfrac{1}{a}] = K\langle\langle X_1, \ldots, X_n\rangle\rangle$$

(the Tate algebra). Set $\mathfrak{X} = \mathrm{Spf}\, V\langle\langle X_1, \ldots, X_n\rangle\rangle$, and observe first that, for any admissible blow-up $\mathfrak{X}' \to \mathfrak{X}$ of , we have $\Gamma(\mathfrak{X}, \mathscr{O}_{\mathfrak{X}}) \cong \Gamma(\mathfrak{X}', \mathscr{O}_{\mathfrak{X}'})$ by GFGA and finiteness theorem. Hence

$$\Gamma(\mathscr{X}, \mathscr{O}_{\mathscr{X}}) = \Gamma(\langle\mathscr{X}\rangle, \mathscr{O}_{\mathscr{X}}^{\mathrm{int}})\left[\tfrac{1}{a}\right] = \left(\varinjlim_{\mathfrak{X}'} \Gamma(\mathfrak{X}', \mathscr{O}_{\mathfrak{X}'})\right)\left[\tfrac{1}{a}\right]$$
$$= V\langle\langle X_1, \ldots, X_n\rangle\rangle\left[\tfrac{1}{a}\right],$$

where the second equality (not only for H^0 but for H^q) is due to [SGA4-2, Exposé VII, Théorème 5.7].

Notice that, as indicated in the above calculation, we often need, in order to show basic results like this, to have fundamental theorems like GFGA theorems and finiteness theorems for formal schemes. In [EGA, **III**], these theorems are proven only in Noetherian cases; but for our purpose, we need them in more general situations. This is one of the reasons why we need to generalize EGA formal geometry.

---

[8] It can be shown, by the way, that the rigid structure sheaf $\mathscr{O}_{\mathscr{X}}$ is coherent as a module over itself; cf. §4.2 below.

2.5 Summary

Figure 4.2 summarizes our approach to rigid geometry. As the figure indicates, our approach has two fundamental backbones, Raynaud's viewpoint of rigid geometry and Zariski's birational geometry.

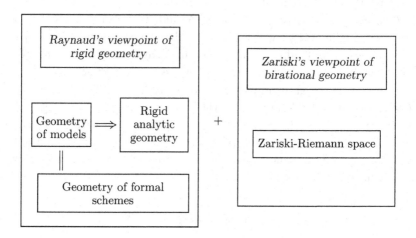

Figure 4.2 Birational approach to rigid geometry

In the course of the explanation of our approach to rigid geometry, we have already indicated several times that a consistent and practical generalization of formal geometry is necessary for full-fledged execution of the program. In the following sections, we will present a part of such attempts.

3 Adhesive rings

3.1 Adic topology

Let A be a ring, $I \subseteq A$ an ideal, and M an A-module. The *I-adic topology*[9] on M is the topology compatible with the additive group structure

---

[9] According to EGA terminology (cf. [EGA, I_{new}, **0**, §7]),

preadic + separated and complete = adic.

This terminology, however, is not frequently adopted nowadays. Also in this paper, we avoid using the terminology 'preadic', and suppose 'adic' does not imply 'separated and complete', except for *adic ring*, which is already widespread to be adically complete (= separated and complete with respect to a preadic topology); cf. the beginning of §4.1.

for which the descending filtration $\{I^n M\}_{n \geq 0}$ gives a fundamental system of open neighborhoods of $0 \in M$. For another ideal $J \subseteq A$, the I-adic topology on M coincides with the J-adic topology if and only if

$$I^n M \subseteq J^m M \subseteq IM$$

holds for some $n, m \geq 0$.

We say that M is I-*adically separated* if $\bigcap_{n \geq 0} I^n M = 0$, or what amounts to the same, the natural map

$$i_M \colon M \longrightarrow \widehat{M} := \varprojlim_{n \geq 0} M/I^n M$$

is injective. We say that M is I-*adically complete* if the map i_M is bijective, that is, M is Hausdorff complete with respect to the I-adic topology.

Let A be a ring, and $I \subseteq A$ a *finitely generated* ideal. Then the I-*adic completion* $\widehat{A} := \varprojlim_{n \geq 0} A/I^n$ is I-adically (that is, $I\widehat{A}$-adically) complete. Let M be an A-module such that M/IM is finitely generated as an A-module. Then the I-adic completion $\widehat{M}$ (defined as above) is I-adically complete, and is finitely generated as an $\widehat{A}$-module.

Consider an A-module M endowed with the I-adic topology by an ideal $I \subseteq A$ (not necessarily finitely generated). If $M \to N$ is a surjective A-linear map, then the quotient topology on N is I-adic. On the other hand, if N is an A-submodule of M, then it is not always the case that the induced topology on N is I-adic. The condition for this to hold is the following:

(∗) for any $n \geq 0$, there exists $m \geq 0$ such that $N \cap I^m M \subseteq I^n N$.

The classical Artin-Rees lemma implies that, if A is Noetherian and M is finitely generated, then the condition is satisfied for any A-submodule $N \subseteq M$.

3.2 Adhesive rings

Let A be a ring, $I \subseteq A$ an ideal, and M an A-module. An element $x \in M$ is said to be a-*torsion* (for $a \in A$) if there exists $n \geq 0$ such that $a^n x = 0$; x is said to be I-*torsion* if it is a-torsion for any $a \in I$. The set of all I-torsion (resp. a-torsion) elements, denoted by $M_{I\text{-tor}}$ (resp. $M_{a\text{-tor}}$), is an A-submodule, called the I-*torsion* (resp. a-*torsion*) *part* of M.

Proposition 3.1 *Let A be a ring, and $I \subseteq A$ a finitely generated ideal. Then the following conditions are equivalent:*

(a) *for any finitely generated A-module M, $M/M_{I\text{-tor}}$ is finitely presented;*

(b) $\mathrm{Spec}\, A \setminus V(I)$ *is a Noetherian scheme, and for any finitely generated A-module M, $M_{I\text{-tor}}$ is finitely generated;*

(c) *for any finitely generated A-module M, and any A-submodule N of M, the I-saturation*

$$\widetilde{N} := \{x \in M \mid \text{for any } a \in I \text{ there exists } n \geq 0 \text{ such that } a^n x \in N\}$$

of N in M is finitely generated.

Proof The equivalence of (a) and (c) follows from [3, Chap. I, §2.8, Lemma 9] and the following facts:

- the I-saturation $\widetilde{N}$ sits in the exact sequence

$$0 \longrightarrow \widetilde{N} \longrightarrow M \longrightarrow (M/N)/(M/N)_{I\text{-tor}} \longrightarrow 0;$$

- for a finitely generated A-module F and a surjective morphism $\phi \colon F \to M/M_{I\text{-tor}}$, $\ker(\phi)$ is I-saturated.

To show (c) $\Rightarrow$ (b), first remark that $M_{I\text{-tor}}$ is nothing but the I-saturation of $\{0\}$ in M, and hence is finitely generated. To show that $\mathrm{Spec}\, A \setminus V(I)$ is a Noetherian scheme, it suffices to show that $A[\frac{1}{a}]$ is a Noetherian ring for any $a \in I$. Let J be an ideal of $A[\frac{1}{a}]$, and J' the pull-back of J by $A \to A[\frac{1}{a}]$. Then J' is easily seen to be I-saturated, and we have $J'A[\frac{1}{a}] = J$. Since J' is finitely generated, so is J.

Finally, we are to show (b) $\Rightarrow$ (c). Let N be a submodule of a finitely generated A-module M. Since I is finitely generated, $\mathrm{Spec}\, A \setminus V(I)$ is quasi-compact, and hence is covered by finitely many affine open subsets of the form $\mathrm{Spec}\, A[\frac{1}{a_i}]$ with $a_i \in I$. Since each $N \otimes_A A[\frac{1}{a_i}]$ is finitely generated, we can find a finitely generated submodule N' of N such that $N' \otimes_A A[\frac{1}{a_i}] \cong N \otimes_A A[\frac{1}{a_i}]$. Since N/N' is I-torsion, we deduce that $\widetilde{N'} = \widetilde{N}$. Thus, we may replace N by N', and hence, may suppose N is finitely generated. Then, by the exact sequence

$$0 \longrightarrow N \longrightarrow \widetilde{N} \longrightarrow (M/N)_{I\text{-tor}} \longrightarrow 0,$$

we deduce that $\widetilde{N}$ is finitely generated. $\qquad\square$

Definition 3.2 We say that A is *I-adically adhesive* if the conditions in 3.1 are satisfied. We say that A is *I-adically universally adhesive* if, for any finite type map $A \to B$, B is IB-adically adhesive.

By definition, if A is I-adically universally adhesive, and B is an A-algebra of finite type, then B is IB-adically universally adhesive.

Proposition 3.3 (1) *If A is I-adically adhesive (resp. I-adically universally adhesive), and $S \subseteq A$ a multiplicative subset, then the localization $S^{-1}A$ is $IS^{-1}A$-adically adhesive (resp. $IS^{-1}A$-adically universally adhesive).*

(2) *If A is I-adically adhesive (resp. I-adically universally adhesive), and B is J-adically adhesive (resp. J-adically universally adhesive), then $A \times B$ is $I \times J$-adically adhesive (resp. $I \times J$-adically universally adhesive).*

(3) *If A is I-adically adhesive, and $A \to B$ is quasi-finite, then B is IB-adically adhesive.*

(4) *Let A be a ring, and $I \subseteq A$ a finitely generated ideal. If a faithfully flat A-algebra B is IB-adically adhesive (resp. IB-adically universally adhesive), then A is I-adically adhesive (resp. I-adically universally adhesive).*

It follows from (3) that, in order to verify that A is I-adically universally adhesive, it suffices to check that the polynomial rings $B = A[X_1, \ldots, X_n]$ are IB-adically adhesive.

Proof By the remark we have just mentioned, the 'universally adhesive' cases in (1), (2), and (4) will follow easily from the corresponding 'adhesive' cases. We thus devote ourselves only to showing the 'adhesive' cases.

(1) Set $B = S^{-1}A$. We are going to check the condition (b) in 3.1. It is easy to see that $\operatorname{Spec} B \setminus V(IB)$ is a Noetherian scheme. Let N be a finitely generated B-module. There exists a finitely generated A-module M such that $M \otimes_A B \cong N$. One verifies easily that $M_{I\text{-tor}} \otimes_A B \cong N_{IB\text{-tor}}$, which is finitely generated, as desired.

(2) We claim that any $I \times J$-torsion free finitely generated $A \times B$-module M is finitely presented, which amounts to checking the condition (a) in 3.1. The $A \times B$-module M is uniquely decomposed as $M = M_A \times M_B$ by a finitely generated A-module M_A and a finitely generated B-module M_B. Moreover, since M is $I \times J$-torsion free, M_A is I-torsion free and M_B is J-torsion free. As M_A is a finitely presented A-module, and M_B is a finitely presented B-module, M is finitely presented, as desired.

(3) The assertion is clear if $A \to B$ is finite. In general, we apply Zariski's Main Theorem ([EGA, **IV**, (18.12.13)]) to reduce to this case, using (1), (2), and (4).

(4) Let M be an finitely generated I-torsion free A-module. We are to show that M is finitely presented. Consider an exact sequence

$$0 \longrightarrow K \longrightarrow A^m \longrightarrow M \longrightarrow 0.$$

Since $K \otimes_A B$ is finitely generated ([3, Chap. I, §2.8, Lemma 9]), we deduce that K is finitely generated by [3, Chap. I, §3.1, Prop. 2]. $\square$

3.3 Some examples

One can consider the notion of I-adically (universally) adhesive rings as a generalization of the notion of Noetherian rings by the following (obvious) facts:

- any Noetherian ring A is I-adically universally adhesive for any ideal $I \subseteq A$;
- a ring A is Noetherian if and only if it is 1-adically adhesive.

More interesting examples are provided by valuation rings. Let V be a valuation ring of non-zero height (that is, V is not a field), and consider a non-zero element $a \in \mathfrak{m}_V \setminus \{0\}$ in the maximal ideal. One considers the a-adic topology on V; recall that, in a valuation ring, every finitely generated ideal is principal.

Proposition 3.4 *In the situation as above, the following conditions are equivalent:*

(a) *V is a-adically adhesive;*
(b) *V is a-adically separated;*
(c) *$V[\frac{1}{a}]$ is a field $(= \mathrm{Frac}(V))$.*

Proof We first show (a) $\Rightarrow$ (b). Suppose that V is a-adically adhesive, and consider the ideal $J = \bigcap_{n \geq 1}(a^n)$. Since J is easily seen to be a-saturated, we deduce that J is finitely generated, and hence is principal, generated by an element $b \in V$. If we write $b = ac$, then c is again an element of $J = (b)$, since it is divisible by any powers of a. If b is non-zero, it would follow that a is a unit in V, and hence we deduce $b = 0$, and thus $J = 0$, as desired.

To show (b) $\Rightarrow$ (c), let $K = \mathrm{Frac}(V)$ be the field of fractions, and take $x \in K \setminus V$. We have $x^{-1} \in \mathfrak{m}_V$. If x^{-1} does not divide any powers a^n $(n \geq 1)$, then x^{-1} must be divisible by all powers of a, and thus we would have $x^{-1} \in J = \bigcap_{n \geq 1}(a^n) = 0$, which is absurd. Hence there exists $n \geq 1$ such that $xa^n \in V$, that is, $x \in V[\frac{1}{a}]$, which shows that $K = V[\frac{1}{a}]$, as desired.

Finally, we show (c) $\Rightarrow$ (a). First notice that the condition (c) implies the following: any non-zero element $b \in V \setminus \{0\}$ divides some power a^n ($n \geq 1$) of a. Hence, for any V-module M, M is torsion free if and only if it is a-torsion free (that is, $M_{a\text{-tor}} = 0$). In particular, any a-torsion free finitely generated V-module M is V-flat, and hence is a free V-module. $\qquad\qquad\square$

3.4 Artin-Rees type property

Let A be a ring, and $I \subseteq A$ a finitely generated ideal. We regard A as a topological ring endowed with the I-adic topology. We say that A has the property **(AP)** (adicness-preserving condition) if the following condition is satisfied:

(AP) for any finitely generated A-module M and any A-submodule $N \subseteq M$, the condition $(*)$ in §3.1 is satisfied, that is, for any $n \geq 0$, there exists $m \geq 0$ such that $N \cap I^m M \subseteq I^n N$.

A slightly stronger condition has been considered in the classical Artin-Rees lemma:

(AR) for any finitely generated A-module M and any A-submodule $N \subseteq M$, there exists a positive integer $c > 0$ such that, for every $n > c$, we have $N \cap I^n M = I^{n-c}(I^c M \cap N)$.

The classical Artin-Rees lemma asserts that, if A is Noetherian, then it satisfies **(AR)**, and hence also **(AP)**. The main assertion in this subsection is the following.

Proposition 3.5 *An I-adically adhesive ring A has the property* **(AP)**.

This is a special case of a more general theorem, which says the following. In general, an I-adically topologized ring A as above is said to have the property **(BT)** (bounded-torsion condition) if for any finitely generated A-module M, its I-torsion part $M_{I\text{-tor}}$ is bounded, that is, there exists $N \geq 0$ such that $I^N M_{I\text{-tor}} = 0$. Clearly, I-adically adhesive rings have this property. Now, the general theorem states that, if A satisfies **(BT)** and is Noetherian outside I (that is, $\operatorname{Spec} A \setminus V(I)$ is Noetherian), then it enjoys **(AP)**.

Here, we omit the proof of this fact (which will be presented in [5] and [7]), and limit ourselves to show the proposition only in the case where I is principal.

Proof of Proposition 3.5 in principal ideal case Set $I = (a)$. Consider the I-saturation $\widetilde{N}$ (cf. Proposition 3.1 (c)) of N in M. One can easily show the equality

$$\widetilde{N} \cap a^n M = a^n \widetilde{N}$$

for $n \geq 0$. Since $\widetilde{N}/N \subseteq (M/N)_{a\text{-tor}}$, and since $(M/N)_{a\text{-tor}}$ is finitely generated, there exists $c > 0$ such that $a^c \widetilde{N} \subseteq N$. Then we have

$$N \cap a^n M = N \cap a^n \widetilde{N} = a^{n-c}(N \cap a^c \widetilde{N}) = a^{n-c}(N \cap a^c M)$$

for any $n > c$. $\qquad\qquad\qquad\qquad\qquad\qquad\qquad\qquad\qquad\qquad\quad$ □

As the above proof shows, A in this particular situation actually satisfies the stronger condition **(AR)**.

Here is a useful consequence of the proposition:

Proposition 3.6 *Let A be an I-adically adhesive ring.*
   (1) *For any finitely generated A-module M, the natural map*

$$M \otimes_A \widehat{A} \longrightarrow \widehat{M}$$

is an isomorphism. In particular, if A is I-adically complete, then every finitely generated A-module is I-adically complete.
   (2) *The natural ring homomorphism $A \to \widehat{A}$ is flat.*

The most important key for the proof of the proposition is the following exactness of the I-adic completion:

Lemma 3.7 *Let A be an I-adically topologized ring by a finitely generated ideal $I \subseteq A$ that verifies the condition **(AP)**. Then, if $L \to M \to N$ is an exact sequence of finitely generated A-modules, the sequence $\widehat{L} \to \widehat{M} \to \widehat{N}$ induced by passage to I-adic completions is again exact.*

This is a special case of [3, Chap. III, §2.12, Lemma 2], and so we omit the proof.

Proof of Proposition 3.6 The assertion (1) in case M is finitely presented can be shown similarly to the proof of [3, Chap. III, §3.4, Theorem 3 (ii)] with the aid of 3.7. In general, since A is Noetherian outside I, M is, so to speak, 'finitely presented outside I'. Then it is an easy exercise to show that there exists a surjective morphism $N \to M$ from a finitely presented A-module N whose kernel K is I-torsion. As the I-torsion K is bounded by our assumption, one readily sees that $K \otimes_A \widehat{A} \to \widehat{K}$ ($\cong K$) is bijective. Now we look at the following commutative diagram with exact rows:

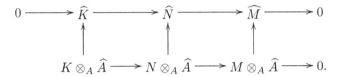

Since the first two vertical arrows are isomorphisms, so is the other one, which concludes the proof of (1). The assertion (2) can be shown from (1) by the same argument as in the proof of [3, Chap. III, §3.4, Theorem 3 (iii)]. $\qquad\square$

By Proposition 3.6 (2), one deduces in particular the following: if A is I-adically adhesive and I-adically complete, then for $f_1, \ldots, f_r \in A$ such that $(f_1, \ldots, f_r) = A$, the natural map

$$A \longrightarrow \prod_{i=1}^{r} A\langle\!\langle f_i^{-1} \rangle\!\rangle$$

is faithfully flat, or equivalently, Zariski localization of the affine formal scheme Spf A is flat.

3.5 Some useful properties

Finally, let us list a few more useful properties that I-adically (universally) adhesive rings have.

First, we have the following *structure sheaf coherency*:

Proposition 3.8 *Let A be an I-adically universally adhesive ring that is I-torsion free. Then any finitely presented A-algebra B is a coherent ring, that is, any finitely generated ideal of B is finitely presented.*

The proof is not difficult, and is left to the reader. From the proposition, it follows in particular that, for any locally of finite presentation scheme X over A as in the proposition, its structure sheaf $\mathscr{O}_X$ is a coherent module over itself.

Next, we have the following *local criterion of flatness*: Let A be a ring with the I-adic topology by a finitely generated ideal $I \subseteq A$, and M an A-module. We write, for each integer $k \geq 0$, $A_k = A/I^{k+1}$ and

$M_k = M/I^{k+1}M$. Then the local criterion of flatness (cf. [3, Chap. III, §5.2, Theorem 1]) states that the following conditions are equivalent:

(a) M *is A-flat;*
(b) $\mathrm{Tor}_1^A(N, M) = 0$ *for any A_0-module N;*
(c) M_0 *is A_0-flat, and* $\mathrm{Tor}_1^A(A_0, M) = 0$;
(d) M_k *is A_k-flat for any $k \geq 0$.*

Now, by an argument similar to that in [3, Chap. III, §5.3], one can prove the following:

Proposition 3.9 *Let $A \to B$ be a ring homomorphism, $I \subseteq A$ a finitely generated ideal of A, and M a finitely generated B-module. Suppose that the ring A (resp. B) is I-adically (resp. IB-adically) adhesive. Suppose moreover that B is I-adically Zariskian (that is, $1 + IB \subseteq B^\times$, this is automatic if B is I-adically complete). Then the above conditions (a), (b), (c), and (d) are equivalent to each other.*

4 Universally adhesive formal schemes

4.1 T.u.a. rings and universally adhesive formal schemes

Let A be an *adic ring*, that is, a topological ring endowed with a *separated and complete* adic topology by an ideal $I \subseteq A$ (cf. [EGA, I$_{\mathrm{new}}$, **0**, §7.1]). We assume that A has a *finitely generated* ideal of definition $I \subseteq A$. Recall that a *topologically finitely generated A-algebra* is an A-algebra isomorphic to one of the form

$$A\langle\!\langle X_1, \ldots, X_r \rangle\!\rangle / \mathfrak{a},$$

where $A\langle\!\langle X_1, \ldots, X_r \rangle\!\rangle$ denotes the so-called *restricted power series ring*, that is, the I-adic completion of the polynomial ring $A[X_1, \ldots, X_r]$, and $\mathfrak{a}$ is an ideal of $A\langle\!\langle X_1, \ldots, X_r \rangle\!\rangle$. In the sequel, unless otherwise clearly stated, topologically finitely generated A-algebras are further assumed to be I-adically complete, or what amounts to the same, the ideal $\mathfrak{a}$ as above is assumed to be *closed*.

Definition 4.1   An adic ring A with a finitely generated ideal of definition $I \subseteq A$ is said to be *topologically universally adhesive*, or *t.u.a.* for short, if, for any $n \geq 0$, the restricted power series ring $A\langle\!\langle X_1, \ldots, X_n \rangle\!\rangle$ is universally adhesive.

If A is a t.u.a. ring, then A itself is I-adically universally adhesive. Hence, as we say in §3.4, the affine formal spectrum Spf A enjoys flat Zariski localization. Moreover, any localization of the form $A\langle\!\langle f^{-1}\rangle\!\rangle$ ($f \in A$) is again t.u.a. Thus, we readily see that, if A is I-torsion free, then the structure sheaf $\mathscr{O}_X$ on the affine formal scheme $X = \text{Spf } A$ is coherent as a module over itself (due to 3.8).

In view of these facts, one arrives at the following definition of a new class of formal schemes:

Definition 4.2 An adic formal scheme X is said to be *universally adhesive* if, for any *of finite type* map of the form Spf $A \to X$ from an affine adic formal scheme, the ring A is universally adhesive.

It can be seen easily that an affine adic formal scheme Spf A is universally adhesive if and only if the ring A is t.u.a. Universally adhesive formal schemes enjoy many of the nice properties that locally Noetherian formal schemes (cf. [EGA, I_{new}, (10.4.2)]) have; for example:

- let $f\colon X \to Y$ be an adic morphism between universally adhesive formal schemes, and suppose that Y has an ideal of definition $\mathscr{I}$; for each $k \geq 0$, set $Y_k = (Y, \mathscr{O}_Y/\mathscr{I}^k)$ and $X_k = (X, \mathscr{O}_X/\mathscr{I}^k\mathscr{O}_X)$, and let $f_k\colon X_k \to Y_K$ be the induced morphism between schemes; then f is locally of finite presentation if and only if f_k is locally of finite presentation for any $k \geq 0$;
- with the notation as above, f is flat (that is, f induces flat maps between stalks of the structure sheaves) if and only if each f_k is flat for $k \geq 0$ (cf. 3.9);
- suppose f is locally of finite presentation, and $\mathscr{O}_Y$ is $\mathscr{I}$-torsion free, then the structure sheaf $\mathscr{O}_X$ of X is coherent as a module over itself (cf. 3.8); in particular, any finitely presented $\mathscr{O}_X$-module is coherent, and vice versa.

4.2 Complete valuation rings

The new notion of complete rings thus introduced would not be quite interesting, unless we have plenty of non-Noetherian examples. In this respect, the following theorem due to Gabber says we actually have a lot of interesting and practical examples:

Theorem 4.3 (O. Gabber) *Let V be a valuation ring of non-zero height, and consider the a-adic topology by $a \in \mathfrak{m}_V \setminus \{0\}$. If V is a-adically complete, then it is a t.u.a. ring.*

Hence, in particular, any topologically of finite type adic ring A over such a valuation ring V is t.u.a., or equivalently, any locally of finite type formal scheme X over Spf V is universally adhesive; notice that, since V is a-torsion free, such a formal scheme X has a coherent structure sheaf (as we have seen in the previous subsection). Since the formal schemes of this form, in particular when V is of height 1, play an important role as formal models of classical rigid geometry (see §2.1), we now have a good chance to generalize Raynaud's picture of rigid geometry.

The proof of the theorem is rather long, referring to a few technically important results; one of them is the following:

Proposition 4.4 *Let A be an adic ring with a finitely generated ideal of definition $I \subseteq A$. Then A is a t.u.a. ring if and only if the restricted power series ring $A\langle\!\langle X_1, \ldots, X_n \rangle\!\rangle$ for each $n \geq 0$ is adhesive.*

The significance of the proposition lies in that the t.u.a.-ness can be checked only by adhesiveness in the definition (4.1), rather than universally adhesiveness, which simplifies substantially the checking.

Although we have to omit the full proof of this result, let us mention that the proof is built up with the following two statements.

The first one is the following fact, which is closely related to the 'descent' type property of adhesiveness:

Lemma 4.5 *Let A be a ring endowed with an adic topology by a finitely generated ideal $I \subseteq A$, and suppose A is Noetherian outside I. Suppose moreover that the following conditions are satisfied:*

(a) *the I-adic completion $\widehat{A}$ is adhesive;*
(b) *$A \to \widehat{A}$ is flat.*

Then A is adhesive.

Second point in the proof of 4.4 is:

Proposition 4.6 *Let A be an adic ring with a finitely generated ideal of definition $I \subseteq A$, and suppose that A is Noetherian outside I. Then the restricted power series ring $A\langle\!\langle X \rangle\!\rangle$ is flat over $A[X]$.*

To show this strong statement, one has to make quite a lot of preparations:

- (due to O. Gabber) an adic ring as in 4.6 satisfies the conditions **(BT)** and **(AP)** (§3.4);

- then one can show that the restricted power series ring $A\langle\!\langle X\rangle\!\rangle$ is flat over A;
- finally, one has the following useful lemma:

Lemma 4.7 (Gluing of flatness) *Let A be a ring, $I \subseteq A$ a finitely generated ideal, and B an A-algebra. Let M be a B-module such that the following conditions are satisfied:*

(a) *B and M are flat over A;*
(b) *M/IM is flat over B/IB, and $\widetilde{M}$ (the associated quasi-coherent sheaf on $\operatorname{Spec} B$) is flat over $\operatorname{Spec} B \setminus V(IB)$.*

Then M is B-flat.

All these details will be discussed with precise proofs in [5] and [7]. Yet one more preparation for the proof of 4.3 is the following (not very difficult) lemma:

Lemma 4.8 *Let A be an a-adically complete V-algebra, and $S \subseteq V$ a multiplicative subset of V such that $S \cap \sqrt{aV} = \emptyset$. Then the localization A_S is a-adically complete, and hence we have the equalities*

$$A_S = A \otimes_V V_S = A\widehat{\otimes}_V V_S$$

up to canonical isomorphisms.

Having all these preparatory results, one is able to show the theorem as follows:

Proof of Theorem 4.3 In view of 4.4, we only need to check that a ring of the form $A = V\langle\!\langle X_1, \ldots, X_n\rangle\!\rangle$ is adhesive. Let M be an a-torsion free finitely generated A-module; we want to show that M is finitely presented.

Take a surjection $A^N \to M$ and consider the exact sequence

$(*)$ $\qquad\qquad 0 \longrightarrow L \longrightarrow A^N \longrightarrow M \longrightarrow 0.$

Then what to prove is that L is finitely generated. This will be shown in two steps.

STEP 1. Suppose the height of V is finite. Since A is a-adically complete, by [11, Theorem 8.4], it is enough to show that L/aL is a finitely generated A/aA-module. This is now equivalent to that M/aM is of finite presentation as an A/aA-module [3, Chap. I, §2.8, Lemma 9]. Since M has no a-torsion, it is V-flat, and hence M/aM is (V/aV)-flat. Now, A/aA is the polynomial ring $A/aA \cong (V/aV)[X_1, \ldots, X_n]$, and $(V/aV)_{red}$ is a valuation ring (possibly of height 0), since $\mathfrak{p} = \sqrt{(a)}$ is a prime ideal. Hence we can apply [13, Théorème (3.4.6)] to conclude that M/aM is of finite presentation as desired.

STEP 2. Now we prove the proposition in general. Let $\mathfrak{p}$ be as above, and set $V' = V_{\mathfrak{p}}$. Then V' is again an a-adically complete valuation ring of, this time, height one. Moreover, by 4.8, we know that $A \otimes_V V'$ is isomorphic to $V'\langle\!\langle X_1, \ldots, X_n \rangle\!\rangle$. Hence one can apply the argument of STEP 1 to conclude that $M \otimes_V V'$ is an $A \otimes_V V'$-module of finite presentation, or equivalently, $L \otimes_V V'$ is a finitely generated $A \otimes_V V'$-module. Hence we can take $x_1, \ldots, x_d \in L$ that generate $L \otimes_V V'$. Now we look at the exact sequence induced by $(*)$

$$(**) \qquad 0 \longrightarrow L/\mathfrak{p}L \longrightarrow (A/\mathfrak{p}A)^N \longrightarrow M/\mathfrak{p}M \longrightarrow 0;$$

note that this is exact, since L is a-saturated (and hence M is V-flat). Since $(**)$ is an exact sequence of modules over a polynomial ring $A/\mathfrak{p}A \cong (V/\mathfrak{p})[X_1, \ldots, X_n]$, and since $M/\mathfrak{p}M$ is flat over $V/\mathfrak{p}$, it follows from [13, Corollaire (3.4.7)] that $L/\mathfrak{p}L$ is a finitely generated $(A/\mathfrak{p}A)$-module. Hence one can take $y_1, \ldots, y_e \in L$ that generates $L/\mathfrak{p}L$.

We finally claim that $x_1, \ldots, x_d, y_1, \ldots, y_e$ generates L as an A-module. Take any $z \in L$. There exist $\alpha_1, \ldots, \alpha_e \in A$ such that

$$z - (\alpha_1 y_1 + \cdots + \alpha_e y_e) \in \mathfrak{p}L.$$

Set $\gamma y = z - (\alpha_1 y_1 + \cdots + \alpha_e y_e)$ ($\gamma \in \mathfrak{p}$, $y \in L$). We can find $\beta_1, \ldots, \beta_d \in A \otimes_V V'$ such that $y = \beta_1 x_1 + \cdots + \beta_d x_d$. Now, since $\mathfrak{p}V' \subseteq V$, $\gamma\beta_i \in A$ for each $i = 1, \ldots, d$, and hence

$$z = \alpha_1 y_1 + \cdots + \alpha_e y_e + (\gamma\beta_1)x_1 + \cdots + (\gamma\beta_d)x_d$$

gives the desired expression by A-linear combination. $\qquad\square$

4.3　Some formal geometry

GFGA comparison theorem

Let A be a universally adhesive ring with respect to a finitely generated ideal $I \subseteq A$, and assume that A is I-torsion free. Consider an A-algebraic space Y locally of finite presentation, and a proper Y-algebraic space $f : X \to Y$ of finite presentation. Let $\widehat{f} : \widehat{X} \to \widehat{Y}$ be the I-adic completion of f. We have the natural commutative diagram

$$
\begin{array}{ccc}
X & \xleftarrow{\ j\ } & \widehat{X} \\
{\scriptstyle f}\downarrow & & \downarrow{\scriptstyle \widehat{f}} \\
Y & \xleftarrow{\ i\ } & \widehat{Y},
\end{array}
$$

which is Cartesian in the category of formal algebraic spaces. It is shown in [7] that the functor $\mathrm{R}f_*$ maps $\mathbf{D}^*_{\mathrm{coh}}(X)$ to $\mathbf{D}^*_{\mathrm{coh}}(Y)$, where $* = $" ", $+, -, \mathrm{b}$.

By means of I-adic completion we have the functor

$$
\mathbf{Mod}_X \longrightarrow \mathbf{Mod}_{\widehat{X}}, \qquad \mathscr{F} \longmapsto \widehat{\mathscr{F}}.
$$

But by 3.6 (1) this functor restricted on coherent $\mathscr{O}_X$-modules is naturally equivalent to the functor

$$
\mathscr{F} \longmapsto j^* \mathscr{F},
$$

which is, by 3.6 (2), exact.

Hence it is reasonable to consider the exact functor

$$
\mathrm{for}\colon \mathbf{Mod}_X \longrightarrow \mathbf{Mod}_{\widehat{X}}, \qquad \mathscr{F} \longmapsto \mathscr{F}^{\mathrm{for}} = j^* \mathscr{F},
$$

which induces an exact functor of triangulated categories

$$
\mathbf{D}^*(X) \longrightarrow \mathbf{D}^*(\widehat{X}),
$$

for $* = $" ", $+, -, \mathrm{b}$. We write the functor thus obtained as

$$
M \longmapsto M^{\mathrm{for}}.
$$

Thus we have the following diagram:

$(*)$

$$
\begin{array}{ccc}
\mathbf{D}^*_{\mathrm{coh}}(X) & \xrightarrow{\ \mathrm{for}\ } & \mathbf{D}^*(\widehat{X}) \\
{\scriptstyle \mathrm{R}f_*}\downarrow & {\scriptstyle \rho}\nearrow & \downarrow{\scriptstyle \mathrm{R}\widehat{f}_*} \\
\mathbf{D}^*_{\mathrm{coh}}(Y) & \xrightarrow[\ \mathrm{for}\ ]{} & \mathbf{D}^*(\widehat{Y}),
\end{array}
$$

where the natural transformation

$$
\rho = \rho_f \colon \mathrm{for} \circ \mathrm{R}f_* \longrightarrow \mathrm{R}\widehat{f}_* \circ \mathrm{for}
$$

(comparison map) is constructed as follows.

As well-known, there exists a canonical natural transformation

$$i^{-1} \circ \mathrm{R}f_* \longrightarrow \mathrm{R}\widehat{f_*} \circ j^{-1}.$$

Indeed, for any object M of $\mathbf{D}(X)$, represent M by a complex $\mathscr{J}^\bullet$ consisting of injective $\mathscr{O}_X$-modules. Then we have the following chain of natural morphisms:

$$i^{-1}\mathrm{R}f_*M \xrightarrow{\sim} i^{-1}f_*\mathscr{J}^\bullet \to \widehat{f_*}j^{-1}\mathscr{J}^\bullet \xleftarrow{\sim} \mathrm{R}\widehat{f_*}j^{-1}M,$$

where the first and the last morphisms are quasi-isomorphisms; note that the last quasi-isomorphy follows from that $j^{-1}\mathscr{J}^\bullet$ gives a flasque resolution of $j^{-1}M$. In view of the fact that the maps i and j are flat (as maps of locally ringed spaces), one can extend the above morphism to

$$(\mathrm{R}f_*M)^{\mathrm{for}} \longrightarrow \mathrm{R}\widehat{f_*}M^{\mathrm{for}};$$

moreover, since the formation of this morphism is canonical, we get the desired natural transformation ρ as above.

In this situation, we show in [7] the following theorem:

Theorem 4.9 (GFGA comparison theorem) *The diagram $(*)$ as above is 2-commutative, that is, the natural transformation ρ gives a natural equivalence for $* =$" ", $+$, $-$, b.*

GFGA existence theorem

Let B be a t.u.a. ring with a finitely generated ideal of definition $I \subseteq B$, and suppose that B is I-torsion free. Let $f\colon X \to Y = \operatorname{Spec} B$ be a proper morphism of algebraic spaces of finite presentation. As before, we have the I-adic completion $\widehat{f}\colon \widehat{X} \to \widehat{Y}$, and the comparison functor $.^{\mathrm{for}}\colon \mathbf{D}^*(X) \longrightarrow \mathbf{D}^*(\widehat{X})$.

Theorem 4.10 (GFGA existence theorem) *In this situation, the comparison functor for bounded derived categories*

$$\mathbf{D}^{\mathrm{b}}_{\mathrm{coh}}(X) \xrightarrow{\mathrm{for}} \mathbf{D}^{\mathrm{b}}_{\mathrm{coh}}(\widehat{X})$$

is an exact equivalence of triangulated categories preserving the canonical cohomology functors.

Finiteness theorem

Theorem 4.11 *Let $f\colon X \to Y$ be a proper morphism of finite presentation between quasi-compact universally adhesive formal schemes, and* suppose locally that $\mathscr{O}_Y$ is $\mathscr{I}$-torsion free, where $\mathscr{I}$ is an ideal of defini-tion. Then the functor $\mathrm{R}f_*$ maps $\mathbf{D}^*_{\mathrm{coh}}(X)$ to $\mathbf{D}^*_{\mathrm{coh}}(Y)$, where $* =$" ", $+, -, \mathrm{b}$.

To show the theorem, we first remark that it is enough (by truncation) to prove only in case $* = \mathrm{b}$; then by a standard argument (induction with respect to amplitudes) one reduce the theorem to the following one.

Theorem 4.12 *Let $f: X \to Y$ be as above. Then, for any coherent $\mathscr{O}_X$-module $\mathscr{F}$, $\mathrm{R}^q f_* \mathscr{F}$ is coherent for any $q \geq 0$.*

A further technical reduction process (involving admissible blow-ups) allows one to assume Y has an *invertible* ideal of definition $\mathscr{I}$. Then the theorem in this case can be shown by an argument similar to that by P. Ullrich[16]. The details will be shown in [7].

5 Rigid geometry

With the topological ring theoretic and formal geometric backgrounds as above, let us now briefly summarize our generalized rigid geometry. Our basic policy of doing it has already been explained in §2, and so the following summary is done according to that thread. Most of the materials and discussions in this section are basically the ones that have been already exhibited in our previous survey [6].

5.1 Coherent rigid spaces

We denote by $\mathbf{AcCFs}^*$ the category whose objects are coherent adic formal schemes that have locally an ideal of definition of finite type, and arrows are adic morphisms. It can be shown that such a formal scheme has in fact a globally defined ideal of definition of finite type.

Admissible blow-ups

Definition 5.1 (Admissible ideal) Let X be an object of $\mathbf{AcCFs}^*$, and $\mathscr{J}$ an ideal of $\mathscr{O}_X$. Then $\mathscr{J}$ is said to be *admissible* if it is an adically quasi-coherent open ideal of finite type.

Here an $\mathscr{O}_X$-module $\mathscr{F}$ is said to be *adically quasi-coherent* if the following conditions are satisfied:

(a) $\mathscr{F}$ is complete with respect to $\mathscr{I}$-adic topology, where $\mathscr{I}$ is an ideal of definition of X;

(b) for any $k \geq 0$, the sheaf $\mathscr{F}_k = \mathscr{F}/\mathscr{I}^{k+1}\mathscr{F}$ is a quasi-coherent sheaf on the scheme $X_k = (X, \mathscr{O}_X/\mathscr{I}^{k+1})$.

Definition 5.2 (Admissible blow-up)   Let X be an object of **AcCFs***, and $\mathscr{J}$ an admissible ideal. The *admissible blow-up* along $\mathscr{J}$ is the morphism of formal schemes

$$X' = \varinjlim_{k \geq 0} \operatorname{Proj}\left(\bigoplus_{n \geq 0} \mathscr{J}^n \otimes \mathscr{O}_{X_k}\right) \longrightarrow X,$$

where $X_k = (X, \mathscr{O}_X/\mathscr{I}^{k+1})$ is the scheme defined as above.

As X' is clearly of finite type over X, X' is again an object of **AcCFs***. Notice that the above definition of admissible blow-ups does not depend on the choice of an ideal of definition $\mathscr{I}$.

Coherent rigid spaces

We can now define rigid spaces in our approach by applying Raynaud's idea.

Definition 5.3 (Coherent rigid spaces)   The category **CRf** of coherent rigid spaces is defined to be the quotient category of **AcCFs*** where all admissible blow-ups are inverted:

$$\mathbf{CRf} = \mathbf{AcCFs}^*/\{\text{admissible blow-ups}\}.$$

We denote the quotient functor **AcCFs*** → **CRf** by

$$X \longmapsto X^{\mathrm{rig}}.$$

For a coherent rigid space $\mathscr{X}$, a *formal model* of $\mathscr{X}$ is defined to be an object X of **AcCFs*** such that $X^{\mathrm{rig}} \cong \mathscr{X}$. A formal model X of $\mathscr{X}$ is said to be *distinguished* if $\mathscr{O}_X$ is $\mathscr{I}$-torsion free, where $\mathscr{I}$ is an ideal of definition of X.

Admissible topology

Definition 5.4   (1) A morphism $\mathscr{U} \to \mathscr{X}$ of coherent rigid spaces is said to be a (*coherent*) *open immersion* if it has as a formal model an open immersion $U \hookrightarrow X$.

(2) Let $\{\mathscr{U}_\alpha \hookrightarrow \mathscr{X}\}$ be a family of open immersions between coherent rigid spaces. We say that the family is a *covering* with respect to the *admissible topology* if it has a finite refinement $\{\mathscr{V}_i \hookrightarrow \mathscr{X}\}$ satisfying the following condition: there exist a formal model X of $\mathscr{X}$ and formal models $V_i \hookrightarrow X$ of $\mathscr{V}_i \hookrightarrow \mathscr{X}$ such that $X = \bigcup V_i$.

The last notion gives rise to a topology on **CRf**, called the admissible topology. The resulting site is denoted by **CRf**$_{ad}$.

5.2 General rigid spaces

The category **AcCFs*** allows "formal birational patching" (cf. §2.1); the following statement is a consequence of the existence of formal birational patching of morphisms.

Proposition 5.5 *Any representable presheaf on* **CRf**$_{ad}$ *is a sheaf.*

This proposition allows a consistent definition of more general rigid spaces.

Definition 5.6 (General rigid spaces) A *general rigid space* is a sheaf $\mathscr{F}$ of sets on the site **CRf**$_{ad}$ such that the following conditions are satisfied:

(a) there exists a surjective map of sheaves

$$\coprod_{\alpha \in L} \mathscr{Y}_\alpha \longrightarrow \mathscr{F},$$

where $\{\mathscr{Y}_\alpha\}_{\alpha \in L}$ is a collection of sheaves represented by coherent rigid spaces;

(b) for $\alpha, \beta \in L$, the map $\mathscr{Y}_\alpha \times_{\mathscr{F}} \mathscr{Y}_\beta \longrightarrow \mathscr{Y}_\alpha$ is isomorphic to the direct limit of a direct system $\{\mathscr{U}_\lambda \to \mathscr{Y}_\alpha\}_{\lambda \in \Lambda}$ of maps between coherent rigid spaces such that all maps in the commutative diagram for $\mu \leq \lambda$

are coherent open immersions.

We denote by **Rf** the category of general rigid spaces. It has **CRf** as a full subcategory. The category **Rf** has a natural topology, constructed and enhanced[10] from the admissible topology of **CRf**, which we also call the *admissible topology*.

---

[10] That is, not only 'coherent open immersions' as defined in 5.4 (1), a 'stretch of coherent open immersions' as in 5.6 (b) are also counted as an open immersion.

5.3 Visualization

Following the argument in §2.4, one has the notion of visualization also
for the rigid spaces in our sense:

Definition 5.7   Let $\mathscr{X} = X^{\mathrm{rig}}$ be a coherent rigid space.
  (1) Define the projective limit

$$\langle \mathscr{X} \rangle = \varprojlim_{X' \to X} X'$$

along all admissible blow-ups of X taken in the category of locally ringed
spaces. Note that, by 2.5, the underlying topological space of $\langle \mathscr{X} \rangle$ is
sober and coherent. The canonical projection map $\langle \mathscr{X} \rangle \to X'$ for any
admissible blow-up X' of X is called the *specialization map*, and is denoted by

$$\mathrm{sp}_{X'} \colon \langle \mathscr{X} \rangle \longrightarrow X'.$$

This is a quasi-compact and closed map.
  (2) The structure sheaf of $\langle \mathscr{X} \rangle$, which is the direct limit of the sheaves
$\mathrm{sp}_{X'}^{-1} \mathscr{O}_{X'}$, is called the *integral structure sheaf*, and is denoted by $\mathscr{O}_{\mathscr{X}}^{\mathrm{int}}$.
  (3) The *rigid structure sheaf* $\mathscr{O}_{\mathscr{X}}$ is the sheaf on $\langle \mathscr{X} \rangle$ defined by

$$\mathscr{O}_{\mathscr{X}} = \varinjlim_{n \geq 0} \mathrm{Hom}_{\mathscr{O}_{\mathscr{X}}^{\mathrm{int}}}(\mathscr{I}^n, \mathscr{O}_{\mathscr{X}}^{\mathrm{int}});$$

here we take an ideal of definition $\mathscr{I}_X$ of X and set $\mathscr{I} = (\mathrm{sp}_X^{-1} \mathscr{I}_X) \mathscr{O}_{\mathscr{X}}^{\mathrm{int}}$.

  Here the definition of $\mathscr{O}_{\mathscr{X}}$ calls for an explanation. It turns out, similarly to 2.11, that the sheaf $\mathscr{O}_{\mathscr{X}}^{\mathrm{int}}$ of local rings is $\mathscr{I}$-valuative, and one
sees that the sheaf $\mathscr{O}_{\mathscr{X}}$ is also a sheaf of local rings.

Definition 5.8 (Zariski-Riemann triple)   We write

$$\mathbf{ZR}(\mathscr{X}) = (\langle \mathscr{X} \rangle, \mathscr{O}_{\mathscr{X}}^{\mathrm{int}}, \mathscr{O}_{\mathscr{X}}),$$

and call it the *Zariski-Riemann triple* associated to the rigid space $\mathscr{X}$.

  One can of course extend the above definition to general rigid spaces
by gluing of topological spaces. The following proposition shows that the
spaces thus obtained are the correct ones to 'visualize' rigid spaces:

Proposition 5.9   *The topos associated to the topological space $\langle \mathscr{X} \rangle$ is
isomorphic to the admissible topos $\mathscr{X}_{\mathrm{ad}}^{\sim}$.*

5.4 Universally adhesive rigid spaces

Although it sometimes happens that one has to really consider the rigid spaces in the fully generalized sense as above, practically more important are the rigid spaces that come locally from universally adhesive formal schemes (4.2):

Definition 5.10 A (general) rigid space $\mathscr{F}$ is called a *universally adhesive* rigid space if it has a covering (as in 5.6) such that each $\mathscr{Y}_\alpha$ ($\alpha \in L$) has a coherent universally adhesive formal scheme as a formal model.

On the other hand, for coherent rigid spaces, one has another way of defining universally adhesiveness; namely, one can define them as the objects in the quotient category

$$\mathbf{AdhCFs}^*/\{\text{admissible blow-ups}\},$$

where **AdhCFs*** denotes the full subcategory of **AcCFs*** consisting of all coherent universally adhesive formal schemes.

One can show that these two possible definitions are equivalent to each other; here is a rough sketch of the proof. Let $\mathscr{F}$ be a coherent rigid space, universally adhesive in the sense of 5.10. Since it is coherent, it can be shown that the covering as in 5.10 has a finite subcovering. Moreover, one can show that each map $\mathscr{Y}_\alpha \to \mathscr{F}$ is represented by a coherent open immersion. Let Y_α be a universally adhesive formal model of $\mathscr{Y}_\alpha$ for each α, and X a formal model of $\mathscr{F}$. By a formal patching argument and careful comparison of topologies, one can replace these formal models by admissible blow-ups such that $\{Y_\alpha\}$ gives a Zariski covering of X; this is done by an argument similar to that in the proof of 2.6. Hence X is universally adhesive, and thus rigid space $\mathscr{F}$ is universally adhesive also in the latter sense.

5.5 GAGA

GAGA functor

Let A be an adic ring with a finitely generated ideal of definition $I \subseteq A$, and set $S = \operatorname{Spec} A$. Let $f\colon X \to U = \operatorname{Spec} A \setminus V(I)$ be a separated U-scheme of finite type. In this situation we consider a *Nagata compactification* of f over S, that is, a commutative diagram of the form

$$
\begin{array}{ccc}
X & \hookrightarrow & \overline{X} \\
{\scriptstyle f}\downarrow & & \downarrow{\scriptstyle \overline{f}} \\
U & \hookrightarrow & S,
\end{array}
$$

where $\overline{f}\colon \overline{X} \to S$ is a proper S-scheme, and $X \hookrightarrow \overline{X}$ is a birational open immersion. Then set

$$Z = (\overline{X} \times_S U) \setminus X,$$
$$\overline{Z} = \text{the closure of } Z \text{ in } \overline{X},$$
$$\widetilde{X} = \overline{X} \setminus \overline{Z}.$$

Let $\widehat{\widetilde{X}} \hookrightarrow \widehat{\overline{X}}$ be the I-adic completion of the open immersion $\widetilde{X} \hookrightarrow \overline{X}$. We have the open immersion

$$(\widehat{\widetilde{X}})^{\mathrm{rig}} \longrightarrow (\widehat{\overline{X}})^{\mathrm{rig}}$$

of coherent rigid spaces.

We define X^{an}, first as a sheaf on the site $\mathbf{Rf}_{\mathscr{S},\mathrm{ad}}$ (the category of general rigid spaces over $\mathscr{S} = (\operatorname{Spf} A)^{\mathrm{rig}}$ endowed with the admissible topology (§5.2)), by

$$X^{\mathrm{an}} = \varinjlim (\widehat{\widetilde{X}})^{\mathrm{rig}},$$

where the inductive limit is taken along the filtered category consisting of all Nagata compactifications as above. Then it can be seen that X^{an} is a quasi-separated rigid space.

Notice that, by the construction, we have always a canonical open immersion

$$(\widehat{\widetilde{X}})^{\mathrm{rig}} \longrightarrow X^{\mathrm{an}}$$

for any Nagata compactification $(X \hookrightarrow \overline{X})$. In particular, if $f\colon X \to U$ is proper, then we have $X^{\mathrm{an}} = (\widehat{\overline{X}})^{\mathrm{rig}}$, which is a coherent rigid space.

A similar construction can be done for morphisms between separated U-schemes of finite type, and thus we get a functor[11]

$$\left\{ \begin{matrix} \text{separated} & \text{schemes} \\ \text{of finite type over } U \end{matrix} \right\} \longrightarrow \left\{ \begin{matrix} \text{rigid spaces over} \\ \mathscr{S} = (\operatorname{Spf} A)^{\mathrm{rig}} \end{matrix} \right\}, \quad X \longmapsto X^{\mathrm{an}}$$

called the *GAGA functor*.

Remark 5.11   We have announced in the workshop that we can extend the GAGA functor to suitable algebraic spaces, due to the following theorem. Here we would like to mention that the GAGA functor for algebraic spaces has been independently constructed by Conrad-Temkin [4].

---

[11] The domain of the functor can be easily extended to the category of all locally of finite type U-schemes.

Theorem 5.12   *Let S be a coherent universally adhesive formal scheme, and $Y \to S$ a formal algebraic space of finite type. Then there exists an admissible blow-up $Y' \to Y$ such that Y' is a formal scheme.*

Remark 5.13   Recently, Fujiwara has shown the following strong Nagata compatification theorem, which allows one to define the GAGA functor for all separated of finite type algebraic spaces with essentially no change in the construction:

Theorem 5.14   *Let Z be a coherent scheme, and $Y \to Z$ a separated of finite type algebraic space over Z. Then there exists a proper Z-scheme $\overline{Y} \to Z$ that admits a dense open immersion $Y \hookrightarrow \overline{Y}$ such that the boundary $\overline{Y} \setminus Y$ is a scheme. Moreover, if Y is a scheme, then $\overline{Y}$ can be chosen to be a scheme.*

Comparison map and comparison functor

We keep the notations as before. One has the canonical morphism

$$\rho_X : (\langle X^{\mathrm{an}}\rangle, \mathscr{O}_{X^{\mathrm{an}}}) \longrightarrow (X, \mathscr{O}_X)$$

of locally ringed spaces. Set-theoretically, this map is roughly obtained as follows. Each point $x \in \langle X^{\mathrm{an}}\rangle$ of the Zariski-Riemann space corresponds, similarly to the classical case as surveyed in §2.4, to a morphism of the form

$$\alpha_x : \operatorname{Spf} V \longrightarrow \langle X^{\mathrm{an}}\rangle$$

(a so-called *rigid point*), where V is an a-adically complete valuation ring ($a \in \mathfrak{m}_V \setminus \{0\}$). Then for any Nagata compactification of X as before, we have the induced map $\operatorname{Spec} V \to \widetilde{X}$, which yields $\operatorname{Spec} V[\frac{1}{a}] \to X$; the last map determines a point of X, which is nothing but the desired point $\rho_X(x)$.

It can be shown that the map ρ_X is flat, that is, induces flat maps between the local rings. Hence we have an exact functor

$$\rho_X^* : \mathbf{Mod}_X \longrightarrow \mathbf{Mod}_{X^{\mathrm{an}}}.$$

In particular, if the adic ring A is t.u.a. and I-torsion free, then one can speak about coherent sheaves on X^{an} (which is a universally adhesive rigid space), and the functor ρ_X^* maps $\mathbf{Coh}_X$ to $\mathbf{Coh}_{X^{\mathrm{an}}}$.

GAGA comparison theorem

Let us assume that the adic ring A is t.u.a., and consider a proper U-morphism

$$f : X \longrightarrow Y$$

of separated and of finite type U-schemes. We have the commutative diagram

$$
\begin{array}{ccc}
X & \xleftarrow{\ \rho_X\ } & (\langle X^{\mathrm{an}}\rangle, \mathscr{O}_{X^{\mathrm{an}}}) \\
\downarrow{\scriptstyle f} & & \downarrow{\scriptstyle f^{\mathrm{an}}} \\
Y & \xleftarrow[\ \rho_Y\ ]{} & (\langle Y^{\mathrm{an}}\rangle, \mathscr{O}_{Y^{\mathrm{an}}})
\end{array}
$$

of locally ringed spaces.

We know that $\mathrm{R}f_*$ maps $\mathbf{D}^*_{\mathrm{coh}}(X)$ to $\mathbf{D}^*_{\mathrm{coh}}(Y)$ for $* = \text{`` ''}, +, -, \mathrm{b}$. On the other hand, the exact functor ρ_X^* induces an exact functor

$$
\mathbf{D}^*(X) \longrightarrow \mathbf{D}^*(X^{\mathrm{an}}), \quad M \longmapsto M^{\mathrm{rig}}.
$$

Then we get the following diagram of triangulated categories:

$$
(*) \qquad
\begin{array}{ccc}
\mathbf{D}^*_{\mathrm{coh}}(X) & \xrightarrow{\ \mathrm{rig}\ } & \mathbf{D}^*_{\mathrm{coh}}(X^{\mathrm{an}}) \;, \\
\downarrow{\scriptstyle \mathrm{R}f_*} & \nearrow^{\rho} & \downarrow{\scriptstyle \mathrm{R}f^{\mathrm{an}}_*} \\
\mathbf{D}^*_{\mathrm{coh}}(Y) & \xrightarrow[\ \mathrm{rig}\ ]{} & \mathbf{D}^*(Y^{\mathrm{an}})
\end{array}
$$

where the natural transformation

$$
\rho = \rho_f \colon \mathrm{rig} \circ \mathrm{R}f_* \longrightarrow \mathrm{R}f^{\mathrm{an}}_* \circ \mathrm{rig}
$$

can be obtained quite similarly to GFGA case (§4.3).

Theorem 5.15 (GAGA comparison theorem) *Suppose $f\colon X \to Y$ is proper. Then the natural transformation ρ gives a natural equivalence; hence the diagram $(*)$ is 2-commutative.*

GAGA existence theorem

We continue with the assumption that A is a t.u.a. ring.

Theorem 5.16 (GAGA existence theorem) *Let $f\colon X \to U$ be a proper U-scheme. Then the comparison functor*

$$
\mathbf{D}^{\mathrm{b}}_{\mathrm{coh}}(X) \xrightarrow{\ \mathrm{rig}\ } \mathbf{D}^{\mathrm{b}}_{\mathrm{coh}}(X^{\mathrm{an}})
$$

is an exact equivalence of triangulated categories.

References

[1] Bosch, S.; Lütkebohmert, W.: *Formal and rigid geometry. I. Rigid spaces*, Math. Ann. **295** (1993), no. 2, 291–317.

[2] Bourbaki, N.: *Elements of mathematics. General topology. Part 1.* Hermann, Paris; Addison-Wesley Publishing Co., Reading, Mass.-London-Don Mills, Ont. 1966.

[3] Bourbaki, N.: *Elements of mathematics. Commutative algebra.* Translated from the French. Hermann, Paris; Addison-Wesley Publishing Co., Reading, Mass., 1972.

[4] Conrad, B.; Temkin, M.: *Non-archimedean analytification of algebraic spaces*, to appear in J. Alg. Geom.

[5] Fujiwara, K.; Gabber, O.; Kato, F.: *On Hausdorff completions of commutative rings in rigid geometry*, preprint.

[6] Fujiwara, K.; Kato, F.: *Rigid geometry and applications*, Advanced Studies in Pure Mathematics **45**, 2006, Moduli Spaces and Arithmetic Geometry (Kyoto, 2004), pp. 327–386.

[7] Fujiwara, K.; Kato, F.: *Foundations of rigid geometry*, in preparation.

[8] Huber, R.: *Continuous valuations*, Math. Z. **212** (1993), no. 3, 455–477.

[9] Huber, R.: *A generalization of formal schemes and rigid analytic varieties*, Math. Z. **217** (1994), no. 4, 513–551.

[10] Johnstone, P.T.: *Stone spaces.* Cambridge Studies in Advanced Mathematics, 3. Cambridge University Press, Cambridge, 1982.

[11] Matsumura, H.: *Commutative ring theory.* Translated from the Japanese by M. Reid. Second edition. Cambridge Studies in Advanced Mathematics, 8. Cambridge University Press, Cambridge, 1989.

[12] Nagata, M.: *A generalization of the imbedding problem of an abstract variety in a complete variety.* J. Math. Kyoto Univ. **3** (1963), 89–102.

[13] Raynaud, M.; Gruson, L.: *Critères de platitude et de projectivité. Techniques de "platification" d'un module.* Invent. Math. **13** (1971), 1–89.

[14] Raynaud, M.: *Géométrie analytique rigide d'après Tate, Kiehl,···.* Table Ronde d'Analyse non archimédienne (Paris, 1972), pp. 319–327, Bull. Soc. Math. France, Mem. No. **39–40**, Soc. Math. France, Paris, 1974.

[15] Tate, J.: *Rigid analytic spaces.* Invent. Math. **12** (1971), 257–289.

[16] Ullrich, P.: *The direct image theorem in formal and rigid geometry.* Math. Ann. **301** (1995), no. 1, 69–104.

[17] Zariski, O.: *The reduction of the singularities of an algebraic surface.* Ann. of Math. (2) **40**, (1939), 639–689.

[18] Zariski, O.: *Local uniformization on algebraic varieties.* Ann. of Math. (2) **41**, (1940), 852–896.

[19] Zariski, O.: *A simplified proof for the resolution of singularities of an algebraic surface.* Ann. of Math. (2) **43**, (1942), 583–593.

[20] Zariski, O.: *The compactness of the Riemann manifold of an abstract field of algebraic functions*, Bull. Amer. Math. Soc. **50**, (1944), 683–691.

[21] Zariski, O.; Samuel, P.: *Commutative algebra.* Volume II. The University Series in Higher Mathematics. D. Van Nostrand Co., Inc., Princeton, N. J.-Toronto-London-New York, 1960.

[EGA] Grothendieck, A., Dieudonné, J.: *Éléments de géométrie algébrique.* Inst. Hautes Études Sci. Publ. Math., no. **4**, **8**, **11**, **17**, **20**, **24**, **28**, **32**, 1961-1967.

[EGA, I_{new}] Grothendieck, A., Dieudonné, J.: *Éléments de géométrie algébrique I.* Die Grundlehren der mathematischen Wissenschaften in Einzeldarstellungen, Band 166, Springer-Verlag, Berlin, Heidelberg, New York, 1971.

[SGA4-2] *Théorie des topos et cohomologie étale des schémas. Tome 2.* Séminaire de Géométrie Algébrique du Bois-Marie 1963–1964 (SGA 4). Dirigé par M. Artin, A. Grothendieck et J. L. Verdier. Avec la collaboration de N. Bourbaki, P. Deligne et B. Saint-Donat. Lecture Notes in Mathematics, Vol. **270**. Springer-Verlag, Berlin-New York, 1972.

5

The Grothendieck ring of varieties

Johannes Nicaise and Julien Sebag

1 Introduction

Since its creation in the middle of the nineties, the theory of motivic integration has been developed in different directions, following a geometric and/or model-theoretic approach. The theory has profound applications in several areas of mathematics, such as algebraic geometry, singularity theory, number theory and representation theory.

A common feature of the different versions of motivic integration is that the integrals take their values in an appropriate *Grothendieck ring*, often the *Grothendieck ring of varieties*. Many applications of motivic integration involve equalities of certain motivic integrals, and hence equalities in the Grothendieck ring of varieties; see, for example, the Batyrev-Kontsevich Theorem [6], which motivated the introduction of motivic integration. Therefore, it is natural to ask for the geometric meaning behind such equalities in the Grothendieck ring.

Unfortunately, the Grothendieck ring of varieties is quite hard to grasp, and little is known about it; many basic and fundamental questions remain unanswered. The central question is arguably the one raised by Larsen and Lunts (see Section 6.2), for which only partial results have been obtained so far.

The present paper is a survey on the Grothendieck ring of varieties. We recall its definition (Section 3), and its main realization maps (Section 4). These realizations constitute the motivic nature of the Grothendieck ring. Besides, we motivate the study of the Grothendieck

Motivic Integration and its Interactions with Model Theory and Non-Archimedean Geometry (Volume I), ed. Raf Cluckers, Johannes Nicaise, and Julien Sebag. Published by Cambridge University Press. © Cambridge University Press 2011.

ring by listing the principal known results, and formulating some challenging open problems (Section 5), which are connected to fundamental questions in algebraic geometry.

Notations, conventions

In this paper, we will tacitly assume that all categories $\mathscr{C}$ under consideration satisfy the property that the isomorphism classes of their objects form a set, which we denote by $\mathfrak{Iso}\mathscr{C}$. If X is an object of $\mathscr{C}$, we denote by $\{X\}$ its class in $\mathfrak{Iso}\mathscr{C}$.

We denote by k a field. We fix an algebraic closure k^a of k, and we denote by k^s the separable closure of k in k^a. We denote by $G_k := \mathrm{Gal}(k^s/k)$ the absolute Galois group of k. We fix a prime number ℓ different from the characteristic p of k. A k-variety is a reduced separated k-scheme of finite type. We denote by Var_k the category of k-varieties, and by (sft/k) the category of separated k-schemes of finite type. These are full subcategories of the category of k-schemes Sch_k. The full subcategory of Var_k whose objects are the smooth projective k-varieties is denoted by PSm_k. For every scheme S, we denote by S_{red} its maximal reduced closed subscheme.

Recall that two connected smooth proper k-varieties X, Y are called *stably birational* if there exist integers $m, n \geq 0$ such that $X \times_k \mathbb{P}_k^m$ and $Y \times_k \mathbb{P}_k^n$ are birational. Stable birationality is an equivalence relation. We say that an integral k-variety X is *uniruled* if there exists an integral algebraic k-variety Y and a dominant, generically finite rational map $Y \times_k \mathbb{P}_k^1 \dashrightarrow X$. The *Kodaira dimension* of a connected smooth proper k-variety X is denoted by $\kappa(X)$.

We denote by $\mathrm{CH}_i(X)$ ($\mathrm{CH}^i(X)$) the Chow group of i-dimensional (i-codimensional) cycles on X modulo rational equivalence. If X and Y are smooth projective k-varieties, a *correspondence* of degree $r \in \mathbb{Z}$ between X and Y is an element of

$$\mathrm{Corr}^r(X, Y) = \bigoplus_{X_i \in \pi_0(X)} \mathrm{CH}^{\dim(X_i)+r}(X_i \times_k Y) \otimes_{\mathbb{Z}} \mathbb{Q}.$$

Here $\pi_0(X)$ denotes the set of connected components of X. If Z is another smooth and projective k-variety, and s is an element of $\mathbb{Z}$, then using the intersection product on $X \times Y \times Z$, one defines a bilinear composition morphism

$$\mathrm{Corr}^r(X, Y) \otimes \mathrm{Corr}^s(Y, Z) \to \mathrm{Corr}^{r+s}(X, Z) : f \otimes g \mapsto g \circ f.$$

2 What is a Grothendieck group?

Grothendieck groups appear in various forms in several branches of mathematics. In this section and the following one, we recall some classical constructions of such objects.

2.1 The Grothendieck group of an additive category

The most basic construction of the Grothendieck group is the following. Let $\mathscr{C}$ be a category that admits finite sums (e.g. an additive category). If X and Y are two objects of $\mathscr{C}$, then we denote by $X \oplus Y$ the sum of X and Y in $\mathscr{C}$.

Definition 2.1 We denote by $K_0^{\mathrm{ad}}(\mathscr{C})$ the quotient of the free abelian group $\mathbb{Z}[\mathfrak{Iso}_\mathscr{C}]$ by the subgroup generated by the elements of the form $\{X \oplus Y\} - \{X\} - \{Y\}$, with X and Y objects in $\mathscr{C}$. We denote by $[X]$ the class of X in $K_0^{\mathrm{ad}}(\mathscr{C})$. We call $K_0^{\mathrm{ad}}(\mathscr{C})$ the (additive) Grothendieck group of the category $\mathscr{C}$.

Example 2.2 *The additive Grothendieck group of $\mathbb{Z}$-modules of finite type.* Let us denote by $\mathrm{Mod}_\mathbb{Z}^{\mathrm{ft}}$ the category of $\mathbb{Z}$-modules of finite type. Let M be an object in $\mathrm{Mod}_\mathbb{Z}^{\mathrm{ft}}$. By the structure theorem for finitely generated modules over principal ideal domains, one knows that there exist prime numbers $p_1, \ldots, p_t$ and integers $r \geq 0$ and $n_1, \ldots, n_t > 0$ such that

$$M \simeq \mathbb{Z}^r \oplus (\mathbb{Z}/(p_1)^{n_1}\mathbb{Z}) \oplus \cdots \oplus (\mathbb{Z}/(p_t)^{n_t}\mathbb{Z}) \, .$$

This decomposition is unique up to permutation of the summands. If we denote by

$$\mathbb{Z}[r, \{(p^n) \,|\, p \text{ prime number}, n \in \mathbb{Z}_{>0}\}]$$

the free abelian group generated by the symbols r and (p^n), then the unique group morphism

$$\mathbb{Z}[r, \{(p^n) \,|\, p \text{ prime number}, n \in \mathbb{Z}_{>0}\}] \to K_0^{\mathrm{ad}}(\mathrm{Mod}_\mathbb{Z}^{\mathrm{ft}}),$$

that sends r to $[\mathbb{Z}]$ and (p^n) to $[\mathbb{Z}/p^n\mathbb{Z}]$ is an isomorphism.

Example 2.3 *The Grothendieck ring of rational Chow motives.* We refer to [2, 39] for a general introduction to the theory of Chow motives. The category $\mathrm{Mot}_{k,\mathbb{Q}}^{\mathrm{eff}}$ of effective Chow motives with rational coefficients over k is a category whose objects are the pairs (X, p), with X an object of PSm_k and p a *projector* on X, i.e., an element in $\mathrm{Corr}^0(X, X)$ such that $p \circ p = p$. Its morphisms are given by

$$\mathrm{Hom}_{\mathrm{Mot}^{\mathrm{eff}}_{k,\mathbb{Q}}}((X, p), (Y, q)) = q \circ \mathrm{Corr}^0(X, Y) \circ p.$$

The object

$$\mathbb{L}_{\mathrm{mot}} := (\mathbb{P}^1_k, p)$$

of $\mathrm{Mot}_{k,\mathbb{Q}}$, with p the class of the cycle

$$\mathbb{P}^1_k \times \{0\} \subset \mathbb{P}^1_k \times_k \mathbb{P}^1_k,$$

is called the *Lefschetz motive*.

The category $\mathrm{Mot}_{k,\mathbb{Q}}$ of Chow motives with rational coefficients over k, is the category whose objects are triples (X, p, m) with X an object of PSm_k, p a projector on X and $m \in \mathbb{Z}$, and whose morphisms are given by

$$\mathrm{Hom}_{\mathrm{Mot}_{k,\mathbb{Q}}}((X, p, m), (Y, q, n)) = q \circ \mathrm{Corr}^{n-m}(X, Y) \circ p.$$

There is a natural full embedding of categories

$$\mathrm{Mot}^{\mathrm{eff}}_{k,\mathbb{Q}} \to \mathrm{Mot}_{k,\mathbb{Q}} : (X, p) \mapsto (X, p, 0).$$

In $\mathrm{Mot}_{k,\mathbb{Q}}$, the Lefschetz motive $\mathbb{L}_{mot}$ is isomorphic to the motive

$$(\mathrm{Spec}(k), \mathrm{id}, -1).$$

The categories $\mathrm{Mot}^{\mathrm{eff}}_{k,\mathbb{Q}}$ and $\mathrm{Mot}_{k,\mathbb{Q}}$ are additive. We define the Grothendieck group of (effective) Chow motives by

$$K_0(\mathrm{Mot}^{\mathrm{eff}}_{k,\mathbb{Q}}) := K_0^{\mathrm{ad}}(\mathrm{Mot}^{\mathrm{eff}}_{k,\mathbb{Q}})$$
$$K_0(\mathrm{Mot}_{k,\mathbb{Q}}) := K_0^{\mathrm{ad}}(\mathrm{Mot}_{k,\mathbb{Q}})$$

The fibered product over k in PSm_k induces a tensor structure on $\mathrm{Mot}^{\mathrm{eff}}_{k,\mathbb{Q}}$ and $\mathrm{Mot}_{k,\mathbb{Q}}$. This tensor structure endows the two Grothendieck groups of motives with a ring structure. The natural full embedding

$$\mathrm{Mot}^{\mathrm{eff}}_{k,\mathbb{Q}} \to \mathrm{Mot}_{k,\mathbb{Q}}$$

is additive and compatible with the tensor product, so it yields a ring morphism

$$\rho : K_0(\mathrm{Mot}^{\mathrm{eff}}_{k,\mathbb{Q}}) \to K_0(\mathrm{Mot}_{k,\mathbb{Q}}).$$

This ring morphism localizes to an isomorphism

$$K_0(\mathrm{Mot}^{\mathrm{eff}}_{k,\mathbb{Q}})([\mathbb{L}_{\mathrm{mot}}]^{-1}) \cong K_0(\mathrm{Mot}_{k,\mathbb{Q}}).$$

The inverse of $[\mathbb{L}_{\mathrm{mot}}]$ in $K_0(\mathrm{Mot}_{k,\mathbb{Q}})$ is the class of the motive $(\mathrm{Spec}(k), \mathrm{id}, 1)$.

2.2 The Grothendieck group of an exact category

Let $\mathscr{C}$ be an exact category, i.e., a strictly full additive subcategory of an abelian category $\mathscr{A}$ that is closed under extensions in $\mathscr{A}$. A sequence

$$X_1 \to X_2 \to X_3$$

in $\mathscr{C}$ is exact iff

$$0 \to X_1 \to X_2 \to X_3 \to 0$$

is a short exact sequence in $\mathscr{A}$.

Definition 2.4 We denote by $K_0^{\mathrm{ex}}(\mathscr{C})$ the quotient of the free abelian group $\mathbb{Z}[\mathfrak{Iso}_\mathscr{C}]$ by the subgroup generated by the elements of the form $\{X\} - \{X'\} - \{X''\}$, with X, X', and X'' objects in $\mathscr{C}$ that fit into an exact sequence $X' \to X \to X''$. We denote by $[X]$ the class of X in $K_0^{\mathrm{ex}}(\mathscr{C})$. We call $K_0^{\mathrm{ex}}(\mathscr{C})$ the (exact) Grothendieck group of $\mathscr{C}$.

Remark 2.5 In the same way, one defines the Grothendieck group of a so-called *triangulated category*, replacing exact sequences by distinguished triangles.

If $\mathscr{C}$ is abelian, then Definition 2.4 implies the following relation for all exact sequences. An exact sequence of length a

$$0 \to X_1 \to \cdots \to X_i \to \cdots \to X_a \to 0$$

decomposes into $a - 2$ short exact sequences, introducing kernels and cokernels. It follows easily that

$$\sum_{i=1}^{a} (-1)^{i-1}[X_i] = 0.$$

Note that an exact category $\mathscr{C}$ is additive, and that there exists a unique morphism of groups

$$K_0^{\mathrm{ad}}(\mathscr{C}) \to K_0^{\mathrm{ex}}(\mathscr{C})$$

sending $[X]$ to $[X]$ for every object X of $\mathscr{C}$. This is not an isomorphism, in general, as is shown by the following example (compare to Example 2.2).

Example 2.6 *The exact Grothendieck group of* $\mathbb{Z}$*-modules of finite type.* The category $(\mathrm{Mod}_{\mathbb{Z}}^{\mathrm{ft}})$ is abelian. For each $n \in \mathbb{Z}_{>0}$, we have a short exact sequence

$$0 \to \mathbb{Z} \xrightarrow{n \cdot} \mathbb{Z} \to \mathbb{Z}/n\mathbb{Z} \to 0$$

in $(\mathrm{Mod}_{\mathbb{Z}}^{\mathrm{ft}})$, where the second arrow is multiplication by n.

This implies that $[\mathbb{Z}/n\mathbb{Z}] = 0$ in $K_0^{\mathrm{ex}}(\mathrm{Mod}_{\mathbb{Z}}^{\mathrm{ft}})$. It follows easily that the morphism

$$\mathbb{Z} \to K_0^{\mathrm{ex}}(\mathrm{Mod}_{\mathbb{Z}}^{\mathrm{ft}}) : a \mapsto a \cdot [\mathbb{Z}]$$

is an isomorphism.

Example 2.7 *The Grothendieck group of locally free* $\mathscr{O}_X$*-modules of finite type.* Let X be a Noetherian scheme, and let $\mathscr{P}_X$ be the category of locally free coherent $\mathscr{O}_X$-modules. This is an exact subcategory of the abelian category of coherent $\mathscr{O}_X$-modules. We denote by $K(X)$ the Grothendieck group $K_0^{\mathrm{ex}}(\mathscr{P}_X)$.

If R is a Noetherian commutative ring, we denote by $\mathscr{P}(R)$ the exact category of finitely generated projective R-modules. In classical algebraic K-theory, one defines the *zeroth group of K-theory* $K_0(R)$ by

$$K_0(R) := K_0^{\mathrm{ad}}(\mathscr{P}(R)).$$

Since a short exact sequence of projective R-modules is split, and the exact categories $\mathscr{P}(R)$ and $\mathscr{P}_{\mathrm{Spec}\,R}$ are equivalent, we have

$$K_0(R) \simeq K_0^{\mathrm{ex}}(\mathscr{P}(R)) \simeq K(\mathrm{Spec}(R)).$$

If F is a field, then the unique morphism

$$K_0(F) \to \mathbb{Z}$$

mapping the class of a finite-dimensional F-vector space V to the dimension of V is an isomorphism.

Example 2.8 *The Grothendieck group of mixed Hodge structures.* See, for instance, [9, 10, 36, 43] for an introduction to Deligne's mixed Hodge theory.

Let n be an integer. A $\mathbb{Q}$*-Hodge Structure* (abbreviated $\mathbb{Q}$-HS) of weight n is a couple $V = (V_{\mathbb{Q}}, F^{\bullet})$ such that

1. $V_{\mathbb{Q}}$ is a $\mathbb{Q}$-vector space of finite dimension;
2. $(F^p, p \in \mathbb{Z})$ is a *finite, decreasing* filtration on $V_{\mathbb{C}} = V_{\mathbb{Q}} \otimes_{\mathbb{Q}} \mathbb{C}$, by $\mathbb{C}$-subspaces (it is called the *Hodge filtration*), such that $F^{n+1-p} \cap \overline{F^p} = \{0\}$ for every $p \in \mathbb{Z}$ (here $\overline{(\cdot)}$ stands for complex conjugation).

A $\mathbb{Q}$-*Mixed Hodge Structure*, abbreviated $\mathbb{Q} - $ MHS, is a triple $V = (V_{\mathbb{Q}}, W_{\bullet}, F^{\bullet})$ such that

1. $V_{\mathbb{Q}}$ is a $\mathbb{Q}$-vector space of finite dimension;
2. $(W_n, n \in \mathbb{Z})$ is a *finite, increasing* filtration on $V_{\mathbb{Q}}$, by $\mathbb{Q}$-subspaces (it is called the *weight filtration*);
3. $(F^p, p \in \mathbb{Z})$ is a *finite, decreasing* filtration on $V_{\mathbb{C}} = V_{\mathbb{Q}} \otimes_{\mathbb{Q}} \mathbb{C}$, by $\mathbb{C}$-subspaces (it is called the *Hodge filtration*);

and such that

$$(gr_n^W V_{\mathbb{Q}}; F^{\bullet} gr_n^W V_{\mathbb{C}})$$

is a pure $\mathbb{Q}$-Hodge structure of weight n, for each $n \in \mathbb{Z}$. A *morphism* of $\mathbb{Q} - $ MHS is a morphism of $\mathbb{Q}$-vector spaces, compatible with the weight filtration $W_{\bullet}$, and (when tensored by $\mathbb{C}$) with the Hodge filtration $F^{\bullet}$. The category $(\mathbb{Q} - $ MHS$)$ of $\mathbb{Q}$-mixed Hodge structures is abelian, and carries a natural tensor product.

We define the Grothendieck group of mixed Hodge structures by

$$K_{\mathbb{Q}H} := K_0^{\text{ex}}(\text{MHS}_{\mathbb{Q}}).$$

The tensor product on $(\mathbb{Q} - $ MHS$)$ defines a ring structure on $K_{\mathbb{Q}H}$. The *Hodge-Deligne polynomial* of a $\mathbb{Q}$-mixed Hodge structure α is given by

$$HD(\alpha; u, v) = \sum_{p,q \in \mathbb{Z}} (\dim_{\mathbb{C}} gr_F^p gr_{p+q}^W \alpha) u^p v^q \qquad \in \mathbb{Z}[u, v, u^{-1}, v^{-1}].$$

There exists a unique ring morphism

$$HD : K_{\mathbb{Q}H} \to \mathbb{Z}[u, v, u^{-1}, v^{-1}]$$

such that $HD([\alpha]) = HD(\alpha; u, v)$ for every object α of $(\mathbb{Q} - $ MHS$)$.

Let X be a separated $\mathbb{C}$-scheme of finite type. For every $i \in \mathbb{N}$, we denote by $H_c^i(X, \mathbb{Q})$ the degree n rational singular cohomology with compact support of $X(\mathbb{C})$ (w.r.t. the complex topology). By a fundamental result of Deligne's, this cohomology space carries a natural $\mathbb{Q} - $ MHS. We define the Hodge characteristic of X by

$$\chi_{Hodge}(X) := \sum_{i \geq 0} (-1)^i [H_c^i(X, \mathbb{Q})] \in K_{\mathbb{Q}H}.$$

The Hodge-Deligne polynomial of X is defined by

$$HD(X; u, v) := HD(\chi_{Hodge}(X)) \qquad \in \mathbb{Z}[u, v]$$

(it is contained in $\mathbb{Z}[u, v]$ because $F^p H_c^i(X, \mathbb{Q}) = H_c^i(X, \mathbb{Q})$ for $p \leq 0$).

Example 2.9 *The Grothendieck group of ℓ-adic Galois representations.* For background, we refer to [23, §6]. A finite-dimensional $\mathbb{Q}_\ell$-vector space endowed with a continuous action of G_k is called a *continuous representation of G_k*, or *ℓ-adic Galois representation*.

We denote by $\mathrm{Rep}_{G_k}\mathbb{Q}_\ell$ the category of ℓ-adic Galois representations of k. It is an abelian category, and its carries a natural tensor structure, given by the tensor product of representations. The tensor structure defines a ring structure on the Grothendieck group

$$K_0(\mathrm{Rep}_{G_k}\mathbb{Q}_\ell) := K_0^{\mathrm{ex}}(\mathrm{Rep}_{G_k}\mathbb{Q}_\ell)$$

by setting $[V \otimes_{\mathbb{Q}_\ell} V'] = [V] \cdot [V']$, for all ℓ-adic Galois representations V, V'.

Let X be a separated k-scheme of finite type. For every $i \geq 0$, we denote by $H_c^i(X \times_k k^s, \mathbb{Q}_\ell)$ the degree i ℓ-adic cohomology space with compact supports of X. This is a finite-dimensional $\mathbb{Q}_\ell$-vector space, naturally endowed with a continuous action of G_k. We define the étale characteristic of X by

$$\chi_{\mathrm{\acute{e}t}}(X) = \sum_{i \geq 0} (-1)^i [H_c^i(X, \mathbb{Q}_\ell)] \in K_0(\mathrm{Rep}_{G_k}\mathbb{Q}_\ell).$$

If the field k is finitely generated, one can refine the construction, replacing $\mathrm{Rep}_{G_k}\mathbb{Q}_\ell$ by the full abelian subcategory $\mathrm{WRep}_{G_k}\mathbb{Q}_\ell$ of *mixed* ℓ-adic Galois representations. Such objects carry a natural weight filtration W_n, $n \in \mathbb{Z}$. By [23, Lemma 6.8.2], $H_c^i(X, \mathbb{Q}_\ell)$ is an object of $\mathrm{WRep}_{G_k}\mathbb{Q}_\ell$ for each i, so that we can consider the mixed étale characteristic

$$\chi_{W\mathrm{\acute{e}t}}(X) := \sum_{i \geq 0} (-1)^i [H_c^i(X, \mathbb{Q}_\ell)] \in K_0(\mathrm{WRep}_{G_k}\mathbb{Q}_\ell).$$

The *weight polynomial* of an object α in $\mathrm{WRep}_{G_k}\mathbb{Q}_\ell$ is given by

$$W(\alpha; T) = \sum_{n \in \mathbb{Z}} [gr_n^W \alpha] T^n \quad \in K_0(\mathrm{Rep}_{G_k}\mathbb{Q}_\ell)[T, T^{-1}]$$

where $W_\bullet \alpha$ denotes the weight filtration on α. There exists a unique ring morphism

$$W : K_0(\mathrm{WRep}_{G_k}\mathbb{Q}_\ell) \to K_0(\mathrm{Rep}_{G_k}\mathbb{Q}_\ell)[T, T^{-1}]$$

such that $W([\alpha]) = W(\alpha; T)$ for every mixed ℓ-adic representation α of k. Composing W with the forgetful morphism

$$K_0(\mathrm{Rep}_{G_k}\mathbb{Q}_\ell)[T, T^{-1}] \to K_0(\mathbb{Q}_\ell)[T, T^{-1}] \cong \mathbb{Z}[T]$$

we find the *Poincaré polynomial*

$$P : K_0(\mathrm{WRep}_{G_k} \mathbb{Q}_\ell) \to \mathbb{Z}[T, T^{-1}].$$

We define the Poincaré polynomial of a separated k-scheme of finite type X by

$$P(X;T) = P(\chi_{W\acute{e}t}(X)) \quad \in \mathbb{Z}[T]$$

(it is contained in $\mathbb{Z}[T]$ for reasons of weight). This definition can be generalized to arbitrary base fields k; see [32, §8] and Example 4.4.

3 The Grothendieck ring of varieties

3.1 Definition and functoriality

Definition 3.1 Let S be a Noetherian scheme. We define the Grothendieck group of S-varieties $K_0(\mathrm{Var}_S)$ as the quotient of the free abelian group generated by the isomorphism classes $\{X/S\}$ of separated S-schemes of finite type X, by the subgroup generated by the elements of the form $\{X/S\} - \{(X-Y)/S\} - \{Y/S\}$, with X a separated S-scheme of finite type and Y a closed subscheme of X. We denote by $[X/S]$ the class of X in $K_0(\mathrm{Var}_S)$.

The Grothendieck ring of varieties does not fit directly into one of the above frameworks; however, we'll see in Section 6.1 that it can be realized as an additive Grothendieck ring of constructible schemes.

We put $\mathbb{L}_S = [\mathbb{A}^1_S/S]$ and $\mathscr{M}_S = K_0(\mathrm{Var}_S)[\mathbb{L}_S^{-1}]$. If the base scheme S is clear from the context, we write $[X]$ and $\mathbb{L}$ instead of $[X/S]$ and $\mathbb{L}_S$. If S is affine, say $S = \mathrm{Spec}(A)$, then we write $K_0(\mathrm{Var}_A)$ and $\mathscr{M}_A$ instead of $K_0(\mathrm{Var}_S)$ and $\mathscr{M}_S$.

There exists a unique ring product on $K_0(\mathrm{Var}_S)$ such that

$$[X/S] \cdot [Y/S] = [(X \times_S Y)/S]$$

for all S-schemes of finite type X and Y. We call the resulting ring the *Grothendieck ring of S-varieties*. The identity element in this ring is the class $[S]$ of the base scheme. When $S = \mathrm{Spec}(k)$, the above definition yields the Grothendieck ring $K_0(\mathrm{Var}_k)$ of k-varieties. This object has been introduced by Grothendieck in a letter to Serre [16, letter of 16/8/1964].

If $f : T \to S$ is a morphism of Noetherian schemes, then there exists a unique ring morphism

$$f^* : K_0(\mathrm{Var}_S) \to K_0(\mathrm{Var}_T)$$

such that $f^*[X] = [X \times_S T]$ for every separated S-scheme X of finite type. It localizes to a ring morphism

$$f^* : \mathscr{M}_S \to \mathscr{M}_T.$$

If $g : S \to U$ is a separated morphism of finite type between Noetherian schemes, then there exists a unique morphism of abelian groups

$$g_! : K_0(\mathrm{Var}_S) \to K_0(\mathrm{Var}_U)$$

such that for every separated S-scheme X of finite type, we have $g_![X] = [X|_U]$ (here $X|_U$ denotes the U-scheme obtained by composing the structural morphism $X \to S$ with the morphism g). Moreover, there exists a unique morphism of abelian groups

$$g_! : \mathscr{M}_S \to \mathscr{M}_U$$

such that

$$g_!([X]\mathbb{L}_S^i) = [X|_U]\mathbb{L}_U^i$$

for every separated S-scheme of finite type X and every integer i. Beware that $g_!$ maps $\mathbb{L}_S^i = [S]\mathbb{L}_S^i$ to $[S|_U]\mathbb{L}_U^i$.

3.2 Scissor relations and piecewise isomorphisms.

The relations imposed in the definition of the Grothendieck ring are called *scissor relations*, because they allow you to cut a variety into locally closed pieces. Let us look at some easy implications. Let S be a Noetherian scheme.

1. For every integer $m > 0$, we have

$$[\mathbb{P}_S^m] = 1 + \mathbb{L} + \mathbb{L}^2 + \cdots + \mathbb{L}^m$$

   in $K_0(\mathrm{Var}_S)$.

2. Let X be a separated S-scheme of finite type. Applying the scissor relations to the closed immersion $X_{red} \to X$, we find that $[X] = [X_{red}]$ in $K_0(\mathrm{Var}_S)$. It follows easily that the base change morphisms $K_0(\mathrm{Var}_S) \to K_0(\mathrm{Var}_{S_{red}})$ and $\mathscr{M}_S \to \mathscr{M}_{S_{red}}$ are ring isomorphisms.

3. A morphism of S-schemes $f : Y \to X$ is called a *piecewise isomorphism* if there exists a finite partition of X into reduced subschemes $X_1, \ldots, X_r$ such that the morphism

$$(Y \times_X X_i)_{red} \to X_i$$

is an isomorphism for each i. In particular, for every partition of X into reduced subschemes $X_1, \ldots, X_r$, the morphism $\sqcup_{i=1}^r X_i \to X$ is a piecewise isomorphism.

We call two S-schemes *piecewise isomorphic* if they can be joint by a chain of piecewise isomorphisms of S-schemes. The scissor relations immediately imply that piecewise isomorphic separated S-schemes of finite type define the same class in the Grothendieck ring of S-varieties.

4. If X is a separated S-scheme of finite type, and C is a constructible subset of X, then we can write C as a finite disjoint union of locally closed subsets $C_1, \ldots, C_r$ of X. We endow each C_i with its induced reduced structure. By the scissor relations, the element

$$[C] := \sum_{i=1}^r [C_i]$$

in $K_0(\mathrm{Var}_S)$ is independent of the chosen decomposition.

5. Let X, Y, and Z be separated S-schemes of finite type. A morphism of S-schemes $f : Y \to X$ is called a *piecewise trivial fibration* with fiber Z if there exists a finite partition of X into subschemes $X_1, \ldots, X_r$ and an isomorphism of X_i-schemes

$$(Y \times_X X_i)_{red} \to (X_i \times_S Z)_{red}$$

for every i. Then, in particular, the S-schemes Y and $X \times_S Z$ are piecewise isomorphic, so that they define the same class in the Grothendieck ring of S-varieties. It follows that $[Y] = [X] \cdot [Z]$ in $K_0(\mathrm{Var}_S)$.

3.3 Bittner's presentation.

If k has characteristic zero, Bittner has given another presentation of $K_0(\mathrm{Var}_k)$ [5, 3.1], which is crucial for the construction of several realization morphisms.

Definition 3.2 Let k be a field. We denote by $K_0^{(bl)}(\mathrm{Var}_k)$ the abelian group given by

- generators: isomorphism classes $[X]_{bl}$ of smooth, projective k-varieties X
- relations: $[\emptyset]_{bl} = 0$, and if Y is a closed subvariety of X, smooth over k, $X' \to X$ is the blow-up of X with center Y, and $E = X' \times_X Y$ is the exceptional divisor, then $[X']_{bl} - [E]_{bl} = [X]_{bl} - [Y]_{bl}$ ("blow-up relations").

The product $[X_1]_{bl} \cdot [X_2]_{bl} = [X_1 \times_k X_2]_{bl}$ defines a ring structure on $K_0^{(bl)}(\mathrm{Var}_k)$. Note that the product is well-defined, since blow-ups commute with flat base change. The ring $K_0^{(bl)}(\mathrm{Var}_k)'$ is defined in the same way, replacing "projective" by "proper". Since a blow-up is an isomorphism over the complement of its center, it follows immediately from the definition that there exist unique ring morphisms

$$\alpha : K_0^{(bl)}(\mathrm{Var}_k) \to K_0(\mathrm{Var}_k)$$
$$\alpha' : K_0^{(bl)}(\mathrm{Var}_k)' \to K_0(\mathrm{Var}_k)$$

mapping $[X]_{bl}$ to $[X]$ for any smooth, projective (resp. proper) k-variety X.

Theorem 3.3 (Bittner [5], Thm. 3.1) *If k has characteristic zero, then the natural ring morphisms*

$$\alpha : K_0^{(bl)}(\mathrm{Var}_k) \to K_0(\mathrm{Var}_k)$$
$$\alpha' : K_0^{(bl)}(\mathrm{Var}_k)' \to K_0(\mathrm{Var}_k)$$

are isomorphisms.

It follows easily from Hironaka's resolution of singularities and the scissor relations that α and α' are surjective. Using Weak Factorization [1], Bittner also proved injectivity.

3.4 Spreading out.

The classical technique of *spreading out* can be formulated at the level of Grothendieck rings of varieties (see also [32]). Let $(I, \leq)$ be a directed set, and let

$$\mathscr{A} = (A_i, \varphi_{i,j} : A_i \to A_j)$$

be a direct system of Noetherian commutative rings, indexed by I. If we denote by A the direct limit of this system in the category of rings, then $\mathrm{Spec}(A)$ is the projective limit of the system

$$(\mathrm{Spec}(A_i), \mathrm{Spec}(\varphi_{i,j}) : \mathrm{Spec}(A_j) \to \mathrm{Spec}(A_i))$$

in the category of schemes. If X is a separated A-scheme of finite type, and i is an element in I, then an A_i-model for X is a separated A_i-scheme of finite type X_i endowed with an isomorphism of A-schemes $X_i \times_{A_i} A \to X$. Likewise, an A_i-model of a morphism of separated A-schemes of finite type $f : X \to Y$ is a morphism of A_i-schemes $f_i : X_i \to Y_i$ such that X_i and Y_i are A_i-models of X, resp. Y, and such that f coincides with the base change of f to A (modulo the identifications $X_i \times_{A_i} A \cong X$ and $Y_i \times_{A_i} A \cong Y$).

For all $i, j \in I$ with $i \leq j$, we consider the base change morphisms

$$\phi_{i,j} : K_0(\mathrm{Var}_{A_i}) \to K_0(\mathrm{Var}_{A_j})$$
$$\psi_{i,j} : \mathcal{M}_{A_i} \to \mathcal{M}_{A_j}$$

as well as

$$\phi_i : K_0(\mathrm{Var}_{A_i}) \to K_0(\mathrm{Var}_A)$$
$$\psi_i : \mathcal{M}_{A_i} \to \mathcal{M}_A$$

We obtain direct systems of rings $(K_0(\mathrm{Var}_{A_i}), \phi_{i,j})$ and $(\mathcal{M}_{A_i}, \psi_{i,j})$ indexed by I, and the morphisms ϕ_i and ψ_i induce ring morphisms

$$\phi : \varinjlim_{i \in I} K_0(\mathrm{Var}_{A_i}) \to K_0(\mathrm{Var}_A)$$
$$\psi : \varinjlim_{i \in I} \mathcal{M}_{A_i} \to \mathcal{M}_A$$

Proposition 3.4 (Spreading out) *The ring morphisms*

$$\phi : \varinjlim_{i \in I} K_0(\mathrm{Var}_{A_i}) \to K_0(\mathrm{Var}_A)$$
$$\psi : \varinjlim_{i \in I} \mathcal{M}_{A_i} \to \mathcal{M}_A$$

are isomorphisms.

Proof Surjectivity follows from the fact that for any separated A-scheme of finite type X, there exist an index $i \in I$ and an A_i-model X_i for X, by [15, IV.8.8.2+8.10.5]. Injectivity follows from the following facts:

If $i \in I$ and U_i and V_i are separated A_i-schemes of finite type, then the canonical map

$$\varinjlim_{j \geq i} \mathrm{Hom}_{A_j}(U_i \times_{A_i} A_j, V_i \times_{A_i} A_j) \to \mathrm{Hom}_A(U_i \times_{A_i} A, V_i \times_{A_i} A)$$

is a bijection [15, IV.8.8.2]. Moreover, if $f_i : U_i \to V_i$ is a morphism of

A_i-schemes such that the induced morphism $f : U_i \times_{A_i} A \to V_i \times_{A_i} A$ is a closed (resp. open) immersion, then there exists an element $j \geq i$ in I such that the morphism $f_j : U_i \times_{A_i} A_j \to V_i \times_{A_i} A_j$ is a closed (resp. open) immersion [15, IV.8.10.5]. $\qquad\square$

In particular, Proposition 3.4 can be applied to the following example. Denote by $\mathscr{A}_k$ the set of finitely generated sub-$\mathbb{Z}$-algebras of the field k, ordered by inclusion. Then k is the limit of the direct system $\mathscr{A}_k$ in the category of rings, and Proposition 3.4 yields isomorphisms

$$\phi : \varinjlim_{B \in \mathscr{A}_k} K_0(\mathrm{Var}_B) \to K_0(\mathrm{Var}_k)$$

$$\psi : \varinjlim_{B \in \mathscr{A}_k} \mathscr{M}_B \to \mathscr{M}_k$$

3.5 The λ-structure and Kapranov's zeta function.

Definition 3.5 Let A be a ring. A λ-*structure* on A is a sequence $(\lambda^i)_{i \in \mathbb{N}}$ of maps $A \to A$ such that:

$$\left\{ \begin{array}{rcl} \lambda^0(x) & = & 1 \\ \lambda^1(x) & = & x \\ & \vdots & \\ \lambda^n(x+y) & = & \sum_{i+j=n} \lambda^i(x)\lambda^j(y) \end{array} \right. ,$$

for all $x, y \in A$.

A ring endowed with a λ-structure is called a λ-*ring*. One can endow $K_0(\mathrm{Var}_k)$ with a λ-structure, as follows. If X is a quasi-projective k-variety, we denote for each $n \in \mathbb{Z}_{>0}$ by $\bigwedge^n X$ the n-th symmetric power of X, i.e.,

$$\bigwedge^n X = \underbrace{(X \times_k X \ldots \times_k X)}_{n \text{ times}} / \Sigma_n$$

with Σ_n the permutation group of $\{1, \ldots, n\}$. Quasi-projectivity of X guarantees that the quotient is representable by a separated k-scheme of finite type. If k is perfect, the set of the k-points of $\bigwedge^n X$ is canonically bijective to the set of effective 0-divisors of X of degree n. We put $\bigwedge^0 X = \mathrm{Spec}(k)$.

The classes of quasi-projective k-varieties generate the Grothendieck group $K_0(\mathrm{Var}_k)$, by the scissor relations and Noetherian induction. One easily shows that there exists a unique λ-structure on $K_0(\mathrm{Var}_k)$ such that

$\lambda^n[X] := [\bigwedge^n X]$ for every $n \geq 0$ and every quasi-projective k-variety X.

Kapranov's *motivic Hasse-Weil zeta function* of an element α of $K_0(\mathrm{Var}_k)$ is defined by

$$\zeta_\alpha(t) := \sum_{i \geq 0} (\lambda^i \alpha) t^i \quad \in K_0(\mathrm{Var}_k)[[t]].$$

If X is a separated k-scheme of finite type, we put $\zeta_X(t) = \zeta_{[X]}(t)$. If k is finite, then, applying the point counting morphism (see Example 4.2 below)

$$N_k : K_0(\mathrm{Var}_k) \to \mathbb{Z} : [X] \mapsto |X(k)|$$

to the coefficients of $\zeta_X(t)$, we obtain the usual Hasse-Weil zeta function of X.

Kapranov asked if the image of $\zeta_X(t)$ in $\mathcal{M}_k[[t]]$ is rational. He proved this in the case where X is a smooth projective curve [24]. Larsen and Lunts gave a necessary and sufficient geometric criterion for rationality of $\zeta_X(t)$ over $K_0(\mathrm{Var}_k)$ when X is a complex surface [28]. This criterion is not always fulfilled, so that $\zeta_X(t)$ is not always rational over $K_0(\mathrm{Var}_k)$. To our knowledge, the question of rationality over $\mathcal{M}_k$ is still open. We refer to [21] for results at the level of Chow motives.

A notion closely related to λ-structures is that of a *power structure*. For its definition on the Grothendieck ring of varieties and some interesting applications, we refer to [18, 19].

3.6 The Grothendieck ring of definable sets

Let $\mathscr{L}$ be a first-order language, and let M be a $\mathscr{L}$-structure. If n is a positive integer, a subset X of M^n is called *definable* if there exist a formula $\varphi(x_1, \ldots, x_n, y_1, \ldots, y_m)$ in $\mathscr{L}$ and a tuple $b \in M^m$ such that

$$X = \{a \in M^n \mid M \vDash \varphi(a, b)\}.$$

A map between two definable sets $X \subset M^n$ and $Y \subset M^p$ is called *definable* if its graph is a definable subset of M^{p+n}. We denote by Def_M the category of definable sets and definable morphisms.

Definition 3.6 We define the Grothendieck group $K_0(\mathrm{Def}_M)$ of the structure M as the quotient of the free abelian goup $\mathbb{Z}[\mathfrak{Iso}_{\mathrm{Def}_M}]$ by the subgroup generated by the elements of the form

$$\{X \cup Y\} + \{X \cap Y\} - \{X\} - \{Y\}$$

with X, Y definable subsets of M^n, for some $n \in \mathbb{N}$. The Cartesian product of definable sets defines a ring structure on $K_0(\mathrm{Def}_M)$, with

$$[X] \cdot [Y] = [X \times Y].$$

A classical example is the following: if $\mathscr{L}$ is the language of ordered fields, and if we consider the structure $\mathbb{R} = (\mathbb{R}, \leq)$, then the singular Euler characteristic w.r.t. the real topology defines an isomorphism $K_0(\mathrm{Def}_{\mathbb{R}}) \cong \mathbb{Z}$.

3.7 The Grothendieck ring of a theory

For notational convenience, we restrict to a particular case. We fix a base ring A. Let $\mathscr{L}$ be the language of rings over A. This means that formulas in $\mathscr{L}$ consist of quantifiers and Boolean combinations of polynomial equalities with coefficients in A. For every formula $\varphi(x_1, \ldots, x_n)$ in $\mathscr{L}$ and every A-algebra B, we denote by $S_\varphi(B)$ the subset of B^n defined by φ.

Let $\mathscr{T}$ be a theory in $\mathscr{L}$. We say that two formulas $\varphi(x_1, \ldots, x_m)$ and $\psi(y_1, \ldots, y_n)$ in $\mathscr{L}$ are $\mathscr{T}$-equivalent, if there exists a third formula $\eta(x_1, \ldots, x_m, y_1, \ldots, y_n)$ such that, for every A-algebra M that is a model of $\mathscr{T}$, the set $S_\eta(M) \subset M^{m+n}$ is the graph of a bijection between $S_\varphi(M) \subset M^m$ and $S_\psi(M) \subset M^n$.

Definition 3.7 The Grothendieck group $K_0(\mathscr{T})$ of the theory $\mathscr{T}$ is the quotient of the free abelian group on $\mathscr{T}$-equivalence classes $[\varphi]$ of formulas φ in $\mathscr{L}$, by the subgroup generated by elements of the form

$$[\varphi \wedge \psi] + [\varphi \vee \psi] - [\varphi] - [\psi]$$

where φ and ψ are formulas in $\mathscr{L}$ with the same sets of free variables.

If φ and ψ are formulas in $\mathscr{L}$ in disjoint sets of free variables, then we put

$$[\varphi] \cdot [\psi] = [\varphi \wedge \psi].$$

This defines a ring structure on $K_0(\mathscr{T})$.

If M is an A-algebra such that M is a model of $\mathscr{T}$, then there exists a unique ring morphism

$$\alpha : K_0(\mathscr{T}) \to K_0(\mathrm{Def}_M)$$

that maps $[\varphi]$ to $[S_\varphi(M)]$, for each formula φ in $\mathscr{L}$. This morphism is not an isomorphism, in general. However, we have the following result.

Proposition 3.8 *Let k be an algebraically closed field, and let* ACF_k *be the theory of algebraically closed fields over k. Then*

$$\alpha : K_0(\mathrm{ACF}_k) \to K_0(\mathrm{Def}_k)$$

is an isomorphism.

Proof Let $\varphi(x_1, \ldots, x_m)$ and $\psi(y_1, \ldots, y_n)$ be formulas in $\mathscr{L}$, and assume that the definable sets $S_\varphi(k) \subset k^m$ and $S_\psi(k) \subset k^n$ are isomorphic in Def_k. To prove that α is an isomorphism, it is enough to show that φ and ψ are ACF_k-equivalent. Let

$$\eta(x_1, \ldots, x_m, y_1, \ldots, y_n)$$

be a formula in $\mathscr{L}$ such that $S_\eta(k) \subset k^{m+n}$ is the graph of a bijection between $S_\varphi(k)$ and $S_\psi(k)$. We'll show that η defines an ACF_k-equivalence between φ and ψ.

There exists a sentence ϕ in $\mathscr{L}$ such that, for each algebraically closed field F containing k, the sentence ϕ is true over F iff $S_\eta(F)$ is the graph of a bijection between $S_\varphi(F)$ and $S_\psi(F)$. By quantifier elimination in ACF_k, we may assume that ϕ has no quantifiers. Then ϕ is simply a Boolean combination of polynomial expressions over k, and since ϕ is true over k, it is true over every field extension of k. Hence, ϕ defines an ACF_k-equivalence between φ and ψ. □

3.8 The modified Grothendieck ring of varieties and the theory of algebraically closed fields

In this section, we compare the Grothendieck ring of varieties over a field k to the Grothendieck ring of the theory ACF_k of algebraically closed fields over k. We will see that they are naturally isomorphic if k has characteristic zero, but that the question is more subtle in positive characteristic.

A morphism of schemes

$$f : X \to Y$$

is called a *universal homeomorphism* if for every morphism of schemes $Y' \to Y$, the morphism

$$f_{Y'} : X \times_Y Y' \to Y'$$

obtained from f by base change is a homeomorphism [15, **IV**.2.4.2]. This property is obviously stable under base change. If f is of finite presentation, then f is a universal homeomorphism iff f is finite, surjective, and purely inseparable [15, **IV**.8.11.6]. We call two schemes X and Y *universally homeomorphic* if there exists a universal homeomorphism from X to Y or from Y to X. If S is a scheme and X and Y are S-schemes, then we call X and Y universally S-homeomorphic if there exists a universal homeomorphism of S-schemes $X \to Y$ or $Y \to X$.

Definition 3.9   Let S be a Noetherian scheme. We denote by I_S^{uh} the ideal in $K_0(\mathrm{Var}_S)$ generated by elements of the form $[X] - [Y]$, where X and Y are universally S-homeomorphic separated S-schemes of finite type. We put

$$K_0^{\mathrm{mod}}(\mathrm{Var}_S) = K_0(\mathrm{Var}_S)/I_S^{uh}$$

and we call this quotient the modified Grothendieck ring of S-varieties.

If Z is a separated S-scheme of finite type, or a constructible subset of such a scheme, then with slight abuse of notation, we denote the image of $[Z]$ in $K_0^{\mathrm{mod}}(\mathrm{Var}_S)$ again by $[Z]$. Likewise, we denote the image of $\mathbb{L}_S$ in $K_0^{\mathrm{mod}}(\mathrm{Var}_S)$ again by $\mathbb{L}_S$, or simply $\mathbb{L}$. We will always clearly indicate in which ring we are working. We denote by $\mathscr{M}_S^{\mathrm{mod}}$ the localization of $K_0^{\mathrm{mod}}(\mathrm{Var}_S)$ with respect to $\mathbb{L}$.

If $S = \mathrm{Spec}(A)$ for some Noetherian ring A, then we also write I_A^{uh}, $K_0^{\mathrm{mod}}(\mathrm{Var}_A)$, and $\mathscr{M}_A^{\mathrm{mod}}$ instead of I_S^{uh}, $K_0^{\mathrm{mod}}(\mathrm{Var}_S)$, and $\mathscr{M}_S^{\mathrm{mod}}$.

Let S, T, and U be Noetherian schemes. If $f : T \to S$ and $g : S \to U$ are morphisms such that g is separated and of finite type, then the ring morphisms

$$f^* : K_0^{\mathrm{mod}}(\mathrm{Var}_S) \to K_0^{\mathrm{mod}}(\mathrm{Var}_T)$$
$$f^* : \mathscr{M}_S^{\mathrm{mod}} \to \mathscr{M}_T^{\mathrm{mod}}$$

and the morphisms of abelian groups

$$g_! : K_0^{\mathrm{mod}}(\mathrm{Var}_S) \to K_0^{\mathrm{mod}}(\mathrm{Var}_U)$$
$$g_! : \mathscr{M}_S^{\mathrm{mod}} \to \mathscr{M}_U^{\mathrm{mod}}$$

are defined similarly as in Section 3.

Proposition 3.10   *If $f : X \to Y$ is a universal homeomorphism of finite type between Noetherian $\mathbb{Q}$-schemes, then f is a piecewise isomorphism.*

Proof   Since $f_{red} : X_{red} \to Y_{red}$ is still a universal homeomorphism [15, IV.2.4.3(vi)], we may assume that X and Y are reduced. By Noetherian induction, it is enough to find a non-empty open subscheme U of Y such that $X \times_Y U \to U$ is an isomorphism. In particular, we may assume that Y is irreducible. Then X is irreducible, because it is homeomorphic to Y. If we denote by η_Y the generic point of Y, then its inverse image in X consists of a unique point η_X, which is the generic point of X. The residue field $\kappa(\eta_X)$ is a purely inseparable extension of the residue field $\kappa(\eta_Y)$ of η_Y. Since these fields have characteristic zero, we see that f induces an isomorphism $\kappa(\eta_X) \cong \kappa(\eta_Y)$, so that the restriction of f to some dense open subset of X is an open immersion. This concludes the proof. $\qquad\square$

Corollary 3.11 *If S is a Noetherian $\mathbb{Q}$-scheme, then I_S^{uh} is the zero ideal, and the projection*

$$K_0(\mathrm{Var}_S) \to K_0^{\mathrm{mod}}(\mathrm{Var}_S)$$

is an isomorphism.

Lemma 3.12 *Let k be a field. Let m and n be elements of $\mathbb{N}$, and let C and D be constructible subsets of $\mathbb{A}_k^m$, resp. $\mathbb{A}_k^n$. Assume that there exists a formula $\eta(x_1, \ldots, x_m, y_1, \ldots, y_n)$ in $\mathscr{L}_k$ such that, for every algebraically closed field L containing k, the set*

$$S_\eta(L) \subset L^{m+n}$$

is the graph of a bijection between $C(L) \subset L^m$ and $D(L) \subset L^n$. Then we have

$$[C] = [D]$$

in $K_0^{\mathrm{mod}}(\mathrm{Var}_k)$.

Proof By quantifier elimination relative to ACF_k, there exists a unique constructible subset E of $\mathbb{A}_k^{m+n}$ such that

$$S_\eta(L) = E(L) \subset L^{m+n}$$

for every algebraically closed field L that contains k. It suffices to show that $[C] = [E]$ in $K_0^{\mathrm{mod}}(\mathrm{Var}_k)$. We denote by

$$\pi : \mathbb{A}_k^{m+n} \to \mathbb{A}_k^m$$

the projection onto the first m coordinates. By Noetherian induction on C, it is enough to prove that there exists a non-empty open subset U of C such that U is locally closed in $\mathbb{A}_k^m$, $V = \pi^{-1}(U) \cap E$ is locally closed in $\mathbb{A}_k^{m+n}$, and the morphism $V \to U$ induced by π is a universal homeomorphism if we endow U and V with the reduced induced structure.

We may assume that C is irreducible and locally closed in $\mathbb{A}_k^m$, and we endow C with its reduced induced structure. Let E' be an integral subscheme of $\mathbb{A}_k^{m+n}$ whose support is contained in E and such that the morphism $\pi' : E' \to C$ induced by π is dominant. Then π' is a quasi-finite morphism of k-varieties. Hence, there exists a non-empty open subset U of C such that, putting $V = E' \times_C U$, the morphism $V \to U$ is finite and surjective. Surjectivity implies, in particular, that $V = E \cap \pi^{-1}(U)$. It follows from our assumptions that $V \to U$ is purely inseparable, so that it is a universal homeomorphism. $\square$

If n is an element of $\mathbb{N}$, and $i_X : X \to \mathbb{A}_F^n$ is an immersion of F-schemes, then there exists a formula $\varphi(x_1, \ldots, x_n)$ such that

$$S_\varphi(L) = X(L) \subset L^n$$

for every field L that contains F. We call such a formula φ an i_X-formula. It is not unique. If Y is a quasi-affine F-variety, then we say that a formula ψ in $\mathscr{L}_F$ is a Y-formula if it is an i_Y-formula for some immersion $i_Y : Y \to \mathbb{A}_F^m$, with $m \in \mathbb{N}$. Again, such a Y-formula is not unique, but its class $[\psi]$ in $K_0(\mathrm{ACF}_F)$ only depends on Y, and not on the choice of the immersion i_Y or the formula ψ.

Proposition 3.13 *Let k be a field of characteristic $p \geq 0$. There exists a unique ring morphism*

$$\beta : K_0(\mathrm{Var}_k) \to K_0(\mathrm{ACF}_k)$$

such that, for every quasi-affine k-variety X, the image of $[X]$ under β is the class in $K_0(\mathrm{ACF}_k)$ of an X-formula φ_X. The morphism β is surjective, and its kernel equals I_k^{uh}, so that β factors through an isomorphism

$$K_0^{\mathrm{mod}}(\mathrm{Var}_k) \to K_0(\mathrm{ACF}_k).$$

Proof   First, we show that $ker(\beta)$ contains the ideal I_k^{uh}. Let $f : X \to Y$ be a universal homeomorphism of k-varieties. We choose a partition $\{Y_1, \ldots, Y_r\}$ of Y into locally closed subsets, and we endow Y_i with its reduced induced structure, for each i. We may assume that Y_i is affine for every i. If we put $X_i = (X \times_Y Y_i)_{red}$, then the morphism $X_i \to Y_i$ is a universal homeomorphism, X_i is affine, and we have

$$[X] = \sum_{i=1}^r [X_i] \quad \text{and} \quad [Y] = \sum_{i=1}^r [Y_i]$$

in $K_0(\text{Var}_k)$. So it is enough to prove that $\beta([X] - [Y]) = 0$ if $f : X \to Y$ is a universal homeomorphism of affine k-varieties. We denote by Γ_f the graph of f in $X \times_k Y$, and we choose closed immersions $i_X : X \to \mathbb{A}_k^m$ and $i_Y : Y \to \mathbb{A}_k^n$, with $m, n \in \mathbb{N}$. These induce a closed immersion $i_f : \Gamma_f \to \mathbb{A}_k^{m+n}$. If we choose an i_X- formula φ_X, an i_Y-formula φ_Y and an i_f-formula φ_f, then φ_f defines an ACF_k-equivalence between φ_X and φ_Y, so that $\beta([X]) = \beta([Y])$.

Hence, the morphism β factors through a ring morphism

$$\gamma : K_0^{\text{mod}}(\text{Var}_k) \to K_0(\text{ACF}_k).$$

We'll show that it is an isomorphism by constructing its inverse. If $\varphi(x_1, \ldots, x_m)$ is a formula in $\mathscr{L}_k$, then by quantifier elimination relative to ACF_k, there exists a unique constructible subset C_φ of $\mathbb{A}_k^m$ such that, for every algebraically closed field L that contains F, the subsets $S_\varphi(L)$ and $C_\varphi(L)$ of L^m coincide. It follows from Lemma 3.12 that the class of C_φ in $K_0^{\text{mod}}(\text{Var}_k)$ only depends on the ACF_k-equivalence class of φ. It is easily seen that there exists a unique ring morphism

$$\delta : K_0(\text{ACF}_k) \to K_0^{\text{mod}}(\text{Var}_k)$$

that maps $[\varphi]$ to $[C_\varphi]$ for every formula φ in $\mathscr{L}_k$. This ring morphism is inverse to γ. $\qquad\square$

It seems likely that, when k is a field of positive characteristic, the projection

$$K_0(\text{Var}_k) \to K_0^{\text{mod}}(\text{Var}_k)$$

is not an isomorphism, but this question is quite difficult to answer, since the standard realization morphisms that one can use to distinguish classes in the Grothendieck ring all factor through $K_0^{\text{mod}}(\text{Var}_k)$ (see Section 4.2).

3.9 Fibrations

One of the most important results in the theory of motivic integration is the *change of variables formula*. This formula is based on the appearance of certain fibrations in the geometry of Greenberg schemes. The following proposition describes how these fibrations behave in the Grothendieck ring of varieties.

Proposition 3.14 *Let S be a Noetherian scheme, and let X, Y, and Z be separated S-schemes of finite type. Let $f : X \to Y$ be a morphism of S-schemes.*

1. *Assume that for every point y of Y, there exists an isomorphism of $k(y)$-schemes*

$$(X \times_Y y)_{\mathrm{red}} \to (Z \times_S y)_{\mathrm{red}}$$

(here $k(y)$ denotes the residue field of y). Then f is a piecewise trivial fibration with fiber Z, so that

$$[X] = [Y] \cdot [Z]$$

in $K_0(\mathrm{Var}_S)$.

2. *Assume that for every perfect field F and every morphism of schemes $\mathrm{Spec}(F) \to Y$, there exists a universal homeomorphism of F-schemes*

$$X \times_Y \mathrm{Spec}(F) \to Z \times_S \mathrm{Spec}(F).$$

Then

$$[X] = [Y] \cdot [Z]$$

in $K_0^{\mathrm{mod}}(\mathrm{Var}_S)$.

Proof We only prove point (2). The proof of point (1) is similar; a special case can be found in [40, 4.2.3].

By Noetherian induction and the scissor relations in the Grothendieck ring, it is enough to find a non-empty open subscheme U of Y such that

$$[X \times_Y U] = [U] \cdot [Z]$$

in $K_0^{\mathrm{mod}}(\mathrm{Var}_S)$. We may assume that Y is affine and integral. We denote by B the ring of regular functions on Y. Let F be the perfect closure of the function field $Frac(B)$ of Y. We know that there exists a universal homeomorphism of F-schemes

$$g : X \times_Y \mathrm{Spec}(F) \to Z \times_S \mathrm{Spec}(F).$$

The B-algebra F is the direct limit of its finitely generated sub-B-algebras. Hence, by [15, **IV**.8.8.2 and **IV**.8.10.5], there exist a finitely generated sub-B-algebra B' of F, and a universal homeomorphism of B'-schemes

$$g' : X \times_Y \mathrm{Spec}(B') \to Z \times_S \mathrm{Spec}(B')$$

such that g is obtained from g' by base change from $\mathrm{Spec}(B')$ to $\mathrm{Spec}(F)$.

Since $\text{Spec}(B') \to Y$ is purely inseparable over the generic point of Y, and the generic point of Y is the projective limit of the dense open subschemes of Y, it follows from [15, **IV**.8.10.5] that there exists a dense open subscheme U of Y such that

$$\text{Spec}(B') \times_Y U \to U$$

is a universal homeomorphism.

Looking at the diagram of universal homeomorphisms of separated S-schemes of finite type

$$X \times_Y (\text{Spec}(B') \times_Y U) \xrightarrow{g' \times_Y id_U} Z \times_S (\text{Spec}(B') \times_Y U)$$

$$\downarrow \qquad\qquad\qquad\qquad\qquad\qquad \downarrow$$

$$X \times_Y U \qquad\qquad\qquad\qquad\qquad Z \times_S U$$

we find that

$$[X \times_Y U] = [U] \cdot [Z]$$

in $K_0^{\text{mod}}(\text{Var}_S)$. $\qquad\qquad\qquad\qquad\qquad\qquad\qquad\qquad$ □

4 Why is the Grothendieck ring of varieties interesting?

Because of its realization maps! These realization morphisms consitute the "motivic nature" of $K_0(\text{Var}_k)$. The Batyrev-Kontsevich Theorem, which initiated the story of motivic integration (see [6] in this volume), is a characteristic example of the power of the joint use of the change of variables formula for motivic integrals and the realization maps of the Grothendieck ring of varieties.

4.1 The notion of additive invariant

Definition 4.1 Let S be a Noetherian scheme, and let A be a group. An *additive A-invariant* of S-varieties is a map

$$\mu : \mathfrak{Iso}_{(\text{sft}/S)} \to A$$

such that

$$\mu(\{X\}) = \mu(\{Y\}) + \mu(\{X \setminus Y\})$$

for each separated S-scheme of finite type X, and each closed subscheme Y of X.

We say that μ is *multiplicative* if, moreover, A is a ring, and

$$\mu(X \times_S Y) = \mu(X) \cdot \mu(Y)$$

for every pair of separated S-schemes of finite type X, Y.

It follows immediately from the definitions that the map

$$\mathfrak{Iso}_{\mathrm{Var}_S} \to K_0(\mathrm{Var}_S) : \{X\} \mapsto [X]$$

is the universal additive (resp. additive and multiplicative) invariant of S-varieties, in the following sense. For each additive A-invariant of k-varieties μ, there exists a unique morphism of groups

$$\widetilde{\mu} : K_0(\mathrm{Var}_k) \to A$$

such that $\widetilde{\mu}([X]) = \mu(X)$ for every separated S-scheme of finite type X. If A is a ring and μ is multiplicative, then $\widetilde{\mu}$ is a ring morphism. The morphism $\widetilde{\mu}$ is called a *realization morphism* of $K_0(\mathrm{Var}_S)$. With slight abuse of notation, we will often denote $\widetilde{\mu}$ by μ. We will now give a list of some important realization morphisms.

Example 4.2 *Counting points.* If k is finite, then there exists a unique ring morphism

$$N_k : K_0(\mathrm{Var}_k) \to \mathbb{Z}$$

such that $N_k([X]) = |X(k)|$ for each object X of (sft/k). It localizes to a ring morphism

$$N_k : \mathscr{M}_k \to \mathbb{Q}.$$

Example 4.3 *Étale realization and Euler characteristic.* Let k be an arbitrary field. By the Künneth formula and the excision exact sequence for ℓ-adic cohomology, the étale characteristic $\chi_{\text{ét}}$ from Example 2.9 is an additive and multiplicative invariant. Hence, there exists a unique ring morphism

$$\chi_{\text{ét}} : K_0(\mathrm{Var}_k) \to K_0(\mathrm{Rep}_{G_k} \mathbb{Q}_\ell)$$

such that $\chi_{\text{ét}}([X]) = \chi_{\text{ét}}(X)$ for each separated k-scheme of finite type X. Since $\chi_{\text{ét}}(\mathbb{A}_k^1) = [\mathbb{Q}_\ell(-1)]$ is invertible in $K_0(\mathrm{Rep}_{G_k} \mathbb{Q}_\ell)$, this morphism localizes to a ring morphism

$$\chi_{\text{ét}} : \mathscr{M}_k \to K_0(\mathrm{Rep}_{G_k} \mathbb{Q}_\ell).$$

Composing this morphism with the forgetful morphism

$$K_0(\mathrm{Rep}_{G_k}\mathbb{Q}_\ell) \to K_0(\mathbb{Q}_\ell) \cong \mathbb{Z}$$

we obtain the *topological Euler characteristic*

$$\chi_{\mathrm{top}} : \mathscr{M}_k \to \mathbb{Z}.$$

For each separated k-scheme of finite type X, we put

$$\chi_{\mathrm{top}}(X) := \chi_{\mathrm{top}}([X]) = \sum_{n \geq 0}(-1)^n \mathrm{dim}_{\mathbb{Q}_\ell}(H_c^n(X \times_k k^s, \mathbb{Q}_\ell)).$$

If $k = \mathbb{C}$, then by the comparison theorems for ℓ-adic and singular cohomology, $\chi_{\mathrm{top}}(X)$ coincides with the singular Euler characteristic of $X(\mathbb{C})$.

The topological Euler characteristic χ_{top} is independent of ℓ. If k is a subfield of $\mathbb{C}$, this follows by comparison with singular cohomology. If k is finite, it follows from the cohomological interpretation of the Hasse-Weil zeta function. For the general case, one applies a spreading out argument.

The étale realization can be generalized to Noetherian base schemes S as follows. Assume that there exists a prime ℓ invertible on S. We denote by $K_0(D_c^b(S, \mathbb{Q}_\ell))$ the Grothendieck ring of the abelian category of constructible ℓ-adic sheaves on S (this is also the Grothendieck ring of the triangulated derived category $D_c^b(S, \mathbb{Q}_\ell)$ of bounded constructible complexes). Then there exists a unique ring morphism

$$\chi_{\mathrm{ét}} : K_0(\mathrm{Var}_S) \to K_0(D_c^b(S, \mathbb{Q}_\ell))$$

that maps the class of a separated S-scheme of finite type X to

$$\sum_{i \geq 0}(-1)^i[R^i f_!(\mathbb{Q}_\ell)]$$

where f denotes the structural morphism $X \to S$. It localizes to a ring morphism

$$\chi_{\mathrm{ét}} : \mathscr{M}_S \to K_0(D_c^b(S, \mathbb{Q}_\ell)).$$

If S is the spectrum of a field k, then the category of constructible ℓ-adic sheaves on S is equivalent to the category of ℓ-adic Galois representations of G_k, and we recover the previous construction.

Example 4.4 *The Poincaré polynomial.* If k is finitely generated, the étale characteristic can be refined to a ring morphism

$$\chi_{W\mathrm{ét}} : K_0(\mathrm{Var}_k) \to K_0(\mathrm{WRep}_{G_k}\mathbb{Q}_\ell).$$

It is the unique ring morphism such that $\chi_{W\acute{e}t}([X]) = \chi_{W\acute{e}t}(X)$ for each separated k-scheme of finite type X (see Example 2.9). Composing with the Poincaré polynomial, we find ring morphisms

$$P : K_0(\mathrm{Var}_k) \to \mathbb{Z}[T] : [X] \mapsto P(X;T)$$
$$P : \mathscr{M}_k \to \mathbb{Z}[T,T^{-1}]$$

such that $P(\mathbb{L}) = T^2$.

The definition of the Poincaré polynomial can be generalized to arbitrary base fields, using a spreading out argument (see [32, §8]). If k is any field, then the Poincaré polynomial is a ring morphism

$$P : K_0(\mathrm{Var}_k) \to \mathbb{Z}[T]$$

satisfying the following properties.

Theorem 4.5 *Let X be a separated k-scheme of finite type, and put $P(X;T) = P([X])$.*

1. *$P(X;T)$ is independent of ℓ,*
2. *$P(X;1) = \chi_{\mathrm{top}}(X)$,*
3. *If X is smooth and proper over k, then*

$$P(X;T) = \sum_{i \geq 0}(-1)^i b_i(X) T^i$$

with $b_i(X) = \dim H^i(X \times_k k^s, \mathbb{Q}_\ell)$ the i-th ℓ-adic Betti number of X,
4. *The degree of $P(X;T)$ equals twice the dimension of X, and the leading coefficient of $P(X;T)$ equals the number of irreducible components of maximal dimension of $X \times_k k^s$.*

For the proofs, we refer to [32, §8].

Example 4.6 *Hodge realization.* If $k = \mathbb{C}$, then the Hodge characteristic χ_{Hodge} from Example 2.8 is an additive and multiplicative invariant. This follows from the fact that the excision sequence and the Künneth isomorphism are compatible with mixed Hodge structures. Hence, if k is a subfield of $\mathbb{C}$, there exists a unique ring morphism

$$\chi_{Hodge} : K_0(\mathrm{Var}_k) \to K_{\mathbb{Q}H}$$

that maps $[X]$ to $\chi_{Hodge}(X \times_k \mathbb{C})$, for each k-variety X. Since $\chi_{Hodge}(\mathbb{A}_k^1) = [\mathbb{Q}(-1)]$ is invertible in $K_{\mathbb{Q}H}$, the morphism χ_{Hodge} localizes to a ring morphism

$$\chi_{Hodge} : \mathscr{M}_k \to K_{\mathbb{Q}H}.$$

Composing with the Hodge-Deligne polynomial, we find ring morphisms

$$HD : K_0(\mathrm{Var}_k) \to \mathbb{Z}[u, v]$$
$$HD : \mathcal{M}_k \to \mathbb{Z}[u, v, u^{-1}, v^{-1}]$$

such that $HD([X]) = HD(X \times_k \mathbb{C}; u, v)$ for each separated k-scheme of finite type. These ring morphisms HD are independent of the embedding of k in $\mathbb{C}$: the group $K_0(\mathrm{Var}_k)$ is generated by the classes of smooth and projective k-varieties X, and for such a variety, we have

$$HD([X]) = \sum_{p,q \geq 0} (-1)^{p+q} h^{p,q}(X) u^p v^q \quad \in \mathbb{Z}[u, v]$$

where

$$h^{p,q}(X) = \dim_k H^q(X, \Omega_X^p)$$

is the (p, q)-th Hodge number of X. It is a purely algebraic invariant, defined without reference to any embedding of k into $\mathbb{C}$.

Now assume that k is an arbitrary field of characteristic zero. The field k is the direct limit of its finitely generated subfields, and all of these can be embedded in $\mathbb{C}$. Applying Proposition 3.4, we see that there exists a unique ring morphism

$$HD : K_0(\mathrm{Var}_k) \to \mathbb{Z}[u, v]$$

such that

$$HD([X]) = \sum_{p,q \geq 0} (-1)^{p+q} h^{p,q}(X) u^p v^q \quad \in \mathbb{Z}[u, v]$$

for each smooth and projective k-variety X. Recall that the Hodge numbers $h^{p,q}(X)$ are algebraic invariants of X, defined by

$$h^{p,q}(X) = \dim_k H^p(X, \Omega_X^q).$$

If k is a subfield of $\mathbb{C}$, the morphism HD coincides with the one constructed above.

For any separated k-scheme of finite type Y, we define its Hodge-Deligne polynomial by

$$HD(Y; u, v) := HD([Y]).$$

Since $HD(\mathbb{A}^1; u, v) = uv$, the morphism HD localizes to a ring morphism

$$HD : \mathcal{M}_k \to \mathbb{Z}[u, v, u^{-1}, v^{-1}].$$

The Hodge polynomial specializes to the Poincaré polynomial and the Euler characteristic: we have $P(Y;T) = HD(Y;T,T)$ and $\chi_{\text{top}}(Y) = HD(Y;1,1)$ (see [32, § 8]).

Example 4.7 *Larsen and Lunts' invariant and the Albanese realization.* We denote by $\mathbb{Z}[\mathfrak{sb}]$ the free abelian group generated by the stable birational equivalence classes of connected smooth projective k-varieties. The following result by Larsen and Lunts is a fundamental tool in the study of the Grothendieck ring of varieties.

Theorem 4.8 (Stably birational realization [27]) *If k has characteristic zero, then there exists a unique isomorphism of abelian groups*

$$\Phi_{SB} : K_0(\text{Var}_k)/\mathbb{L}K_0(\text{Var}_k) \to \mathbb{Z}[\mathfrak{sb}]$$

that maps the class $[X]$ of a connected smooth projective variety X to the equivalence class of X in $\mathbb{Z}[\mathfrak{sb}]$. Here $\mathbb{L}K_0(\text{Var}_k)$ denotes the ideal of $K_0(\text{Var}_k)$ generated by $\mathbb{L}$.

As explained in [27, 2.4 and 2.7], the existence of Φ_{SB} follows immediately from Theorem 3.3, and the fact that it is an isomorphism follows easily from Weak Factorization [1]. In [27] it was assumed that k is algebraically closed, but this is not necessary [25, p. 28].

Corollary 4.9 (Albanese realization [37]) *Assume that k has characteristic zero, denote by AV_k the monoid of isomorphism classes of abelian varieties over k, and by $\mathbb{Z}[\text{AV}_k]$ the associated monoid ring. There exists a unique ring morphism*

$$Alb : K_0(\text{Var}_k) \to \mathbb{Z}[\text{AV}_k]$$

that sends the class $[X]$ of a smooth, projective, connected k-variety X to the isomorphism class of its Albanese $Alb(X)$ in $\mathbb{Z}[\text{AV}_k]$.

In particular, if A, B are abelian varieties over k, then $[A] = [B]$ in $K_0(\text{Var}_k)$ iff $A \cong B$.

Proof The Albanese is invariant under stably birational equivalence, and commutes with products. $\square$

Note that $Alb(\mathbb{L}) = 0$, so that Alb does not localize to a non-trivial realization of $\mathcal{M}_k$.

Example 4.10 *Chow motives.* For every smooth and projective variety X over k, we denote by $M(X)$ the effective Chow motive (X, id) associated to X in $\text{Mot}^{\text{eff}}_{k,\mathbb{Q}}$. With slight abuse of notation, we'll use the same notation for its image $(X, \text{id}, 0)$ in $\text{Mot}_{k,\mathbb{Q}}$.

Theorem 4.11 (Gillet-Soulé [12], Guillen-Navarro Aznar [17], Bittner [5]) *Assume that k has characteristic zero. There exist unique ring morphisms*

$$\chi_{\text{mot}}^{\text{eff}} : K_0(\text{Var}_k) \to K_0(\text{Mot}_{k,\mathbb{Q}}^{\text{eff}})$$
$$\chi_{\text{mot}} : \mathcal{M}_k \to K_0(\text{Mot}_{k,\mathbb{Q}})$$

such that, for any smooth and projective k-variety X, $\chi_{\text{mot}}^{\text{eff}}(X)$ (resp. $\chi_{\text{mot}}(X)$) is the class of $M(X)$ in $K_0(\text{Mot}_{k,\mathbb{Q}}^{\text{eff}})$ (resp. $K_0(\text{Mot}_{k,\mathbb{Q}})$).

The question about the existence of such a morphism $\chi_{\text{mot}}^{\text{eff}}$ was raised already by Grothendieck in a letter to Serre [16, letter of 16/8/1964]; he also asked how far the morphism $\chi_{\text{mot}}^{\text{eff}}$ is from being bijective. The morphism $\chi_{\text{mot}}^{\text{eff}}$ is not injective: isogeneous abelian varieties have isomorphic Chow motives with rational coefficients, while, if k has characteristic zero, the classes of two abelian varieties in $K_0(\text{Var}_k)$ coincide iff the varieties are isomorphic, because of the existence of the Albanese realization (Corollary 4.9). By a more involved argument, Ekedahl has shown that there exists a pair of isogenous abelian k-varieties A and B with different classes in $\mathcal{M}_k$ (see the proof of [11, 3.5]). This shows that χ_{mot} is not injective either. (In [32, 7.9], the first author has shown in a different way that χ_{mot} is not injective if k is the field of Laurent series $\mathbb{C}((t))$. The proof uses the *motivic Serre invariant*; see Example 4.12 below). Nothing seems to be known about the surjectivity of χ_{mot} and $\chi_{\text{mot}}^{\text{eff}}$, but it seems plausible that these morphisms are not surjective.

Theorem 4.11 still holds if we replace Chow motives with rational coefficients by Chow motives with integer coefficients [12, Thm. 4]. By Theorem 3.3, we only have to check that Chow motives with integer coefficients satisfy the blow-up relations. For rational coefficients, this was proven in [17, 5.1], but the same proof holds for $\mathbb{Z}$-coefficients (see [3, 0.1.3] for a computation of the Chow groups). If we work with $\mathbb{Z}$-coefficients, we do not know if χ_{mot} and $\chi_{\text{mot}}^{\text{eff}}$ are injective.

Example 4.12 *The motivic Serre invariant.* If K is a complete discretely valued field of characteristic zero, with perfect residue field k, then the first author has constructed in [32] a ring morphism

$$S : K_0(\text{Var}_K)/(\mathbb{L} - 1) \to K_0^{\text{mod}}(\text{Var}_k)/(\mathbb{L} - 1)$$

which maps the class $[X]$ of a smooth and proper K-variety X to the *motivic Serre invariant* of X. This motivic Serre invariant is the class of the special fiber of a weak Néron model of X over the ring of integers of K, and was first introduced in [30]. We refer to [33] for background.

4.2 Modified Grothendieck ring of varieties and realization morphisms

In this section, we'll prove that some realization morphisms from Sections above factor through the modified Grothendieck ring.

Proposition 4.13 *Let S be a Noetherian scheme, and let X and Y be separated S-schemes of finite type such that there exists a universal homeomorphism of S-schemes*

$$f : X \to Y.$$

We denote by g_X and g_Y the structural morphisms from X, resp. Y, to S.

1. *If S is the spectrum of a finite field k, then $X(k)$ and $Y(k)$ have the same cardinality. In particular, the point counting realization*

$$N_k : K_0(\mathrm{Var}_k) \to \mathbb{Z}$$

   factors through a ring morphism

$$N_k : K_0^{\mathrm{mod}}(\mathrm{Var}_k) \to \mathbb{Z}.$$

2. *Let ℓ be a prime invertible on S. The natural morphism*

$$R(g_Y)_!(\mathbb{Q}_\ell) \to R(g_X)_!(\mathbb{Q}_\ell)$$

   in $D_c^b(S, \mathbb{Q}_\ell)$ induced by f is an isomorphism. In particular, the étale realization

$$\text{ét} : K_0(\mathrm{Var}_S) \to K_0(D_c^b(S, \mathbb{Q}_\ell))$$

   factors through a ring morphism

$$\text{ét} : K_0^{\mathrm{mod}}(\mathrm{Var}_S) \to K_0(D_c^b(S, \mathbb{Q}_\ell))$$

   and, if S is the spectrum of a field, then the étale realization

$$\text{ét} : K_0(\mathrm{Var}_k) \to K_0(\mathrm{Rep}_{G_k} \mathbb{Q}_\ell)$$

   factors through a ring morphism

$$\text{ét} : K_0^{\mathrm{mod}}(\mathrm{Var}_k) \to K_0(\mathrm{Rep}_{G_k} \mathbb{Q}_\ell).$$

3. *If S is the spectrum of a field k, then the Poincaré polynomials $P(X;T)$ and $P(Y;T)$ are equal. In particular, the Poincaré realization*

$$P : K_0(\text{Var}_k) \to \mathbb{Z}[T]$$

factors through a ring morphism

$$P : K_0^{\text{mod}}(\text{Var}_k) \to \mathbb{Z}[T].$$

Proof (1) Since f is a universal homeomorphism and k is perfect, the map

$$f(k) : X(k) \to Y(k)$$

is a bijection.

(2) By Grothendieck's spectral sequence for the composition of the functors $f_! = f_*$ and $(g_Y)_!$, it suffices to show that

$$f_* \mathbb{Q}_\ell \cong \mathbb{Q}_\ell,$$
$$R^j f_*(\mathbb{Q}_\ell) = 0 \text{ for } j > 0.$$

The first isomorphism follows immediately from the fact that f is a universal homeomorphism. The second follows from finiteness of f.

(3) We may assume that k has characteristic $p > 0$, since otherwise, there is nothing to prove. By [15, **IV**.8.8.2 and **IV**.8.10.5], there exist a finitely generated sub-$\mathbb{F}_p$-algebra A of k, and a universal homeomorphism

$$f' : X' \to Y'$$

of separated A-schemes of finite type, such that $f : X \to Y$ is obtained from f' by base change from $\text{Spec}(A)$ to $S = \text{Spec}(k)$. We denote by $g_{X'}$ and $g_{Y'}$ the structural morphisms from X' and Y' to $\text{Spec}A$.

If we denote by η the generic point of $\text{Spec}(A)$ and by $k(\eta)$ its residue field, then the Poincaré polynomial $P(X;T)$, resp. $P(Y;T)$, is equal to the Poincaré polynomial of the separated $k(\eta)$-scheme of finite type $X' \times_A k(\eta)$, resp. $Y' \times_A k(\eta)$ [32, 8.12]. Since taking the Poincaré polynomial of fibers defines a constructible function on the base $\text{Spec}(A)$ [32, 8.12], it suffices to look at the fibers over closed points of $\text{Spec}(A)$. Hence, we may assume that k is a finite field. In this case, the result follows immediately from point (2), by definition of the Poincaré polynomial of a variety over a finite field [32, 8.1]. □

Corollary 4.14   *If k is a field, then the étale Euler characteristic*

$$\chi_{\text{top}} : K_0(\text{Var}_k) \to \mathbb{Z}$$

factors through a ring morphism

$$\chi_{\text{top}} : K_0^{\text{mod}}(\text{Var}_k) \to \mathbb{Z}.$$

4.3 Direct consequences for the geometry of varieties

Thanks to the realization maps, we can extract some geometric information from the class of a k-variety in the Grothendieck ring of varieties. Let X and Y be separated k-schemes of finite type over k, and assume that $[X] = [Y]$ in $K_0(\text{Var}_k)$.

1. If k has characteristic zero, then X and Y have the same Hodge polynomial. For arbitrary k, the k-schemes X and Y have the same Poincaré polynomial and the same ℓ-adic Euler characteristic. If X and Y are smooth and proper, then they have the same ℓ-adic Betti numbers.

2. By Theorem 4.5, we know that X and Y have the same dimension, and that $X \times_k k^s$ and $Y \times_k k^s$ have the same number of irreducible components of maximal dimension. In particular, $[X] = 0$ in $K_0(\text{Var}_k)$ iff X is empty.

3. Assume that k has characteristic zero, and that X and Y are connected, smooth and proper. By Theorem 4.8, X and Y are stably birational. They have isomorphic Albanese varieties and isomorphic geometric fundamental groups (the geometric fundamental group is invariant under stable birational equivalence).

4. Recall that the *index* $\delta(X/k)$ is the greatest common divisor of the degrees over k of the closed points of X. The integer $\nu(X/k)$ is the minimum of the degrees over k of the closed points of X. If k has characteristic zero, and X, Y are smooth and projective over k, then

$$\text{CH}_0(X) \simeq \text{CH}_0(Y), \quad \delta(X/k) = \delta(Y/k), \quad \nu(X/k) = \nu(Y/k)$$

(see [29, Corollary 6]). In particular, X has a rational point (equivalently, $\nu(X/k) = 1$) if and only if Y has a rational point.

5. Assume that k has characteristic zero, and that X and Y are smooth and projective over k. In [13, Conjecture 2.5], Göttsche conjectured that the classes of two Chow motives in $K_0(\text{Mot}_{k,\mathbb{Q}}^{\text{eff}})$ are equal iff the motives are isomorphic. He noted that this conjecture would follow from the Beilinson-Murre Conjectures. Assuming Göttsche's

conjecture, we see that X and Y have isomorphic rational Chow groups.

Remark 4.15 Points (1), (2) and (5) remain valid if we only assume that $[X] = [Y]$ in $\mathcal{M}_k$ (for (5), see [32, § 2.3]).

5 Basic properties and questions

In spite of its central role in the theory of motivic integration, and its independent interest, the Grothendieck ring of varieties has recieved little attention so far, and several fundamental questions remain unanswered. We denote by k a field.

5.1 What we know about the Grothendieck ring of varieties

The subring $\mathbb{Z}[\mathbb{L}]$. The ring morphism

$$\mathbb{Z}[U] \to \mathcal{M}_k$$

that maps U to $\mathbb{L}$ is injective. To see this, note that the composition of this morphism with the Poincaré polynomial

$$P : \mathcal{M}_k \to \mathbb{Z}[T, T^{-1}]$$

maps U to T^2. As a consequence, the subring $\mathbb{Z}[\mathbb{L}]$ of $K_0(\mathrm{Var}_k)$ is isomorphic to $\mathbb{Z}[U]$. In particular, $K_0(\mathrm{Var}_k)$ has characteristic zero.

The cardinality of the Grothendieck ring. Since $K_0(\mathrm{Var}_k)$ is generated, as a group, by the classes of affine k-varieties, it is not hard to see that its cardinality is bounded by $\max\{|\mathbb{N}|, |k|\}$. If k has characteristic zero, then we know that non-isomorphic elliptic k-curves have distinct classes in $K_0(\mathrm{Var}_k)$ (Corollary 4.9). Looking at their j-invariants, we find that the set of isomorphism classes of elliptic k-curves has cardinality $|k|$, so that

$$|K_0(\mathrm{Var}_k)| = |k|$$

if k has characteristic zero.

The Grothendieck ring admits large algebraically independent families. Recall that we say that a family $(x_i)_{i \in I}$ of elements of $K_0(\mathrm{Var}_k)$ is *algebraically independent* if the morphism

$$\mathbb{Z}[(t_i)_{i \in I}] \to K_0(\mathrm{Var}_k) : t_i \mapsto x_i$$

is injective. There exist large algebraically independent families in $K_0(\text{Var}_k)$.

If k is finite, then Naumann [31, Thm. 12] has constructed a family of $|\mathbb{N}|$ smooth, projective, geometrically connected k-curves whose classes in $K_0(\text{Var}_k)$ are algebraically independent. In [31, Thm. 13] he proved that there exists an infinite family of elliptic curves with algebraically independent classes over a number field. In [26, 5.8], Scanlon and Krajíček prove that $K_0(\text{Var}_{\mathbb{C}})$ admits an algebraically independent family of cardinality $|\mathbb{R}|$.

If k has characteristic zero, then Liu and the second author have shown in [29, 4.4] that any family of abelian k-varieties with no common isogeny factor is algebraically independent over $\mathbb{Z}$.

The Grothendieck ring of varieties is not noetherian. Let k be a field of characteristic zero. In [29, Corollary 4], Liu and the second author prove that $K_0(\text{Var}_k)$ is not noetherian. The argument is based on the birational simplification technique, which we recall below (Theorem 6.2).

The Grothendieck ring of varieties is not a domain. Let k be a field of characteristic zero.

In [37], Poonen constructs abelian k-varieties A and B such that $A \times_k A$ and $B \times_k B$ are isomorphic, while A and B are not isomorphic. Since $[A] = [B]$ in $K_0(\text{Var}_k)$ iff $A \cong B$ (Corollary 4.9) this implies that $[A]-[B]$ and $[A]+[B]$ are zero-divisors in $K_0(\text{Var}_k)$. Recently, Ekedahl has shown in [11, 3.5] that for the abelian varieties A and B in Poonen's example, we also have $[A] \neq [B]$ in $\mathcal{M}_k$, so that $[A] - [B]$ and $[A] + [B]$ are zero-divisors in $\mathcal{M}_k$.

In [25], Kollár proves that for any smooth projective conic C over k without rational point, the element $[C] - [\mathbb{P}^1_k]$ is a zero-divisor in $K_0(\text{Var}_k)$. Indeed, we have

$$[C] \cdot ([C] - [\mathbb{P}^1_k]) = 0$$

in $K_0(\text{Var}_k)$, and $[C] \neq [\mathbb{P}^1_k]$ by point (4) in Section 4.3.

Let A be an abelian k-variety that admits a non-trivial torsor X over k. Then $[A] \neq [X]$ in $K_0(\text{Var}_k)$, by point (4) in Section 4.3. Since $[X]([X] - [A]) = 0$ in $K_0(\text{Var}_k)$, we see that $[X]$ is a zero-divisor in $K_0(\text{Var}_k)$. If k is a complete discretely valued field of characteristic zero with algebraically closed residue field, and A has good reduction, then the first author showed in [32, 7.10] that $[X] \neq [A]$ in $\mathcal{M}_k$, so that $[X]$ is also a zero-divisor in $\mathcal{M}_k$. The proof uses the *motivic Serre invariant* (see [33]).

Now let k be any field that is not separably closed, and let k' be a Galois extension of k of degree $d > 1$. It is not hard to see that $[\mathrm{Spec}(k')]^2 = d \cdot [\mathrm{Spec}(k')]$ in $K_0(\mathrm{Var}_k)$, while $[\mathrm{Spec}(k')] \neq d$ in $\mathscr{M}_k$. This shows that $K_0(\mathrm{Var}_k)$ and $\mathscr{M}_k$ are not domains. See [38, 3.5] and [32, 2.12].

5.2 What we don't know about the Grothendieck ring of varieties

Zero divisors. It is not known if $K_0(\mathrm{Var}_k)$ or $\mathscr{M}_k$ have zero-divisors if k is a separably closed field of positive characteristic.

Is $\mathbb{L}$ a zero-divisor? In several theories of motivic integration, the integrals take their values in the localized Grothendieck ring $\mathscr{M}_k$ (or an appropriate completion). It is not known if the localization morphism

$$K_0(\mathrm{Var}_k) \to \mathscr{M}_k$$

is injective, i.e., if $\mathbb{L}$ is a zero-divisor in $K_0(\mathrm{Var}_k)$. Any element of $K_0(\mathrm{Var}_k)$ can be written as $[X] - [Y]$, with X and Y varieties over k, so the question is whether $\mathbb{L}[X] = \mathbb{L}[Y]$ implies $[X] = [Y]$. This question can be interpreted as a piecewise analog of *Zariski's cancellation problem*. The cancellation problem asks whether $X \times_k \mathbb{A}_k^1 \cong Y \times_k \mathbb{A}_k^1$ implies $X \cong Y$. It has a negative answer in general (see e.g. Danielewski's famous counterexample [8]). There also exist irreducible complex varieties X and Y such that $X \times_{\mathbb{C}} \mathbb{A}_{\mathbb{C}}^1$ and $Y \times_{\mathbb{C}} \mathbb{A}_{\mathbb{C}}^1$ are birational while X and Y are not [4].

But, to our understanding, the known counterexamples to the cancellation problem are not counterexamples to its piecewise analog. In particular, as noted by the second author in [41], this kind of problem raises the following question: is it possible to find two irreducible k-varieties X and X' such that $X \times_k \mathbb{A}_k^1 \cong X' \times_k \mathbb{A}_k^1$ but X and X' are not birational? An answer to this question would refine the understanding of Zariski's problem. If k is an algebraically closed field of characteristic zero, it follows from Lemma 9 in [29] that there are no such varieties X and X' of dimension two.

Is the completion morphism injective? In the theory of motivic integration, in order to obtain a sufficiently large class of measurable functions on arbitrary varieties, it is often necessary to replace the localized Grothendieck ring $\mathscr{M}_k$ by its *dimensional completion* $\widehat{\mathscr{M}_k}$. For each $m \in \mathbb{N}$, we denote by $F^m \mathscr{M}_k$ the subgroup of $\mathscr{M}_k$ generated by objects of the form $[X]\mathbb{L}^{-i}$, with X a k-variety and i an integer such that $\dim(X) + m \leq i$. We denote by $\widehat{\mathscr{M}_k}$ the completion of $\mathscr{M}_k$ w.r.t. the dimensional filtration $F^\bullet$. Since

$$F^m \mathscr{M}_k \cdot F^n \mathscr{M}_k \subset F^{m+n} \mathscr{M}_k$$

for all m, $n \in \mathbb{N}$, the completion $\widehat{\mathscr{M}_k}$ carries a natural ring structure. It is not known if the completion morphism

$$\mathscr{M}_k \to \widehat{\mathscr{M}_k}$$

is injective, i.e., if $\cap_{m\in\mathbb{N}} F^m \mathscr{M}_k = 0$. We denote by $\overline{\mathscr{M}}_k$ the image of $\mathscr{M}_k$ in $\widehat{\mathscr{M}_k}$. If k has characteristic zero, then the Hodge-Deligne polynomial

$$HD : \mathscr{M}_k \to \mathbb{Z}[u, v, u^{-1}, v^{-1}]$$

factors through $\overline{\mathscr{M}}_k$, since HD sends elements of $F^m \mathscr{M}_k$ to Laurent polynomials of total degree $\leq -2m$, and Laurent polynomials of degree $-\infty$ are zero. Likewise, for an arbitrary field k, the Poincaré polynomial P factors through $\overline{\mathscr{M}}_k$. In particular, the topological Euler characteristic χ_{top} factors through $\overline{\mathscr{M}}_k$.

What about purely inseparable morphisms in $K_0(\text{Var}_k)$? Assume that k has positive characteristic. If X and Y are separated k-schemes of finite type, and $f : X \to Y$ is a universal homeomorphism, is there any relation between $[X]$ and $[Y]$ in $K_0(\text{Var}_k)$? Is the projection

$$K_0(\text{Var}_k) \to K_0^{\text{mod}}(\text{Var}_k)$$

an isomorphism? The most basic question in this context is the following: if k is imperfect and k' is a finite purely inseparable extension of k, do we have $[\text{Spec}(k')] \neq 1$ in $K_0(\text{Var}_k)$?

The question of Larsen and Lunts. Arguably, the most fundamental question about the Grothendieck ring of varieties is the one raised by Larsen and Lunts. Solving this question would also mean a major step in the study of the questions formulated above. Larsen and Lunts' question will be discussed in the following section.

6 Piecewise geometry and the question of Larsen and Lunts

6.1 Piecewise algebraic geometry

Let k be a field, and consider the category Sch_k of k-schemes. Let Sch_k^o be the full subcategory of Sch_k whose objects are the k-schemes X such that for each point x of X, the local ring $\mathcal{O}_{X,x}$ is a field. Equivalently, the objects of Sch_k^o are the zero-dimensional reduced k-schemes. In [15, IV], Grothendieck suggested the following construction.

Proposition 6.1 ([34, 35]) *The embedding*

$$i \; : \mathit{Sch}_k^o \to \mathit{Sch}_k$$

has a right adjoint, denoted by $(\cdot)^{\mathrm{cons}}$.

We refer to [29, § 2] for a detailed account. It is shown there that two k-schemes X and Y are *piecewise isomorphic* (Section 3.2) iff X^{cons} and Y^{cons} are isomorphic in Sch_k^o. This result allows us to realize the Grothendieck group of varieties as a Grothendieck group in the sense of Section 2.1. We denote by CVar_k is the essential image of Var_k under the functor $(\cdot)^{\mathrm{cons}}$. The category CVar_k admits finite sums (given by the disjoint union of schemes in Sch_k^o), so that we can consider the additive Grothendieck group $K_0^{\mathrm{ad}}(\mathrm{CVar}_k)$. There exists a unique group morphism

$$K_0(\mathrm{Var}_k) \to K_0^{\mathrm{ad}}(\mathrm{CVar}_k)$$

that sends $[X]$ to $[X^{\mathrm{cons}}]$, for any k-variety X. By [29, 2.14], this is an isomorphism.

6.2 The question of Larsen and Lunts

One has no complete, geometric interpretation of the equality of classes in $K_0(\mathrm{Var}_k)$. The most natural guess is the following question of Larsen and Lunts [27, 1.2].

Let X and Y be k-varieties such that $[X] = [Y]$ in $K_0(\mathrm{Var}_k)$. Is it true that X and Y are piecewise isomorphic?

Note that it suffices to find isomorphic non-empty open subvarieties of X_{red} and Y_{red} : a Noetherian induction argument then yields a positive answer to Larsen and Lunts' question.

An equivalent way of formulating Larsen and Lunts' question is the following. The *Grothendieck semi-group* of k-varieties $K_0(\mathrm{Var}_k)^+$ is the abelian semi-group generated by isomorphism classes $\{X\}$ of k-varieties X, with the scissor relation $\{X\} = \{Y\} + \{X \setminus Y\}$ for every k-variety X and every closed subvariety Y of X. Two k-varieties have the same class in $K_0(\mathrm{Var}_k)^+$ iff they are piecewise isomorphic. The Grothendieck group $K_0(\mathrm{Var}_k)$ is the abelian group associated to $K_0(\mathrm{Var}_k)^+$. The question of Larsen and Lunts has an affirmative answer iff the natural morphism of semi-groups

$$K_0(\mathrm{Var}_k)^+ \to K_0(\mathrm{Var}_k)$$

is injective.

As observed by the second author in [41], if k has characteristic zero, then resolution of singularities allows to reduce the question of Larsen and Lunts to the following equivalent form:

Let X and Y be smooth projective k-varieties such that $[X] = [Y]$ in $K_0(\mathrm{Var}_k)$. Are the connected components of X and of Y pairwise birational?

Even for the particular case where X and Y are connected, no answer is known. We know that X and Y are stably birational, and that they have the same dimension (Section 4.3). In [4], Beauville, Colliot-Thélène, Sansuc and Swinnerton-Dyer have constructed projective smooth varieties that are stably rational (i.e., stably birational to a point), but not rational. This means that stably birational equivalence does not imply birational equivalence, even for varieties of the same dimension. Therefore, we can not deduce directly from Theorem 4.8 that X and Y are birational.

6.3 The work of Liu and Sebag

The only systematic study of the question of Larsen and Lunts so far is [29]. The results in that paper are based on Larsen-Lunts' invariant (Theorem 4.8) and the following birational cancellation theorem.

Theorem 6.2 ([29]) *Let k be a field, and let X, Y be integral k-varieties of the same dimension. Assume that there are geometrically integral k-varieties W, Z such that one of the following conditions is satisfied:*

1. *X or Y is non-uniruled, and W, Z are rationally chain connected;*
2. *the characteristic of k is zero, X, Y, W, Z are projective and smooth,*

$$\kappa(X) \geq 0 \text{ or } \kappa(Y) \geq 0, \quad \text{and } \kappa(W) = \kappa(Z) = -\infty.$$

Assume moreover that there exists a birational map

$$f : X \times_k W \dashrightarrow Y \times_k Z.$$

Then there exists a unique birational map $g : X \dashrightarrow Y$ *such that the diagram*

$$
\begin{array}{ccc}
X \times_k W & \xrightarrow{\ f\ } & Y \times_k Z \\
\downarrow{\scriptstyle p_1} & & \downarrow{\scriptstyle q_1} \\
X & \xdashrightarrow{\ \ g\ \ } & Y
\end{array}
$$

commutes (the vertical arrows are the projections on the first factors).

One knows:

Theorem 6.3 ([29],[41]) *Let* k *be a field; let* X *and* Y *be* k-*varieties such that* $[X] = [Y]$ *in* $K_0(\mathrm{Var}_k)$.

1. *Assume that* X *and* Y *are zero-dimensional. If* k *has characteristic zero, or* k *is a finite field, or* k *is algebraically closed, then* X *is isomorphic to* Y.

2. *Assume that* k *is algebraically closed of characteristic zero. If* X *and* Y *are* k-*varieties such that* $\dim X \leq 1$, *then* X *is piecewise isomorphic to* Y.

3. *Assume that* k *is algebraically closed of characteristic zero. If* X *and* Y *are smooth projective* k-*varieties of dimension two, then* X *is piecewise isomorphic to* Y.

4. *Assume that* k *is algebraically closed of characteristic zero. If* X *and* Y *are* k-*varieties of dimension two, and the irreducible components of* X *of dimension 2 are non-ruled or rational surfaces, then* X *is piecewise isomorphic to* Y.

5. *Assume that* k *has characteristic zero. If* X *and* Y *are geometrically irreducible* k-*curves, then* X *is piecewise isomorphic to* Y.

6. *Assume that* k *has characteristic zero. Suppose that one of the following assumptions holds:*

   1. *the field* k *is algebraically closed and* X *contains only finitely many rational curves,*
   2. $X \times_k k^a$ *does not contain any rational curve (e.g. if* X *is a subvariety of an abelian variety).*

   Then X *is piecewise isomorphic to* Y.

Remark 6.4 The results in [42] allow to prove that, if X, Y are complex Calabi-Yau varieties of dimension at most 4 such that $[X] = [Y]$ in $K_0(\mathrm{Var}_k)$, then X and Y are piecewise isomorphic.

We should also mention the work of Kollár [25] and the generalizations in Hogadi [22]. They show that two products of Severi-Brauer varieties of dimension ≤ 2, defined over a field k of characteristic zero, have the same class in $K_0(\mathrm{Var}_k)$ iff they are birational.

The work of Larsen-Lunts and Liu-Sebag illustrates the crucial role of the existence of rational curves on varieties for the geometric understanding of $K_0(\mathrm{Var}_k)$. The strategy of birational cancellation developed in [29] is blocked by the existence of stably rational varieties which are not rational [4]. However, this does not mean that such a strategy cannot work. If X is a stably rational k-variety X that is not rational, one should prove that its class $[X] \in K_0(\mathrm{Var}_k)$ cannot be equal to the class of a rational k-variety. This seems to be a challenging problem.

6.4 Some consequences

We indicate some consequences of the above results. We refer to [29] for details (see also [41] for the first point). Let k be a field of characteristic zero, and let X and X' be integral k-varieties such that $[X] = [X']$ in $K_0(\mathrm{Var}_k)$.

1. If X or X' is non-uniruled, then they are birational.
2. If X, X' are connected, smooth and projective, then they have the same Kodaira dimension.
3. Assume that X is smooth, projective and connected, of dimension $d \geq 2$, ruled over a smooth projective connected k-variety D of dimension $d-1$. Then there exist a finite number of smooth projective connected k-varieties C_i, $i \in I$, of dimension at most $d - 2$ such that

$$[X] = \mathbb{L}([D] + \sum_{i \in I} \varepsilon_i [C_i]) + [D],$$

   with $\varepsilon_i \in \{-1, 1\}$.
4. As a partial converse, one has the following property. Assume that X is smooth, projective and connected. Assume that there exist an element a in $K_0(\mathrm{Var}_k)$ and a smooth projective connected k-variety Y, with $\dim Y < \dim X$, such that $[X] = [Y] + a\mathbb{L}$. Then $\kappa(X) = -\infty$.

6.5 Related problems

In this paragraph, we indicate some problems, already mentioned in [41], which are related to Larsen and Lunts' question.

Rational points and Grothendieck ring of varieties. A positive answer to the question of Larsen and Lunts would imply a positive answer to the following one.

Question 6.5 Let k be a field of characteristic zero. Let X be a k-variety. Are the following assertions equivalent?
i) X has a rational point;
ii) there exists a k-variety Y such that $[X] = [Y] + 1$ in $K_0(\mathrm{Var}_k)$.

Note that $i) \Rightarrow ii)$ is trivial, but its converse is quite challenging. One can prove the following partial results (see [41]).

- If X is a connected smooth proper k-variety, then X has a rational point iff there exists a connected smooth proper k-variety $\overline{Y}$ and a smooth divisor D on $\overline{Y}$ such that $Y := \overline{Y} \backslash D$ satisfies the relation $[X] = 1 + [Y]$ in $K_0(\mathrm{Var}_k)$.
- If X is connected smooth proper, non-ruled, k-surface then X has a rational point iff there exists a connected smooth proper k-variety $\overline{Y}$ and a simple normal crossings divisor D on $\overline{Y}$ such that $Y := \overline{Y} \backslash D$ satisfies the relation $[X] = 1 + [Y]$ in $K_0(\mathrm{Var}_k)$.
- If X is a geometrically irreducible smooth k-curve, then X has a rational point if and only if there exists a geometrically irreducible smooth k-curve Y such that $[X] = 1 + [Y]$ in $K_0(\mathrm{Var}_k)$.

Gromov's question, the complement problem and the Grothendieck ring of varieties. The *complement problem* in affine geometry asks the following. Let $N \geq 1$ be an integer; let $f, g \in \mathbb{C}[\underline{x}] := \mathbb{C}[x_1, \ldots, x_N]$ be two *irreducible* polynomials. Assume that there exists a $\mathbb{C}$-isomorphism $\mathbb{C}[\underline{x}]_f \to \mathbb{C}[\underline{x}]_g$. *Is it true that the hypersurfaces $V(f)$ and $V(g)$ of $\mathbb{A}_{\mathbb{C}}^N$ are $\mathbb{C}$-isomorphic?* In [14], Gromov asks the following related question:

Let two algebraic varieties X and Y admit embeddings to a third one, say $X \hookrightarrow Z$ and $Y \hookrightarrow Z$, such that the complements $Z \backslash X$ and $Z \backslash Y$ are [...] isomorphic. How far are X and Y from being birationally equivalent? When does there exist a constructible bijection $X \to Y$?

In our framework, Gromov's question can be rephrased as follows (see also [29, 41]):

Question 6.6 Let k be a field of characteristic zero. Let X be a k-variety, and let U, V be closed subvarieties of X. Assume that $X \backslash U$ and $X \backslash V$ are isomorphic. Does this imply that U and V are piecewise isomorphic?

Of course, if Larsen-Lunts' question admits a positive answer, the last question does too, since U and V define the same class in $K_0(\mathrm{Var}_k)$. Thanks to the results explained above, we can conclude in some cases. For example, if k is algebraically closed, and X is integral of dimension 2 or X does not contain a rational curve, then Question 6.6 admits a positive answer.

References

[1] Abramovich, D.; Karu, K.; Matsuki, K.; Włodarczyk, J. *Torification and factorization of birational maps.* J. Amer. Math. Soc. 15 (2002), no. 3, 531–572.

[2] André, Y. *Une introduction aux motifs (motifs purs, motifs mixtes, périodes).* Panoramas et Synthèses, 17. Société Mathématique de France, Paris, 2004.

[3] Beauville, A. *Variétés de Prym et jacobiennes intermédiaires.* Ann. Sci. École Norm. Sup. (4) 10 (1977), no. 3, 309–391.

[4] Beauville, A.; Colliot-Thélène, J.-L.; Sansuc, J.-J.; Swinnerton-Dyer, P. *Variétés stablement rationnelles non rationnelles.* Ann. of Math. (2) 121 (1985), no. 2, 283–318.

[5] Bittner, F. *The universal Euler characteristic for varieties of characteristic zero.* Compos. Math. 140 (2004), no. 4, 1011–1032.

[6] Blickle, M. *A short course on geometric motivic integration,* this volume.

[7] Cluckers, R.; Loeser, F. *Constructible motivic functions and motivic integration.* Invent. Math. 173 (2008), no. 1, 23–121.

[8] W. Danielewski. *On a cancellation problem and automorphism groups of affine algebraic varieties.* Preprint, 1989.

[9] Deligne, P. *Théorie de Hodge II.* Inst. Hautes Études Sci. Publ. Math. No. 40 (1971), 5–57.

[10] Deligne, P. *Théorie de Hodge III.* Inst. Hautes Études Sci. Publ. Math. No. 44 (1974), 5–77.

[11] T. Ekedahl. *The Grothendieck group of algebraic stacks.* Preprint, arXiv:0903.3143v2.

[12] Gillet, H.; Soulé, C. *Descent, motives and K-theory.* J. Reine Angew. Math. 478 (1996), 127–176.

[13] Göttsche, L. *On the motive of the Hilbert scheme of points on a surface.* Math. Res. Lett. 8 (2001), no. 5-6, 613–627.

[14] Gromov, M. *Endomorphisms of symbolic algebraic varieties.* J. Eur. Math. Soc. (JEMS) 1 (1999), no. 2, 109–197.

[15] Grothendieck, A.; Dieudonné J. *Éléments de Géométrie Algébrique* (EGA) Publ. Math. IHES, **4, 8, 11, 17, 20, 24, 28, 32**, 1960–1967

[16] *Correspondance Grothendieck-Serre.* Edited by Pierre Colmez and Jean-Pierre Serre. Documents Mathématiques (Paris), 2. Société Mathématique de France, Paris, 2001.

[17] Guillén, F.; Navarro Aznar, V.. *Un critère d'extension des foncteurs définis sur les schémas lisse.* Publ. Math. Inst. Hautes Études Sci. No. 95 (2002), 1–91.

[18] S.M. Gusein-Zade, I. Luengo and A. Melle-Hernández. *A power structure over the Grothendieck ring of varieties.* Math. Res. Lett. 11(1):49–57, 2004.

[19] S.M. Gusein-Zade, I. Luengo and A. Melle-Hernández. *Power structure over the Grothendieck ring of varieties and generating series of Hilbert schemes of points.* Michigan Math. J. 54(2):353–359, 2006.

[20] Hartshorne, R. *Algebraic geometry.* Graduate Texts in Mathematics, No. 52. Springer-Verlag, New York-Heidelberg, 1977.

[21] Heinloth, F. *A note of functional equations for zeta functions with values in Chow motives.* Ann. Inst. Fourier, 57 (2007), no. 6, 1927–1945.

[22] Hogadi, A. *Products of Brauer-Severi surfaces.* Proc. Amer. Math. Soc. 137 (2009), no. 1, 45–50.

[23] Jannsen, U. *Mixed motives and algebraic K-theory.* With appendices by S. Bloch and C. Schoen. Lecture Notes in Mathematics, 1400. Springer-Verlag, Berlin, 1990.

[24] Kapranov, M. *The elliptic curve in the S-duality theory and Eisenstein series for Kac-Moody groups.* Preprint, arXiv:math/0001005.

[25] Kollár, J. *Conics in the Grothendieck ring.* Adv. Math. 198 (2005), no. 1, 27–35.

[26] Krajíček, J.; Scanlon, T. *Combinatorics with definable sets: Euler characteristics and Grothendieck rings.* Bull. Symbolic Logic 6 (2000), no. 3, 311–330.

[27] Larsen, M.; Lunts, V. A. *Motivic measures and stable birational geometry.* Mosc. Math. J. **3** (2003), no. 1, 85–95.

[28] Larsen, M.; Lunts, V. A. *Rationality criteria for motivic zeta functions.* Compos. Math. 140 (2004), no. 6, 1537–1560.

[29] Liu, Q.; Sebag, J. *The Grothendieck ring of varieties and piecewise isomorphisms*, to appear in Math. Z.

[30] Loeser, F.; Sebag, J. *Motivic integration on smooth rigid varieties and invariants of degenerations*, Duke Math. Journal, Vol 119 (2003), no. 2, 315–344.

[31] Naumann, N. *Algebraic independence in the Grothendieck ring of varieties.* Trans. Amer. Math. Soc. 359 (2007), no. 4, 1653–1683

[32] Nicaise, J. *A trace formula for varieties over a discretely valued field*, J. Reine Angew. Math. 650 (2011), 193–238.

[33] Nicaise, J.; Sebag, J. *Motivic invariants of rigid varieties, and applications to complex singularities*, in this volume.

[34] Olivier, J.-P. *Anneaux absolument plats universels et épimorphismes à buts réduits*, Séminaire Samuel. Algèbre commutative, **2** (1967-1968), Exposé No. 6. Secrétariat mathématique, Paris 1967-1968.

[35] Olivier, J.-P. *Le foncteur $T^{-\infty}$. Globalisation du foncteur T*, Séminaire Samuel. Algèbre commutative, **2** (1967-1968), Exposé No 9. Secrétariat mathématique, Paris 1967-1968.

[36] Peters, C. A. M.; Steenbrink, J. H. M. *Mixed Hodge structures*. Ergebnisse der Mathematik und ihrer Grenzgebiete. 3. Folge. A Series of Modern Surveys in Mathematics, 52. Springer-Verlag, Berlin, 2008.

[37] Poonen, B. *The Grothendieck ring of varieties is not a domain*. Math. Res. Lett. 9 (2002), no. 4, 493–497.

[38] Rökaeus, K. *The computation of the classes of some tori in the Grothendieck ring of varieties*. Preprint, arXiv:0708.4396.

[39] Scholl, A. J. *Classical motives*. In *Motives, Seattle, 1991*, volume 55 of *Proc. Symp. Pure Math.*, pages 163–187. Amer. Math. Soc., 1994.

[40] Sebag, J. *Intégration motivique sur les schémas formels*. Bull. Soc. Math. France **132** (2004), no. 1, 1–54.

[41] Sebag, J. *Variations on a question of Larsen and Lunts*. Proc. Am. Math. Soc., vol. 138 (2010), no. 4, 1231–1242.

[42] Sebag, J. *Variétés K-équivalentes et isomorphismes par morceaux*. to appear in Archiv Math. 94 (2010), no. 3, 207–219.

[43] Voisin, C. *Hodge theory and complex algebraic geometry I,II*. Volumes 76 and 77 of *Cambridge Studies in Advanced Mathematics*, Cambridge University Press, 2002 and 2003.

6

A short course on geometric motivic integration

Manuel Blickle[a]

1 The invention of motivic integration

Motivic integration was introduced by Kontsevich [30] to prove the following result conjectured by Batyrev: Let

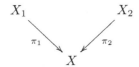

be two crepant resolutions of the singularities of X, which itself is a complex projective Calabi-Yau variety with at worst canonical Gorenstein singularities. For the purpose of these notes we call a normal projective variety X of dimension n Calabi-Yau if the canonical divisor K_X is trivial. Crepant (as in *non discrepant*) means that the pullback of the canonical divisor class on X is the canonical divisor class on X_i, *i.e.* the discrepancy divisor $E_i = K_{X_i} - \pi_i^* K_X$ is numerically equivalent to zero. In this situation Batyrev showed, using p-adic integration, that X_1 and X_2 have the same betti numbers $h_i = \dim H^i(\_, \mathbb{C})$. This led Kontsevich to invent *motivic integration* to show that X_1 and X_2 even have the same Hodge numbers $h^{i,j} = \dim H^i(\_, \Omega^j)$.

This problem was motivated by the *topological mirror symmetry test* of string theory which asserts that if X and X^* are a mirror pair of

[a] These notes are based on several lecture series of the author held in the period between 2002 and 2005 in Stockholm, Helsinki, and Salt Lake City. In consequence, I do not discuss developments that took place after 2005.

smooth Calabi-Yau varieties then they have mirrored Hodge numbers

$$h^{i,j}(X) = h^{n-i,j}(X^*).$$

As the mirror of a smooth Calabi-Yau might be singular, one cannot restrict to the smooth case and the equality of Hodge numbers actually fails in this case. Therefore Batyrev suggested, inspired by string theory, that one should look instead at the Hodge numbers of a crepant resolution, if such exists. The independence of these numbers from the chosen crepant resolution is Kontsevich's result. This makes the *stringy Hodge numbers* $h^{i,j}_{st}(X)$ of X, defined as $h^{i,j}(X')$ for a crepant resolution X' of X, well defined. This leads to a modified mirror symmetry conjecture, asserting that the stringy Hodge numbers of a mirror pair are equal [3].

Batyrev's conjecture is now Kontsevich's theorem and the simplest form to phrase it might be:

Theorem 1.1 (Kontsevich)   *Birationally equivalent smooth Calabi-Yau varieties have the same Hodge numbers.*

Proof   The idea now is to assign to any variety a *volume* in a suitable ring $\hat{\mathscr{M}}_k$ such that the information about the Hodge numbers is retained. The following diagram illustrates the construction of $\hat{\mathscr{M}}_k$:

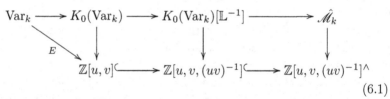

$$(6.1)$$

The diagonal map is the (compactly supported) Hodge characteristic, which on a smooth projective variety X is given by $E(X) = \sum (-1)^{i+j} \dim H^i(X, \Omega^j_X) u^i v^j$. In general it is defined via mixed Hodge structures [7, 8, 9] (or alternatively using the construction in [4] which employs weak factorization theorem of Włodarczyk [1] to reduce the definition of E to the case of X smooth and projective), satisfies $E(X \times Y) = E(X)E(Y)$ for all varieties X, Y and has the property that for $Y \subseteq X$ a closed k-subvariety one has $E(X) = E(Y) + E(X - Y)$. Therefore the Hodge characteristic factors through the *naive Grothendieck ring* $K_0(\mathrm{Var}_k)$ which is the universal object with the latter property. $K_0(\mathrm{Var}_k)$ is the free abelian group on the isomorphism classes $[X]$ of k-varieties subject to the relations $[X] = [X - Y] + [Y]$ for Y a closed subvariety of X. The product is given by $[X][Y] = [X \times_k Y]$. The symbol

$\mathbb{L}$ denotes the class of the affine line $[\mathbb{A}_k^1]$. This explains the left triangle of the diagram.

The bottom row of the diagram is the composition of a localization (inverting uv) and a completion with respect to negative degree. $\hat{\mathcal{M}}_k$ is constructed analogously, by first inverting $\mathbb{L}^{-1} = [\mathbb{A}_k^1]$ (a pre-image of uv) and then completing appropriately (negative dimension). Whereas the bottom maps are injective (easy exercise), the map $K_0(\mathrm{Var}_k) \rightarrow \hat{\mathcal{M}}_k$ is most likely not injective. The need to work with $\hat{\mathcal{M}}_k$ instead of $K_0(\mathrm{Var}_k)$ arises in the setup of the integration theory and will become clear later.

Clearly, by construction it is now enough to show that birationally equivalent Calabi-Yau varieties have the same *volume*, i.e. the same class in $\hat{\mathcal{M}}_k$. This is achieved via the all important *birational transformation rule* of motivic integration. Roughly it asserts that for a proper birational map $\pi : Y \rightarrow X$ the class $[X] \in \hat{\mathcal{M}}_k$ is an *expression* in Y and $K_{Y/X}$ only:

$$[X] = \int_Y \mathbb{L}^{-\mathrm{ord}\, K_{Y/X}}$$

To finish off the proof let X_1 and X_2 be birationally equivalent Calabi-Yau varieties. We resolve the birational map to a Hironaka hut:

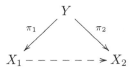

By the Calabi-Yau assumption we have $K_{X_i} \equiv 0$ and therefore $K_{Y/X_i} \equiv K_Y - \pi_i^* K_{X_i} \equiv K_Y$. Hence the divisors K_{Y/X_1} and K_{Y/X_2} are numerically equivalent. This numerical equivalence implies in fact an equality of divisors $K_{Y/X_1} = K_{Y/X_2}$ since, again by the Calabi-Yau assumption, $\dim H^0(X, K_Y) = \dim H^0(X_i, \mathcal{O}_{X_i}) = 1$. By the transformation rule, $[X_1]$ is an expression depending only on Y and $K_{Y/X_1} = K_{Y/X_2}$. The same is true for $[X_2]$ and thus we have $[X_1] = [X_2]$ as desired. □

We conclude this introduction by pointing out that the condition that X_1 and X_2 have a common resolution Y such that K_{Y/X_1} is numerically equivalent K_{Y/X_1} is called K–equivalence. We showed above that two birational Calabi–Yau varieties are K–equivalent. For mildly singular X_i (say canonical) it can be derived from the negativity lemma [29, Lemma 3.39] that K–equivalence implies actual equality of divisors K_{Y/X_1} and

K_{Y/X_1}. Hence the Calabi–Yau assumption was not essential to conclude this (but provides a simple argument).

Remark 1.2   There is now a proof by Ito [27] of this result using p-adic integration, thus continuing the ideas of Batyrev who proved the result for Betti numbers using this technique. Furthermore the recent weak factorization theorem of Włodarczyk [1] allows for a proof avoiding integration of any sort.

These notes were started during a working seminar at MSRI during the year of 2003 and took shape in the course of the past 2 years while I was giving introductory lectures on the subject. They have taken me way too much time to finish and would not have been finished at all if it weren't for the encouragement of many people: Thanks goes to all the participants of the seminar on motivic integration at MSRI (2002/2003), of the Schwerpunkt Junioren Tagung in Bayreuth (2003) and the patient listeners of the mini-courses at KTH, Stockholm (2003), the University of Helsinki (2004) and the Vigre graduate course in Salt Lake City (2005). Special thanks goes to Karen Smith for encouragement to start this project and to Julia Gordon for numerous comments, suggestions and careful reading. Finally I want to thank the referee for a careful reading and many valuable comments.

2   Geometric motivic integration

We now assume that k is algebraically closed and of characteristic zero. In fact, there is one point (see section 4.1) where we will assume that $k = \mathbb{C}$ in order to avoid some technicalities which arise if the field is not uncountable. Thus the reader may choose to replace k by $\mathbb{C}$ whenever it is comforting. We stress that there are significant (though manageable) obstacles one has to overcome if one wants to (a) work with singular spaces or (b) with varieties defined over fields which are not uncountable or not algebraically closed. Or put differently: The theory develops naturally (for an algebraic geometer), and easily, in the smooth case over $\mathbb{C}$, as we hope to demonstrate below. In order to transfer this intuition to any other situation, nontrivial results and extra care is necessary.

All the results in these notes appeared in the papers of Denef and Loeser, Batyrev and Looijenga. Our exposition is particularly influenced by Looijenga [32] and Batyrev [2]. Also Craw [6] was very helpful as a first reading. We also recommend the articles of Hales [24] and Veys

[42], both explain the connection to p-adic integration in detail, which we do not discuss in these notes at all. The above mentioned references are also a great source to learn about the various different applications the theory had to date. We will discuss none of them except for certain applications to birational geometry.

We will now introduce the building blocks of the theory. These are:

1. The value ring of the measure: $\hat{\mathscr{M}}_k$, a *localized and completed Grothendieck ring*.
2. A domain of integration: $\mathscr{J}_\infty(X)$, *the space of formal arcs over X*.
3. An algebra of measurable sets of $\mathscr{J}_\infty(X)$ and a measure: *cylinders/ stable sets and the virtual euler characteristic*.
4. An interesting class of measurable/integrable functions: *Contact order of an arc along a divisor*.
5. A change of variables formula: *Kontsevich's birational transformation rule*.

These basic ingredients appear with variations in all versions of motivic integration (one could argue: of any theory of integration).

2.1 The value ring of the motivic measure

Here we already gravely depart from any previous (classical?) theory of integration since the values of our measure do not lie in $\mathbb{R}$. Instead they lie in a huge ring, constructed from the Grothendieck ring of varieties by a process of localization and completion. This ingenious choice is a key feature of the theory.

We start with the naive Grothendieck ring of the category of varieties over k. Alternatively, the Grothendieck ring of finite type schemes over k leads to the same object because $X - X_{\text{red}} = \emptyset$, however note that the finite type assumption is crucial here: If one would allow non finite type schemes, $K_0(\text{Var}_k)$ would be zero. To see this let Y be any k–scheme and let X be an infinite disjoint union of copies of Y. Then $[X] + [Y] = [X]$ and therefore $[Y] = 0$. The Grothendieck ring of varieties is the ring $K_0(\text{Var}_k)$ generated by the isomorphism classes of all finite type k–varieties and with relation $[X] = [Y] + [X - Y]$ for a closed k-subvariety $Y \subseteq X$, that is such that the inclusion $Y \subseteq X$ is defined over k. The square brackets denote the image of X in $K_0(\text{Var}/k)$. The product structure is given by the fiber product, $[X] \cdot [Y] = [X \times_k Y]$ (which should be thought of as $[(X \times_k Y)_{\text{red}}]$) if k is not algebraically closed). The symbol $\mathbb{L}$ is reserved for the class of the affine line $[\mathbb{A}_k^1]$ and $1 = 1_k$ denotes $\operatorname{Spec} k$. Thus, for example, $[\mathbb{P}^n] = \mathbb{L}^n + \mathbb{L}^{n-1} + \ldots + \mathbb{L} + 1$.

Roughly speaking the map $X \mapsto [X]$ is robust with respect to chopping up X into a disjoint union of locally closed subvarieties, see the beautiful article of Hales [24] which emphasizes precisely this point of $K_0(\operatorname{Var}_k)$ being a *scissor group*. By using a stratification of X by smooth subvarieties, this shows that $K_0(\operatorname{Var}_k)$ is generated by the classes of smooth varieties. Even more is true: In [4] it is shown that $K_0(\operatorname{Var}_k)$ is the abelian group generated by smooth projective varieties subject to a class of relations which arise from blowing up at a smooth center: If Z is a smooth subvariety of X, then the relation is $[X] - [Z] = [\operatorname{Bl}_Z X] - [E]$, where E is the exceptional divisor of the blowup.

Exercise 2.1 Show that the map $Y \mapsto [Y]$ for Y a closed subvariety of X naturally extends to the algebra of constructible subsets of X.

Exercise 2.2 Let $Y \longrightarrow X$ be a piecewise trivial fibration with constant fiber Z. This means one can write $X = \bigsqcup X_i$ as a finite disjoint union of locally closed subsets X_i such that over each X_i one has $f^{-1}X_i \cong X_i \times Z$ and f is given by the projection onto X_i. Show that in $K_0(\operatorname{Var}_k)$ one has $[Y] = [X] \cdot [Z]$.

Dimension

There is a natural notion of dimension of an element of $K_0(\operatorname{Var}_k)$. We say that $\tau \in K_0(\operatorname{Var}_k)$ is *d–dimensional* if there is an expression in $K_0(\operatorname{Var}_k)$

$$\tau = \sum a_i [X_i]$$

with $a_i \in \mathbb{Z}$ and k-varieties X_i of dimension $\leq d$, and if there is no expression like this with all $\dim X_i \leq d - 1$. The dimension of the class of the empty variety is set to be $-\infty$. It is easy to verify (exercise!) that the map

$$\dim : K_0(\operatorname{Var}_k) \longrightarrow \mathbb{Z} \cup \{-\infty\}$$

satisfies $\dim(\tau \cdot \tau') \leq \dim \tau + \dim \tau'$ and $\dim(\tau + \tau') \leq \max\{\dim \tau, \dim \tau'\}$ with equality in the latter if $\dim \tau \neq \dim \tau'$.

A basic observation one should make is that $\dim[V] = \dim V$. By definition of $\dim[V]$ we clearly have $\dim[V] \leq \dim V$. Suppose now that we have a strict inequality, i.e. we can write $[V] = \sum a_i [V_i]$ with $\dim V_i < \dim V$. Applying the (compactly supported) Hodge characteristic (Hodge-Deligne polynomial) to this equality we get

$E(V) = \sum a_i E(V_i)$. But this is impossible since $E(V)$ is a polynomial of degree $2 \dim V$ whereas each term on the right hand side has degree $2 \dim V_i < 2 \dim V$.

The dimension can be extended to the localization $\mathcal{M}_k \overset{\text{def}}{=} K_0(\text{Var}_k) \times [\mathbb{L}^{-1}]$ simply by demanding that $\mathbb{L}^{-1}$ has dimension -1. From the preceding discussion it follows also that the dimension of $\tau \in \mathcal{M}_k$ is equal to $1/2$ the degree of the Hodge characteristic $E(\tau) \in \mathbb{Z}[u, v, (uv)^{-1}]$.

To obtain the ring $\hat{\mathcal{M}}_k$ in which the desired measure will take values we further complete $\mathcal{M}_k$ with respect to the filtration induced by the dimension. The nth filtered subgroup is

$$\mathcal{F}^n(\mathcal{M}_k) = \{ \tau \in \mathcal{M}_k \mid \dim \tau \leq n \}.$$

This gives us the following maps which will be the basis for constructing the sought after motivic measure:

$$\text{Var}_k \longrightarrow K_0(\text{Var}_k) \xrightarrow{\text{invert } \mathbb{L}} \mathcal{M}_k \xrightarrow{\wedge} \hat{\mathcal{M}}_k.$$

We will somewhat ambiguously denote the image of $X \in \text{Var}_k$ in any of the rings to the right by $[X]$. Note that the dimension on $\mathcal{M}_k$ extends to a dimension function $\dim : \hat{\mathcal{M}}_k \to \mathbb{Z} \cap \{\pm\infty\}$ which again has the property that for any $\tau \in \hat{\mathcal{M}}_k$ the dimension $\dim \tau$ is equal to $1/2$ the degree of the power series $E(\tau) \in \mathbb{Z}[u, v, (uv)^{-1}]^{\wedge}$, see diagram (6.1).

It is important to point out here that it is unknown whether the completion map $\_^{\wedge}$ is injective, *i.e.* whether its kernel, $\bigcap \mathcal{F}^d(K_0(\text{Var}_k)[\mathbb{L}^{-1}])$, is zero. It is also unknown whether the localization is injective.

Remark 2.3 In Looijenga [32], the dimension function on $\mathcal{M}_k$ is called the virtual dimension. As described by Batyrev, composing the dimension $\dim : \mathcal{M}_k \to \mathbb{Z} \cup \{-\infty\}$ with the exponential $\mathbb{Z} \subseteq \mathbb{R} \xrightarrow{\exp(\_)} \mathbb{R}_+$ and by further defining $\emptyset \mapsto 0$ we get a map

$$\delta_k : \mathcal{M}_k \to \mathbb{R}_{+,0}$$

which is a *non-archimedean norm*. That means the following properties hold:

1. $\delta_k(A) = 0$ if $A = 0 = [\emptyset]$ in $\mathcal{M}_k$.
2. $\delta_k(A + B) \leq \max\{ \delta_k(A), \delta_k(B) \}$.
3. $\delta_k(A \cdot B) \leq \delta_k(A) \cdot \delta_k(B)$.

The ring $\hat{\mathcal{M}}_k$ is then the completion with respect to this norm, and therefore $\hat{\mathcal{M}}_k$ is complete in the sense that all Cauchy sequences uniquely converge. The condition (2) is stronger than the one used in the definition of an archimedean norm. This non-archimedean ingredient makes the notion of convergence of sums conveniently simple; a sum converges if and only if the sequence of summands converges to zero.

If there was an equality in condition (3) the norm would be called *multiplicative*. It is unknown whether δ is multiplicative on $\mathscr{M}_k$. However, Poonen [37] shows that $K_0(\text{Var}_k)$ contains zero divisors, thus δ restricted to $K_0(\text{Var}_k)$ is *not* multiplicative on $K_0(\text{Var}_k)$.

Remark 2.4 In [37] Poonen shows that $K_0(\text{Var}\,/k)$ is not a domain in characteristic zero. It is expected though that the localization map is not injective and that $\mathscr{M}_k$ is a domain and that the completion map $\mathscr{M}_k \longrightarrow \hat{\mathscr{M}}_k$ is injective. But recently Naumann [36] found in his dissertation zero-divisors in $K_0(\text{Var}_k)$ for k a finite field and these are non-zero even after localizing at $\mathbb{L}$ – thus for a finite field $\mathscr{M}_k$ is not a domain. For infinite fields (*e.g.* k algebraically closed) the above questions remain open.

Exercise 2.5 Convergence of series in $\hat{\mathscr{M}}_k$ is rather easy. For this observe that a sequence of elements $\tau_i \in \mathscr{M}_k$ converges to zero in $\hat{\mathscr{M}}_k$ if and only if the dimensions $\dim \tau_i$ tend to $-\infty$ as i approaches ∞. Show that a sum $\sum_{i=0}^{\infty} \tau_i$ converges if and only the sequence of summands converges to zero.

Exercise 2.6 Show that in $\hat{\mathscr{M}}_k$ the equality $\sum_{i=0}^{\infty} \mathbb{L}^{-ki} = \frac{1}{1-\mathbb{L}^{-k}}$ holds.

2.2 The arc space $\mathscr{J}_\infty(X)$

Arc spaces were first studied seriously by Nash [35] who conjectured a tight relationship between the geometry of the arc space and the singularities of X, see Ishii and Kollar [26] for a recent exposition of Nash's ideas in modern language. Recent work of Mustaţă [33] supports these predictions by showing that the arc spaces contain information about singularities, for example rational singularities of X can be detected by the irreducibility of the jet schemes for complete intersections. In subsequent investigations he and his collaborators show that certain invariants of birational geometry, such as the log canonical threshold of a pair, for example, can be read off from the dimensions of certain components of the jet schemes, see [34, 14, 16] and Section 5 where we will discuss some of these results in detail.

Let X be a (smooth) scheme of finite type over k of dimension n. An m-jet of X is an order m infinitesimal curve in X, *i.e.* it is a morphism

$$\vartheta : \operatorname{Spec} k[t]/t^{m+1} \longrightarrow X.$$

The set of all m-jets carries the structure of a scheme $\mathscr{J}_m(X)$, called the mth *jet scheme*, or space of truncated arcs. It's characterizing property is that it is right adjoint to the functor $\_\!\_ \times \operatorname{Spec} k[t]/t^{m+1}$. In other words,

$$\operatorname{Hom}(Z \times \operatorname{Spec} k[t]/t^{m+1}, X) = \operatorname{Hom}(Z, \mathscr{J}_m(X))$$

for all k-schemes Z, *i.e.* $\mathscr{J}_m(X)$ is the scheme which represents the contravariant functor $\operatorname{Hom}(\_\!\_ \times \operatorname{Spec} k[t]/t^{m+1}, X)$, see Greenberg [18, 19] or [5]. In particular this means that the k–valued points of $\mathscr{J}_m(X)$ are precisely the $k[t]/t^{m+1}$–valued points of X. The so called Weil restriction of scalars, *i.e.* the natural map $k[t]/t^{m+1} \longrightarrow k[t]/t^m$, induces a map $\pi_{m-1}^m : \mathscr{J}_m(X) \longrightarrow \mathscr{J}_{m-1}(X)$ and composition gives a map $\pi^m : \mathscr{J}_m(X) \longrightarrow \mathscr{J}_0(X) = X$. As upper indices are often cumbersome we define $\eta_m = \pi^m$ and $\varphi_m = \pi_{m-1}^m$.

Taking the inverse limit of the resulting system of *affine morphisms* yields the definition of the *infinite jet scheme*, or the *arc space*

$$\mathscr{J}_\infty(X) = \varprojlim \mathscr{J}_m(X).$$

Its k-points are the limit of the k-valued points $\operatorname{Hom}(\operatorname{Spec} k[t]/t^{m+1}, X)$ of the jet spaces $\mathscr{J}_m(X)$. Therefore they correspond to the formal curves (or arcs) in X, that is to maps $\operatorname{Spec} k[\![t]\!] \longrightarrow X$ since $\varprojlim \operatorname{Hom}(R, k[t]/t^{m+1}) \cong \operatorname{Hom}(R, \varprojlim k[t]/t^{m+1}) = \operatorname{Hom}(R, k[\![t]\!])$. There are also maps $\pi_m : \mathscr{J}_\infty(X) \longrightarrow \mathscr{J}_m(X)$ again induced by the truncation map $k[\![t]\!] \longrightarrow k[\![t]\!]/t^{m+1}$. If there is danger of confusion we sometimes decorate the projections π with the space. The following picture should help to remember the notation.

$$\mathscr{J}_\infty(X) \xrightarrow{\ \pi_a\ } \mathscr{J}_a(X) \xrightarrow{\ \pi_b^a\ } \mathscr{J}_b(X) \xrightarrow[\eta_b]{\ \pi^b\ } X \qquad (6.2)$$

are the maps induced by the natural surjections

$$k[\![t]\!] \longrightarrow k[\![t]\!]/t^{a+1} \longrightarrow k[\![t]\!]/t^{b+1} \longrightarrow k .$$

Example 2.7 Let $X = \operatorname{Spec} k[x_1, \dots, x_n] = \mathbb{A}^n$. Then, on the level of k-points one has

$$\mathscr{J}_m(X) = \{ \vartheta : k[x_1, \dots, x_n] \longrightarrow k[\![t]\!]/t^{m+1} \}$$

Such a map ϑ is determined by its values on the x_i's, *i.e.* it is determined by the coefficients of $\vartheta(x_i) = \sum_{j=0}^m \vartheta_i^{(j)} t^j$. Conversely, any choice of coefficients $\vartheta_i^{(j)}$ determines a point in $\mathscr{J}_m(\mathbb{A}^n)$. Choosing coordinates $x_i^{(j)}$ of $\mathscr{J}_m(X)$ with $x_i^{(j)}(\vartheta) = \vartheta_i^{(j)}$ we see that

$$\mathscr{J}_m(X) \cong \operatorname{Spec} k[x_1^{(0)}, \ldots, x_n^{(0)}, \ldots\ldots, x_1^{(m)}, \ldots, x_n^{(m)}] \cong \mathbb{A}^{n(m+1)}.$$

Furthermore observe that, somewhat intuitively, the truncation map $\pi^m : \mathscr{J}_m(X) \longrightarrow X$ is induced by the inclusion

$$k[x_1, \ldots, x_n] \hookrightarrow k[x_1^{(0)}, \ldots, x_n^{(0)}, \ldots\ldots, x_1^{(m)}, \ldots, x_n^{(m)}]$$

sending x_i to $x_i^{(0)}$.

Exercise 2.8 Let $Y \subseteq \mathbb{A}^n$ be a reduced hypersurface given by the vanishing of one equation $f = 0$. Show that $\mathscr{J}_m(Y) \subseteq \mathscr{J}_m(\mathbb{A}^n)$ is given by the vanishing of $m + 1$ equations $f^{(0)}, \ldots, f^{(m)}$ in the coordinates of $\mathscr{J}_m(\mathbb{A}^n)$ described above. (Observe that $f^{(0)} = f(x^{(0)})$ and $f^{(1)} = \sum \frac{\partial}{\partial x_i} f(x^{(0)}) x_i^{(1)}$). Show that

1. $\mathscr{J}_m(Y)$ is pure dimensional if and only if $\dim \mathscr{J}_m(Y) = (m+1)(n-1)$, in which case $\mathscr{J}_m(Y)$ is a complete intersection.
2. $\mathscr{J}_m(Y)$ is irreducible if and only if $\dim(\pi_Y^m)^{-1}(Y_{\mathrm{Sing}}) < (m+1)(n-1)$.

Similar statements hold if Y is locally a complete intersection.

The existence of the jet schemes in general (that is to show the representability of the functor defined above) is proved, for example, in [5]. From the very definition one can easily derive the following étale invariance of jet schemes, which, together with the example of $\mathbb{A}^n$ above gives us a pretty good understanding of the jet schemes of a smooth variety.

Proposition 2.9 *Let $X \longrightarrow Y$ be étale, then $\mathscr{J}_m(X) \cong \mathscr{J}_m(Y) \times_Y X$.*

Proof We show the equality on the level of the corresponding functors of points

$$\operatorname{Hom}(\_, \mathscr{J}_m(X)) \cong \operatorname{Hom}(\_ \times_k \operatorname{Spec} \frac{k[\![t]\!]}{t^{m+1}}, X)$$

and

$$\operatorname{Hom}(\_, \mathscr{J}_m(Y) \times_Y X) = \operatorname{Hom}(\_, \mathscr{J}_m(Y)) \times \operatorname{Hom}(\_, X)$$

$$= \operatorname{Hom}(\_ \times_k \operatorname{Spec} \frac{k[\![t]\!]}{t^{m+1}}, Y) \times \operatorname{Hom}(\_, X).$$

For this let Z be a k–scheme and consider the diagram

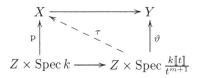

to see that a Z-valued m-jet $\tau \in \mathrm{Hom}(Z \times_k \mathrm{Spec}\, \frac{k[\![t]\!]}{t^{m+1}}, X)$ of X induces a Z-valued m-jet $\vartheta \in \mathrm{Hom}(Z \times_k \mathrm{Spec}\, \frac{k[\![t]\!]}{t^{m+1}}, Y)$ and a map $p \in \mathrm{Hom}(Z, X)$. Virtually by definition of formally étaleness [22, Definition (17.1.1)] for the map from X to Y, the converse holds also, *i.e.* ϑ and p together induce a unique map τ as indicated. $\square$

Using this étale invariance of jet schemes the computation carried out for $\mathbb{A}^n$ above holds locally on any smooth X. Thus we obtain:

Proposition 2.10 *Let X be a smooth k-scheme of pure dimension n. Then $\mathscr{J}_m(X)$ is an $\mathbb{A}^{nm}$–bundle over X. In particular $\mathscr{J}_m(X)$ is smooth of pure dimension $n(m+1)$. In the same way, $\mathscr{J}_{m+1}(X)$ is an $\mathbb{A}^n$–bundle over $\mathscr{J}_m(X)$.*

Note that this is not true for a singular X as can be seen already by looking at the tangent bundle $TX = \mathscr{J}_1(X)$ which is well-known to be a bundle if and only if X is smooth. In fact, over a singular X the jet schemes need not even be irreducible nor reduced and can also be badly singular.

2.3 An algebra of measurable sets

The prototype of a measurable subset of $\mathscr{J}_\infty(X)$ is a *stable set*. They are defined just right so that they receive a natural volume in $\mathscr{M}_k$. Recall that the constructible subsets of a scheme Y form the smallest algebra of sets containing the closed sets in Zariski topology.

Definition 2.11 A subset $A \subseteq \mathscr{J}_\infty(X)$ is called *stable* if there is $m_0 \in \mathbb{N}$ such that for all $m \geq m_0$, $A_m \overset{\text{def}}{=} \pi_m(A)$ is a constructible subset of $\mathscr{J}_m(X)$, $A = \pi_m^{-1}(A_m)$ and the map

$$\pi_m^{m+1} : A_{m+1} \longrightarrow A_m \quad \text{is a } \mathbb{A}^n\text{–bundle.} \qquad (6.3)$$

In this situation we also say that A is *stable at level* m_0.

For any $m \gg 0$ we define the *volume* of the stable set A by

$$\mu_X(A) = [A_m] \cdot \mathbb{L}^{-nm} \in \mathscr{M}_k.$$

That this is independent of m is ensured by condition (6.3) which implies that $[A_{m+1}] = [A_m] \cdot \mathbb{L}^n$.

Remark 2.12   Reid [38], Batyrev [2], and Looijenga [32] use this definition which gives the volume $\mu_X(\mathscr{J}_m(X)) \in \mathscr{M}_k$ of X virtual dimension n. Denef, Loeser [12], and Craw [6] use an additional factor $\mathbb{L}^{-n}$ to give $\mu_X(\mathscr{J}_m(X))$ virtual dimension 0. It seems to be essentially a matter of taste which definition one uses. Just keep this in mind while browsing through different sources in the literature to avoid unnecessary confusion.

Assuming that X is smooth one uses Proposition 2.10 to show that the collection of stable sets forms an algebra of sets, which means that $\mathscr{J}_\infty(X)$ is stable and with A and A' stable the sets $\mathscr{J}_\infty(X) - A$ and $A \cap A'$ are also stable. The smoothness of X furthermore warrants that so called *cylinder sets* are stable (a cylinder is a set $A = \pi_m^{-1}B$ for some constructible $B \subseteq \mathscr{J}_m(X)$). Thus in the smooth case condition (6.3) is superfluous whereas in general it is absolutely crucial and we have that in the smooth case the stable sets are precisely the cylinders.

In fact, a main technical point in defining the motivic measure on singular varieties is to show that the class of stable sets can be enlarged to an algebra of *measurable* sets which contains the cylinders. In particular $\mathscr{J}_\infty(X)$ is then measurable. This is achieved as one would expect by declaring a set measurable if it is approximated in a suitable sense by stable sets. This is essentially carried out in [32], though there are some inaccuracies; but everything should be fine if one works over $\mathbb{C}$ and makes some adjustments following [2, Appendix]. Of course Denef and Loeser also set up motivic integration over singular spaces [11] but their approach differs from the one discussed here in the sense that they assign a volume to the *formula* defining a constructible set rather than to the set of (k-rational) points itself. Thus they elegantly avoid any problems which arise if k is small. To avoid any of these technicalities we assume until the end of this section that X is smooth over the complex numbers $\mathbb{C}$.

2.4 The measurable function associated to a subscheme

From an algebra of measurable sets there arises naturally a notion of measurable function. Since we did not carefully define the measurable sets — we merely described the prototypes — we will for now only discuss an important class of measurable functions.

Let $Y \subseteq X$ be a subscheme of X defined by the sheaf of ideals I_Y. To Y one associates the function

$$\mathrm{ord}_Y : \mathscr{J}_\infty(X) \longrightarrow \mathbb{N} \cup \{\infty\}$$

sending an arc $\vartheta : \mathscr{O}_X \to k[\![t]\!]$ to the order of vanishing of ϑ along Y, *i.e.* to the supremum of all e such that ideal $\vartheta(I_Y)$ of $k[\![t]\!]$ is contained in the ideal (t^e). Equivalently, $\mathrm{ord}_Y(\vartheta)$ is the supremum of all e such that the map

$$\mathscr{O}_X \xrightarrow{\ \vartheta\ } k[\![t]\!] \longrightarrow k[\![t]\!]/t^e$$

sends I_Y to zero. Note that this map is nothing but the truncation $\pi_{e-1}(\vartheta) \in \mathscr{J}_{e-1}(X)$ of ϑ (setting $\mathscr{J}_{-1}(X) = X$ and $\pi_{-1} = \pi_0 = \pi$ to avoid dealing with the case $e = 0$ separately). For a $(e-1)$-jet $\gamma \in \mathscr{J}_{e-1}(X)$ to send I_Y to zero means precisely that $\gamma \in \mathscr{J}_{e-1}(Y)$.

Thus we can rephrase this by saying that $\mathrm{ord}_Y(\vartheta)$ is the supremum of all e such that the truncation $\pi_{e-1}(\vartheta)$ lies in $\mathscr{J}_{e-1}(Y)$. Now it is clear that

$$\mathrm{ord}_Y(\vartheta) \neq 0 \Leftrightarrow \pi(\vartheta) \in Y,$$
$$\mathrm{ord}_Y(\vartheta) \geq s \Leftrightarrow \pi_{s-1}(\vartheta) \in \mathscr{J}_{s-1}(Y) \text{ and}$$
$$\mathrm{ord}_Y(\vartheta) = \infty \Leftrightarrow \vartheta \in \mathscr{J}_\infty(Y).$$

The functions ord_Y just introduced are examples of measurable functions which come up in the applications of motivic integration. For a function to be measurable, one requires that the level sets $\mathrm{ord}_Y^{-1}(s)$ are measurable sets, *i.e.* stable sets (or at least suitably approximated by stable sets). For this consider the set $\mathrm{ord}_Y^{-1}(\geq s) = \{\vartheta \in \mathscr{J}_\infty(X) \mid \mathrm{ord}_Y(\vartheta) \geq s\}$ consisting of all arcs in X which vanish of order *at least* s along Y. By what we just observed $\mathrm{ord}_Y^{-1}(\geq s) = \pi_{s-1}^{-1}(\mathscr{J}_{s-1}(Y))$ is a cylinder. Therefore, the level set $\mathrm{ord}_Y^{-1}(s)$ is also a cylinder equal to

$$\mathrm{ord}_Y^{-1}(\geq s) - \mathrm{ord}_Y^{-1}(\geq s+1) = \pi_{s-1}^{-1}\mathscr{J}_{s-1}(Y) - \pi_s^{-1}\mathscr{J}_s(Y).$$

The exception is $s = 0$ in which case $\mathrm{ord}_Y^{-1}(\geq 0) = \pi^{-1}(X) = \mathscr{J}_\infty(X)$ and $\mathrm{ord}_Y^{-1}(0) = \pi^{-1}(X) - \pi^{-1}(Y)$.

Note that the level set at infinity,

$$\mathrm{ord}_Y^{-1}(\infty) = \bigcap \mathrm{ord}_Y^{-1}(\geq s) = \mathscr{J}_\infty(Y)$$

on the other hand is *not* a cylinder. What this means is that for an arc $\vartheta \in \mathscr{J}_\infty(X)$ to lie in $\mathscr{J}_\infty(Y)$ is a condition that cannot be checked on any truncation. Exercise! Still, since $\mathrm{ord}_Y^{-1}(\infty)$ it is the decreasing intersection of the cylinders $\mathrm{ord}_Y^{-1}(\geq s+1) = \pi_s^{-1}\mathscr{J}_s(Y)$ its alleged volume should be obtained as the limit of the volumes of these cylinders. The volume of $\pi_s^{-1}\mathscr{J}_s(Y)$ is $[\mathscr{J}_s(Y)] \cdot \mathbb{L}^{-ns}$. The dimension of this element of $\mathscr{M}_k$ is $\leq \dim \mathscr{J}_s(Y) - ns$. If Y is smooth of pure dimension $n - c$, the we have seen that $\dim \mathscr{J}_s(Y) = (n-c)(s+1)$. Therefore, $\mathscr{J}_\infty(Y)$ is the intersection of cylinder sets whose volumes have dimension $\leq (n-c)(s+1)-ns = (n-c)-cs$. For increasing s these dimensions tend to negative infinity. But recall that in the ring $\mathscr{M}_k$ this is exactly the condition of convergence to zero. Thus the only sensible assignment of a volume to $\mathrm{ord}_Y^{-1}(\infty)$ is zero. This argument used that Y is smooth to show that the dimension of $\mathscr{J}_m(Y)$ grows significantly slower than the dimension of $\mathscr{J}_m(X)$. This holds in general for singular Y and we phrase it as a proposition whose proof however is postponed until Section 4.

Proposition 2.13 *Let $Y \subseteq X$ be a nowhere dense subvariety of X, then $\mathscr{J}_\infty(Y)$ is measurable and has measure $\mu_X(\mathscr{J}_\infty(Y))$ equal to zero.*

To make this idea into a rigorous theory one has to define a larger class of measurable subsets of $\mathscr{J}_\infty(X)$ and this turns out to be somewhat subtle. In section 4 we will outline this briefly – but since, no matter what, the measure of $\mathrm{ord}_Y^{-1}(\infty)$ will be zero, we will move on at this point and start to integrate.

2.5 Definition and computation of the motivic integral

As before let X be a smooth k-scheme and Y a subscheme. We define the *motivic integral* of $\mathbb{L}^{-\mathrm{ord}_Y}$ on X as

$$\int_{\mathscr{J}_\infty(X)} \mathbb{L}^{-\mathrm{ord}_Y} \, d\mu_X = \sum_{s=0}^{\infty} \mu(\mathrm{ord}_Y^{-1}(s)) \cdot \mathbb{L}^{-s}.$$

Observe that the level set at infinity is already left out from this summation as it has measure zero.

Note that the sum on the right does converge since the virtual dimension of the summands approaches negative infinity. Where the fact that for stable sets $A \subseteq B$ we have $\dim \mu(A) \leq \dim \mu(B)$ applied to $\mathrm{ord}_Y^{-1}(s) \subseteq \mathscr{J}_\infty(X)$ gives that the dimension of $\mu(\mathrm{ord}_Y^{-1}(s)) \cdot \mathbb{L}^{-s}$ is less or equal to $n - s$. The notion of convergence in the ring $\mathscr{M}_k$ is such that this alone is enough to ensure the convergence of the sum. Thus it is justified to call $\mathbb{L}^{-\mathrm{ord}_Y}$ integrable with integral as above.

It is useful to calculate at least one example. For $Y = \emptyset$ one has $\mathrm{ord}_Y \equiv 0$ and thus we get

$$\int_{\mathscr{J}_\infty(X)} \mathbb{L}^{-\mathrm{ord}_Y} d\mu_X = \mu(\mathrm{ord}_Y^{-1}(0)) = [X]$$

where we used that X is smooth. A less trivial example is Y a smooth divisor in X. Then $\mathscr{J}_s(Y)$ is a $\mathbb{A}^{(n-1)s}$-bundle over Y. The level set is $\mathrm{ord}_Y^{-1}(s) = \pi_{s-1}^{-1}\mathscr{J}_{s-1}(Y) - \pi_s^{-1}\mathscr{J}_s(Y)$ and, using that $[\mathscr{J}_s(Y)] = [Y] \cdot \mathbb{L}^{(n-1)s}$, its measure is

$$[\mathscr{J}_{s-1}(Y)] \cdot \mathbb{L}^{-n(s-1)} - [\mathscr{J}_s(Y)] \cdot \mathbb{L}^{-ns} = [Y](\mathbb{L} - 1)\mathbb{L}^{-s}.$$

The integral of ord_Y^{-1} is therefore

$$\begin{aligned}
\int_{\mathscr{J}_\infty(X)} \mathbb{L}^{-\mathrm{ord}_Y} d\mu_X &= [X - Y] + \sum_{s=1}^{\infty}[Y](\mathbb{L} - 1)\mathbb{L}^{-s} \cdot \mathbb{L}^{-s} \\
&= [X - Y] + [Y](\mathbb{L} - 1)\mathbb{L}^{-2}\sum_{s=0}^{\infty}\mathbb{L}^{-2s} \\
&= [X - Y] + [Y](\mathbb{L} - 1)\frac{1}{\mathbb{L}^2(1 - \mathbb{L}^{-2})} \\
&= [X - Y] + [Y](\mathbb{L} - 1)(\mathbb{L}^2 - 1)^{-1} \\
&= [X - Y] + \frac{[Y]}{\mathbb{L} + 1} = [X - Y] + \frac{[Y]}{[\mathbb{P}^1]}
\end{aligned} \qquad (6.4)$$

Note the appearance of a geometric series in line 3 which is typical for these calculations (cf. Exercise 2.6). In fact, the motivic volumes of a wide class of measurable subsets (namely, the *semi-algebraic subsets* of Denef and Loeser in [11]) belong to the ring generated by the image of $\mathscr{M}_k$ under the completion map, and the sums of geometric series with denominators $\mathbb{L}^j$, $j > 0$ [10, Corollary 5.2].

Exercise 2.14   Compute in a similar fashion the motivic integrals $\mathbb{L}^{-\mathrm{ord}_Y}$ where Y is as follows.

a. Y is a smooth subscheme of codimension c in X.
b. $Y = aD$ where D is a smooth divisor and $a \in \mathbb{N}$.
c. $Y = D_1 + D_2$ where the D_i's are smooth and in normal crossing.
d. $Y = a_1 D_1 + a_2 D_2$ with D_i as above and a_i positive integers.

These computations are a special case of a formula which explicitly computes the motivic integral over $\mathbb{L}^{-\mathrm{ord}_Y}$ where Y is an effective divisor with normal crossing support.

Proposition 2.15 *Let $Y = \sum_{i=1}^{s} r_i D_i$ $(r_i > 0)$ be an effective divisor on X with normal crossing support and such that all D_i are smooth. Then*

$$\int_{\mathscr{J}_\infty(X)} \mathbb{L}^{-\operatorname{ord}_Y} d\mu_X = \sum_{J \subseteq \{1,\dots,s\}} [D_J^\circ] (\prod_{j \in J} \frac{\mathbb{L} - 1}{\mathbb{L}^{r_j+1} - 1})$$

$$= \sum_{J \subseteq \{1,\dots,s\}} \frac{[D_J^\circ]}{\prod_{j \in J} [\mathbb{P}^{r_j}]}$$

$D_J = \bigcap_{j \in J} D_j$ *(note that $D_\emptyset = X$) and $D_J^\circ = D_J - \bigcup_{j \notin J} D_j$.*

The proof of this is a computation entirely similar to (though significantly more complicated than) the one carried out in (6.4) above; for complete detail see either Batyrev [2, Theorem 6.28] or Craw [6, Theorem 1.17]. We suggest doing it as an exercise using the following lemma.

Lemma 2.16 *For $J \subseteq \{1, \dots, s\}$ redefine $r_i = 0$ if $i \notin J$. Then*

$$\mu(\cap \operatorname{ord}_{D_j}^{-1}(r_j)) = [D_J^0](\mathbb{L} - 1)^{|J|} \mathbb{L}^{-\sum r_j}$$

where $D_J^0 = \bigcap_{j \in J} D_j - \bigcup_{j \notin J} D_J$ and $|J|$ denotes the cardinality of J.

Exercise 2.17 Show that for $t \geq \max\{r_i\}$ one has locally an isomorphism of the k-points

$$\pi_t(\cap_i \operatorname{ord}_{D_i}^{-1}(r_i)) \cong D_{\operatorname{supp} r}^0 \times (k - \{0\})^{|\operatorname{supp} r|} \times k^{nt - \sum r_i}$$

where $\operatorname{supp} r$ denotes the set $\{i | r_i \neq 0\}$. Show that this statement implies the lemma. To prove the preceding statement reduce to the case that $X = U \subseteq \mathbb{A}^n$ is an open subvariety of $\mathbb{A}^n$ and $\sum D_i$ is given by the vanishing of $x_1 \cdot \ldots \cdot x_s$ where $x_1, \dots, x_n$ is a local system of coordinates (use Proposition 2.9). Then finish this case using the description of the arc space of $\mathbb{A}^n$ as given in Example 2.7.

The explicit formulas of Proposition 2.15 and Lemma 2.16 are one cornerstone underlying many applications of motivic integration. The philosophy one employs is to encode information in a motivic integral, then using the transformation rule of the next section the computation of this integral can be reduced to the computation of an integral over $\mathbb{L}^{-\operatorname{ord}_Y}$ for Y a normal crossing divisor. In this case the above formula gives the answer. Thus we shall proceed to the all important birational transformation rule for motivic integrals.

3 The transformation rule

The power of the theory stems from a formula describing how the motivic integral transforms under birational morphisms:

Theorem 3.1   *Let* $X' \xrightarrow{f} X$ *be a proper birational morphism of smooth k-schemes and let D be an effective divisor on X (or D a closed subscheme of X), then*

$$\int_{\mathscr{J}_\infty(X)} \mathbb{L}^{-\operatorname{ord}_D} d\mu_X = \int_{\mathscr{J}_\infty(X')} \mathbb{L}^{-\operatorname{ord}_{f^{-1}D + K_{X'/X}}} d\mu_{X'}.$$

As the relative canonical sheaf $K_{X'/X}$ is defined by the Jacobian ideal of f, this should be thought of as the change of variables formula for the motivic integral.

As a warmup for the proof, we verify the transformation rule first in the special case of blowing up a smooth subvariety and $D = \emptyset$. Let $X' = \operatorname{Bl}_Y X$ be the blowup of X along the smooth center Y of codimension c in X. Then by [25, Exercise II.8.5] the relative canonical divisor is $K_{X'/X} = (c-1)E$, where E is the exceptional divisor of the blowup. Then, using Proposition 2.15 in its simplest incarnation we compute

$$\int_{\mathscr{J}_\infty(X')} \mathbb{L}^{-\operatorname{ord}_{K_{X'/X}}} d\mu_{X'} = \int_{\mathscr{J}_\infty(X')} \mathbb{L}^{-\operatorname{ord}_{(c-1)E}} d\mu_{X'}$$

$$= [X' - E] + \frac{[E]}{[\mathbb{P}^c]}$$

$$= [X - Y] + [Y] = [X],$$

where we used that E is a $\mathbb{P}^c$–bundle over Y (by definition of blowup) and therefore $[E] = [Y][\mathbb{P}^c]$.

The induced map on the arc space

The proper birational map f induces a map $f_\infty = \mathscr{J}_\infty(f) : \mathscr{J}_\infty(X') \to \mathscr{J}_\infty(X)$. The first task will be to show that away from a set of measure zero f_∞ is a bijection (of sets). Let $\Delta \subseteq X'$ contain the locus where f is not an isomorphism. If X and X' are both smooth we can take $\Delta = \operatorname{Supp}(K_{X'/X})$, for example. We show that every arc $\gamma : \operatorname{Spec} k[\![t]\!] \to X$, which does not entirely lie in $f(\Delta)$ uniquely lifts to an arc in X'. For

illustration consider the diagram

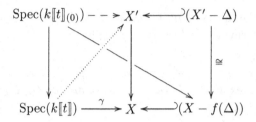

Observe that by assumption the generic point $\operatorname{Spec} k[\![t]\!]_{(0)}$ of $\operatorname{Spec} k[\![t]\!]$ does lie in $X - f(\Delta)$ and since f is an isomorphism over $X' - \Delta$ it thus lifts to X' uniquely (dashed arrow). Now the valuative criterion for properness (see [25, Chapter 2, Theorem 4.7]) yields the unique existence of the dotted arrow. Thus the map $f_\infty : (\mathscr{J}_\infty(X') - \mathscr{J}_\infty(\Delta)) \to (\mathscr{J}_\infty(X) - \mathscr{J}_\infty(f(\Delta)))$ is a bijection of k-valued points. Since $\mathscr{J}_\infty(\Delta)$ has measure zero it can be safely ignored and we will do so in the following.

Exercise 3.2 Let $f : X' \to X$ be a proper birational map (of smooth varieties). Then for every m the map $f : \mathscr{J}_m(X') \to \mathscr{J}_m(X)$ is surjective.

Proof of transformation rule The level sets $C'_e = \operatorname{ord}^{-1}_{K_{X'/X}}(e)$ partition $\mathscr{J}_\infty(X')$, and cutting into even smaller pieces according to order of contact along $f^{-1}(D)$ we define

$$C'_{e,k} = C'_e \cap \operatorname{ord}^{-1}_{f^{-1}D}(k) \text{ and } C_{e,k} = f_\infty(C'_{e,k})$$

to get the following partitions (up to measure zero by 3) of the arc spaces

$$\mathscr{J}_\infty(X') = \bigsqcup C'_{e,k} \text{ and } \mathscr{J}_\infty(X) = \bigsqcup C_{e,k}.$$

The essence of the proof of the transformation rule is captured by the following two crucial facts.

(a). $C_{e,k}$ are stable sets for all e, k.

(b). $\mu_X(C_{e,k}) = \mu_{X'}(C'_{e,k}) \cdot \mathbb{L}^{-e}$.

Two different proofs of these facts will occupy the remainder of this section. Using these facts, the transformation rule is a simple calculation:

$$\int_{\mathscr{J}_\infty(X)} \mathbb{L}^{-\operatorname{ord}_D} d\mu_X = \sum_k \mu(\operatorname{ord}_D^{-1}(k))\mathbb{L}^{-k} = \sum_k \left(\sum_e \mu(C_{e,k}) \right) \mathbb{L}^{-k}$$

$$= \sum_{e,k} \mu(C'_{e,k})\mathbb{L}^{-e}\mathbb{L}^{-k}$$

$$= \sum_t \left(\sum_{e+k=t} \mu(C'_{e,k}) \right) \mathbb{L}^{-t}$$

$$= \sum_t \mu(\operatorname{ord}_{f^{-1}D+K_{X'/X}}^{-1}(t))\mathbb{L}^{-t}$$

$$= \int_{\mathscr{J}_\infty(X')} \mathbb{L}^{-\operatorname{ord}_{f^{-1}D+K_{X'/X}}} d\mu_{X'}$$

Besides facts (a) and (b) one uses that if a cylinder B is written as a disjoint union of cylinders B_i then the measure $\mu(B) = \sum \mu(B_i)$. To check that this is correct one has to use the precise definition of the motivic measure which we have avoided until now. See the Section 4 for more details on this. $\qquad\square$

We first treat properties (a) and (b) in a special case, namely the blowup at a smooth center. In the applications of motivic integration to birational geometry which we discuss below, we can always put ourselves in the favorable situation that the birational map in consideration is a sequence of blowing ups along smooth centers, hence already this simple version goes very far. Furthermore, using the Weak Factorization Theorem of [1] one can make a general proof of the transformation rule by reducing to this case. However, despite the adjective *weak* in the Weak Factorization Theorem it is a very deep and difficult result and its use in the proof of the Transformation rule is overkill. Therefore we give in Appendix A.2 an essentially elementary (though at the first reading somewhat technical) proof following [32].

3.1 Images of cylinders under birational maps.

We start with some basic properties of the behavior of cylinders under birational morphisms, see [13]. These will be useful also in Section 5.

Proposition 3.3 *Let* $f : X' \longrightarrow X$ *be a proper birational map of smooth varieties. Let* $C' = (\pi_m^{X'})^{-1}(B') \subseteq \mathscr{J}_\infty(X')$ *be a cylinder such that* B' *is a union of fibers of* f_m, *then* $C \stackrel{\mathrm{def}}{=} f_\infty(C') = (\pi_m^X)^{-1}(f_m B')$ *is a cylinder in* $\mathscr{J}_\infty(X)$.

Proof   Clearly it is enough to show that $C = \pi_m^{-1}(B)$ since $B = f_m(B')$ is constructible (being the image of a constructible set under a finite type morphism). The nontrivial inclusion is $\pi_m^{-1}(B) \subseteq C$. Let $\gamma \in \pi_m^{-1}(B) \subseteq C$ and consider for every $p \geq m$, the cylinder

$$D_p \stackrel{\text{def}}{=} (\pi_p^{X'})^{-1}(f_p^{-1}(\pi_p^X(\gamma)))$$

which is non empty by exercise 3.2. Clearly, $D_p \supseteq D_{p+1}$ which implies that the D_p form a decreasing sequence of nonempty cylinders. By Proposition 4.4 the intersection of all the D_p is nonempty. Now let $\gamma' \in \bigcap D_p$ and clearly $f_\infty \gamma' = \gamma$. Furthermore, since B' is a union of fibers of f_m we have $C' \supseteq D_m$ and hence $C' \supseteq D_p$ for all $p \geq m$. Hence $\pi_m(\gamma') \in B'$.   □

Exercise 3.4   Let $f : X' \to X$ be a proper birational map of smooth varieties. Then $f_\infty : \mathscr{J}_\infty(X') \to \mathscr{J}_\infty(X)$ is surjective.

The key technical result for blowup at smooth center.
We now proceed to showing the key technical result used in the proof of the transformation formula.

Theorem 3.5 (Denef, Loeser, [11])   *Let $f : X' \to X$ be a proper birational morphism of smooth varieties. Let $C'_e = \text{ord}_{K_{X'/X}}^{-1}(e)$ where $K_{X'/X}$ is the relative canonical divisor and let $C_e \stackrel{\text{def}}{=} f_\infty C'_e$. Then, for $m \geq 2e$*

(a'). the fiber of f_m over a point $\gamma_m \in \pi_m C_e$ lies inside a fiber of π_{m-e}^m.
(a). $\pi_m(C'_e)$ is a union of fibers of f_m.
(b). $f_m : \pi_m C'_e \to \pi_m C_e$ is piecewise trivial $\mathbb{A}^e$-fibration.

Corollary 3.6   *With the notation as in the theorem, C_e is stable at level $m \geq 2e$ and $[C'_e] = [C_e]\mathbb{L}^e$.*

Proof of Theorem 3.5   First note (a') implies (a) since C'_e is stable at level e. Furthermore using the following Lemma 3.7 it is enough to show that a fiber of f_m over a point in C_e is an affine space $\mathbb{A}^e$.

We give the proof here only in the case that $X' = \text{Bl}_Y X \to X$ where Y is a smooth subvariety of X, for the general argument see Appendix A.2 below. Since X is smooth there is an étale morphism $\varphi : X \to \mathbb{A}^n$ such that $\varphi(Y)$ is given by the vanishing of the first $n - c$ coordinates on $\mathbb{A}^n$ and $\varphi^{-1}(\varphi(Y)) = Y$. By the étale invariance of jet schemes (Proposition 2.9) one can hence assume that $Y = \mathbb{A}^{n-c} \subseteq \mathbb{A}^n = X$. To fully justify this we point out that for any sheaf of ideals I on

$\mathbb{A}^n$ one has $\mathrm{Bl}_{I\mathcal{O}_X} X \cong X \times_{\mathbb{A}^n} \mathrm{Bl}_I \mathbb{A}^n$ since $\varphi : X \longrightarrow \mathbb{A}^n$ is flat which implies that $I^m \mathcal{O}_X \cong \varphi^*(I^m)$ and then the equation follows from the definition of the blowup in terms of the proj of the Rees algebra.

To further simplify notation (and notation only) we assume that $n = 3$ and $c = 3$, that is we only have to consider the blowup of the origin in $\mathbb{A}^3$. By definition

$$X' = \mathrm{Bl}_0 \mathbb{A}^3 \subseteq \mathbb{A}^3 \times \mathbb{P}^2$$

is given by the vanishing of the 2×2 minors of the matrix

$$\begin{pmatrix} x_0 & x_1 & x_2 \\ y_0 & y_1 & y_2 \end{pmatrix}$$

where (x_0, x_1, x_2) and (z_0, z_1, z_2) are the coordinates on $\mathbb{A}^3$ and the homogeneous coordinates of $\mathbb{P}^2$ respectively. Due to the local nature of our question it is enough to consider an affine patch of X', say the one determined by $z_0 = 1$. The equations of the minors then reduce to $x_1 = x_0 z_1$ and $x_2 = x_0 z_2$ such that on this patch the map $X' \longrightarrow X$ is given by the inclusion of polynomial rings

$$k[x_0, x_0 z_1, x_0 z_2] \xrightarrow{f} k[x_0, z_1, z_2].$$

The exceptional divisor E is hence given by the vanishing of x_0. The relative canonical divisor $K_{X'/X}$ is equal to $2E$ since $\det(\mathrm{Jac}(f))$ is easily computed to be x_0^2.

Now let $\gamma' \in C'_e$. In our local coordinates, $\gamma'_m = \pi_m(\gamma')$ is uniquely determined by the three truncated powerseries

$$\gamma'_m(x_0) = t^{e/2} \sum_{i=0}^{m-e/2} a_i t^i \quad \text{with } a_0 \neq 0,$$

$$\gamma'_m(z_1) = \sum_{i=0}^{m} b_i t^i, \quad \text{and}$$

$$\gamma'_m(z_2) = \sum_{i=0}^{m} c_i t^i,$$

where the special shape of the first one comes from the condition that γ' has contact order with $K_{X'/X} = 2E$ precisely equal to e (we only have to consider e which are divisible by $c - 1 = 2$ since otherwise C'_e is empty). Its image $\gamma_m = f_m(\gamma'_m) = \gamma'_m \circ f_m$ is analogously determined by the three truncated powerseries

$$\gamma_m(x_0) = t^{e/2} \sum_{i=0}^{m-e/2} a_i t^i \quad \text{with } a_0 \neq 0$$

$$\gamma_m(x_0 z_1) = \gamma'_m(x_0)\gamma'_m(z_1) = t^{e/2} \sum_{i=0}^{m-e/2} a_i t^i \sum_{i=0}^{m} b_i t^i \quad \mod t^{m+1} \quad (6.5)$$

$$\gamma_m(x_0 z_2) = \gamma'_m(x_0)\gamma'_m(z_1) = t^{e/2} \sum_{i=0}^{m-e/2} a_i t^i \sum_{i=0}^{m} c_i t^i \quad \mod t^{m+1}$$

Expanding the product of the sums in the last two equations of (6.5) we observe that (due to the occurrence of $t^{e/2}$) the coefficients $b_{m-\frac{e}{2}+1}, \dots,$ b_m are not visible in $\gamma_m(x_0 z_1)$ since they only appear as coefficients of some t^k for $k > m$. Analogously, $\gamma_m(x_0 z_2)$ does not depend on $c_{m-\frac{e}{2}+1}, \dots, c_m$. Conversely, given

$$\gamma_m(x_0 z_1) = t^{e/2} \sum_{i=0}^{m-\frac{e}{2}} \beta_i t^i$$

and knowing all a_i's (which equally show up in γ'_m and γ_m) we can inductively recover the b_i's:

$$b_0 = (\beta_0)/a_0 \qquad \text{(note: } a_0 \neq 0\text{)}$$
$$b_1 = (\beta_1 - a_1 b_0)/a_0$$
$$b_2 = (\beta_2 - (a_2 b_0 + a_1 b_1))/a_0$$

$$\vdots$$

$$b_t = (\beta_t - (a_t b_0 + a_{t-1} b_1 + \dots + a_1 b_{t-1}))/a_0$$

and this works until $t = m - \frac{e}{2}$ since a_t is known for $t \leq m - \frac{e}{2}$. The analogous statements of course hold also for the c_i's.

Summing up these observations we see that the fiber of f_m over γ_m is an affine space of dimension $e = 2 \cdot \frac{e}{2}$, namely it is spanned by the last $\frac{e}{2}$ of the coefficients b_i and c_i. This proves part (b). Furthermore, any two γ'_m and γ''_m mapping via f_m to γ_m only differ in these last $\frac{e}{2}$ coefficients, hence they become equal after further truncation to level $m - e$. Even to level $m - \frac{e}{2}$ in this case of a blowup of a point in $\mathbb{A}^3$. In general, for a blowup of c codimensional smooth center it is truncation to level $m - \frac{e}{c-1}$ which suffices. Hence uniformly it is truncation to level $m - e$ which works. This shows (a') and the proof is finished. $\square$

Lemma 3.7 *If $\varphi : V \longrightarrow W$ is a morphism of finite type schemes such that all fibers $\varphi^{-1}(x) \cong \mathbb{A}^e \times k(x)$, then φ is a piecewise trivial $\mathbb{A}^e$-fibration.*

Proof We may assume that W is irreducible. Then the fiber over the generic point η of W is by assumption isomorphic to $\mathbb{A}^e$. This means that there is an open subset $U \subseteq W$ such that $f^{-1}(U) \cong \mathbb{A}^e \times U$. This isomorphism $\mathbb{A}^e(k(\eta)) \cong \varphi^{-1}(\eta) = V \times_W \operatorname{Spec} k(\eta)$ is defined via some finitely many rational functions on W. For any $U \subseteq W$ such that these are regular we get $\varphi^{-1}(U) \cong \mathbb{A}^e \times U$. Now φ restricted to the complement of U is a map of the same type but with smaller dimensional base and we can finish the argument by induction. Even though we did not make this explicit in the proof of Theorem 3.5 the statement about the fibers being equal to $\mathbb{A}^e$ holds for all fibers (not only fibers over closed points). $\square$

Exercise 3.8 Use Proposition 3.3 and 3.5 to show that if $C \subseteq \mathscr{J}_\infty(X')$ is a cylinder, then the closure of $f_\infty(C) \subseteq \mathscr{J}_\infty(X)$ is a cylinder, where $f : X' \longrightarrow X$ is a proper birational morphism.

3.2 Proof of transformation rule using Weak Factorization

With the proof given so far we have the transformation formula available for a large class of proper birational morphisms, namely the ones which are obtained as a sequence of blowups along smooth centers. So in particular we have the result for all resolutions of singularities, which is the only birational morphism we will consider in our applications later.

Let me finish by outlining how using the Weak Factorization Theorem one can make a full proof out of this. Let us first recall the statement:

Theorem 3.9 (Weak Factorization Theorem [1]) *Let $\varphi : X' \dashrightarrow X$ be a birational map between smooth complete varieties over k of characteristic zero. Then φ can be factored*

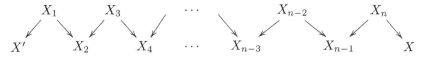

such that all indicated maps are a blowup at a smooth center. Furthermore, there is an index i such that the rational maps to X' to the left ($X' \dashleftarrow X_j$ for $j \leq i$) and the rational maps to X to the right ($X_j \dashrightarrow X$ for $j \geq i$) of that index are in fact regular maps.

One should point out that the second part of the Theorem about the regularity of the maps is crucial in many of its applications, and particularly in the application that follows next.

Proof of Theorem 3.1   So far we have proved the Transformation rule for the blowup along a smooth center. Given a proper birational morphism of smooth varieties $f : X' \to X$ we can factor it into a chain as in the Weak Factorization Theorem. In particular, for each birational map in that chain, the Transformation rule holds. The second part of the Factorzation Theorem together with the assumption that $X' \to X$ is a morphism (and $X \dashrightarrow X'$ is not a morphism unless $X' \cong X$) implies that the index j in the Factorization Theorem is $\leq n-1$ such that $X_{n-1} \to X$ is also a morphism. Part (b) of the following Exercise shows that for this morphism $X_{n-1} \to X$ the Transformation rule holds. Now the shorter chain ending with $X_{n-2} \to X$ is again a chain such that for each map the Transformation rule holds, by part (a) of that exercise. By induction we can conclude that the transformation rule holds for f itself. $\square$

Exercise 3.10   Suppose one is given a commuting diagram of proper birational morphisms

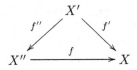

(a). If the Transformation rule holds for f'' and f, then it also holds for $f' = f'' \circ f$.

(b). If the Transformation rule holds for f'' and f', then it also holds for f.

(c). The same as (a) and (b) but with "the Transformation rule" replaced by "the conclusions of Theorem 3.5". (This part is more difficult than the others).

(d). Using (c) and the Weak Factorization Theorem produce a proof of Theorem 3.5 building on the case of the blowup at a smooth center we considered above.

4 Brief outline of a formal setup for the motivic measure

In this section we fill in some details that were brushed over in our treatment of motivic integration so far. First this concerns some basic properties and the well-definedness of the motivic measure and integral.

4.1 Properties of the motivic measure

For simplicity we still assume that X is a smooth $\mathbb{C}$–variety. Recall that we defined for a stable set $C = \pi_m^{-1}(B)$ a volume by setting

$$\mu_X(C) = [B]\mathbb{L}^{-nm} \in \mathcal{M}_k.$$

It is easy to verify that on stable sets (*i.e.* cylinders if X is smooth) the measure is additive on finite disjoint unions. Furthermore, for stable sets $C \subseteq C'$ one has $\dim \mu_X(C) \leq \dim \mu_X(C')$. We begin with a rigorous definition of what is a measurable set extending the above definition.

Definition 4.1 A subset $C \subseteq \mathscr{J}_m(X)$ is called *measurable* if for all $n \in \mathbb{N}$ there is a stable set C_n and stable sets $D_{n,i}$ for $i \in \mathbb{N}$ such that

$$C \Delta C_n \subseteq \bigcup_{i \in \mathbb{N}} D_{n,i}$$

and $\dim \mu(D_{n,i}) \leq -n$ for all i. Here $C \Delta C_n = (C \setminus C_n) \cup (C_n \setminus C)$ denotes the symmetric difference of two sets. In this case the *volume* of C is

$$\mu_X(C) = \lim_{n \to \infty} \mu_X(C_n) \in \mathcal{M}_k.$$

This limit converges and is independent of the C_n's.

The key point in proving the claims in the definition is the so called Baire property of constructible subsets of a $\mathbb{C}$–variety which crucially uses the fact that $\mathbb{C}$ is uncountable, see [21, Corolaire 7.2.6].

Remark 4.2 In [32] a more restrictive definition of measurable is used, namely he requires that $\dim D_{n,i} \leq -(n+i)$. This has the advantage that one does not require the field to be uncountable to conclude the well defined-ness of the measure. Essentially, he uses that if $D \subseteq \bigcup D_i$ are cylinders with $\lim \dim D_i = -\infty$, then D is already contained the union of finitely many of the D_i. This is true if k is infinite, and, if k is uncountable, even true without the assumption on the D_i. The advantage of Looijenga's setup is that one is not bound to an uncountable field, but unfortunately I was not able to verify that in his setup $\mathscr{J}_\infty(Y)$ is measurable and has zero volume.

Proposition 4.3 *Let* $K_1 \supseteq K_2 \supseteq K_3 \supseteq \ldots$ *be an infinite sequence of nonempty constructible subsets of a* $\mathbb{C}$*–variety* X*. Then* $\bigcap_i K_i$ *is nonempty.*

For cylinder sets this implies the following Proposition, which will be used below in a version which asserts that a cylinder C which is contained in the union of countably many cylinders C_i is already contained in the union of finitely many of these.

Proposition 4.4 *Let* $C_1 \supseteq C_2 \supseteq C_3 \supseteq \ldots$ *be an infinite sequence of nonempty cylinders in* $\mathscr{J}_\infty(X)$ *where* X *is a smooth* $\mathbb{C}$*–variety. Then* $\bigcap_i C_i$ *is nonempty.*

Proof By definition of a cylinder and Chevalley's theorem $\pi_0(C_i)$ is a constructible subset of X. Thus we can apply Proposition 4.3 to the sequence $\pi_0(C_1) \supseteq \pi_0(C_2) \supseteq \ldots$ to obtain an element $x_0 \in \bigcap \pi_0(C_i)$. Now consider the sequence of cylinders $C_i' \stackrel{\text{def}}{=} C_i \cap \pi_0^{-1}(x_0)$ and repeat the argument for the sequence of constructible sets $\pi_1(C_1') \supseteq \pi_1(C_2') \supseteq \pi_1(C_2') \supseteq \ldots$ to obtain an element $x_1 \in \bigcap \pi_1(C_i')$. Repeating this procedure we successively lift $x_0 \in X$ to $x_1 \in \mathscr{J}_1(X)$, $x_2 \in \mathscr{J}_2(X)$ and so forth. The limit of these x_i gives an element $x \in \mathscr{J}_\infty(X)$ which is in the intersection of all C_i. Thus this intersection is nonempty. $\square$

On a possibly singular $\mathbb{C}$–variety the statement (and proof) is true if "cylinder" is replaced by "stable set" above. If X is a k–variety with k at most countable these statements might be false. Nevertheless, if one views the constructible sets (resp. cylinders) not as subsets of the k–rational points but rather as certain sub–arrangements (something weaker than a subfunctor) of the functor of points represented by X the above is essentially true. This is the point of view of Denef and Loeser and is carried out in [12].

Justification of Definition 4.1 For both claims it suffices to show $\dim(\mu_X(C_i) - \mu_X(C_j')) \le -i$ for all $j \ge i$, where the prime indicates a second set of defining data as is in the definition. We have

$$C_i - C_j' \subseteq (C\Delta C_i) \cup (C\Delta C_j') \subseteq \bigcup_m C_{i,m} \cup \bigcup_m C_{j,m}'.$$

Since $(C_i - C_j')$ and all terms on the right are cylinders the previous proposition applies and $C_i - C_j'$ is contained in finitely many of the cylinders of the right hand side. This implies that $\dim \mu_X(C_i - C_j') \leq -i$. The same applies to $\dim \mu_X(C_j' - C_i)$. Using that $C_i = (C_i \cap C_j') \cup (C_i - C_j')$ and $C_j' = (C_i \cap C_j') \cup (C_j' - C_i)$ and that μ_X is additive on finite disjoint unions of cylinders we get

$$
\begin{aligned}
\dim(\mu_X(C_i) - \mu_X(C_j')) &= \dim(\mu_X(C_i - C_j') - \mu_X(C_j' - C_i)) \\
&\leq \max\left\{\dim \mu_X(C_i - C_j'), \dim \mu_X(C_j' - C_i)\right\} \\
&\leq -i.
\end{aligned}
$$

For once this shows that the $\mu_X(C_i)$ form a Cauchy sequence, thus the limit exists as claimed. Secondly it immediately follows that this limit does not depend on the chosen data. $\qquad\square$

We summarize some basic properties of the measure and measurable sets.

Proposition 4.5 *The measurable sets form an algebra of sets. μ_X is additive on disjoint unions, thus μ_X is a pre–measure in classical terminology. If C_i are a infinite disjoint sequence of measurable sets such that*

$$
\lim_{i \to \infty} \mu_X(C_i) = 0
$$

then $C = \bigcup C_i$ is measurable and $\mu_X(C) = \sum \mu_X(C_i)$.

Proof The verification of all parts is quite easy. As an example we only show that the complement of a measurable set is also measurable and leave the rest as an exercise.

If C is measurable then cylinders C_i and $C_{i,j}$ can be chosen with the properties as in Definition 4.1. The complements $C_i^c = \mathscr{J}_\infty(X) - C_i$ are cylinders. Since $C^c \Delta C_j^c = C \Delta C_j$ it follows at once that the complement of C is measurable. $\qquad\square$

The following proposition was one of the missing ingredients for the setup of motivic integration we outlined so far. It ensures that our typical functions $\mathbb{L}^{-\operatorname{ord}_Y}$ are in fact measurable (the missing part was the level set at infinity $\mathscr{J}_\infty(Y)$ which we owe the proof that it is measurable with measure zero).

Proposition 4.6 *Let $Y \subseteq X$ be a locally closed subvariety. Then $\mathscr{J}_\infty(Y)$ is a measurable subset of $\mathscr{J}_\infty(X)$ and if $\dim Y < \dim X$ the volume $\mu_X(\mathscr{J}_\infty(Y))$ is zero.*

Proof In [2, Proposition 6.22] Batyrev claims that with the previous results one can reduce the proof to the case that Y is a smooth divisor, where it is easily verified – we discussed this on page 202. Unfortunately I was not able to follow Batyrevs argument, thus the somewhat not so self contained proof is included here. For another proof see Proposition 5.7. The proof of this result relies on a fundamental result of Greenberg [20] which, for our purpose is best phrased as follows:

Proposition 4.7 *Let Y be a variety. Then there exists a positive integer $c \geq 1$ such that*

$$\pi_{\lfloor \frac{m}{c} \rfloor} \mathscr{J}_\infty(Y) = \pi^m_{\lfloor \frac{m}{c} \rfloor} \mathscr{J}_m(Y)$$

for all $m \gg 0$.

This, in particular, implies that the image $\pi_n(\mathscr{J}_\infty(Y))$ is a constructible subset of $\mathscr{J}_n(Y)$. It can be shown that its dimension is just the expected one, namely $\dim \pi_n(\mathscr{J}_\infty(Y)) = (n+1) \dim Y$, see [12, Lemma 4.3]. From these observations we obtain a bound for the dimension of $\mathscr{J}_m(Y)$ as follows. We can work locally and may assume that $Y \subseteq X$ for a smooth X. Then we have

$$\begin{aligned}
\dim \mathscr{J}_m(Y) &\leq \dim(\pi^m_{\lfloor \frac{m}{c} \rfloor} \mathscr{J}_m(Y)) + (m - \lfloor \tfrac{m}{c} \rfloor) \dim X \\
&= \dim(\pi_{\lfloor \frac{m}{c} \rfloor} \mathscr{J}_\infty(Y)) + (m - \lfloor \tfrac{m}{c} \rfloor) \dim X \\
&= (\lfloor \tfrac{m}{c} \rfloor + 1) \dim Y + (m+1) \dim X - (\lfloor \tfrac{m}{c} \rfloor + 1) \dim X \\
&= (m+1) \dim X - (\lfloor \tfrac{m}{c} \rfloor + 1)(\dim X - \dim Y).
\end{aligned}$$

Thus, if the codimension of Y in X is greater or equal to 1, then $\dim \mathscr{J}_m(Y) \mathbb{L}^{-m \dim X}$ approaches $-\infty$ as m approaches ∞. This implies that $\mathscr{J}_\infty(Y)$ is measurable that its measure $\mu_X(\mathscr{J}_\infty(Y))$ is zero. □

Comparison with Lebesgue integration

To guide one's intuition a comparison of the motivic measure with more classical measures such as the Lebesgue measure or p–adic measures is sometimes helpful. We discuss here the similarities with the Lebesgue measure on $\mathbb{A}^n$ since this is wellknown. The analogy with p–adic measures is even more striking, see for example the discussions in [42] and [23, 24].

For convenience we consider the case that $X = \mathbb{A}^n$ in which case we identify the k–points of $\mathscr{J}_\infty(\mathbb{A}^n_k)$ with n–tuples of power series with coefficients in k. That is we identify $\mathscr{J}_\infty(\mathbb{A}^n_k) \cong (k[\![t]\!])^n = \mathbb{A}^n_{k[\![t]\!]}$. Since

$k[\![t]\!]$ is a discrete valuation domain we can use the valuation to define a norm on $(k[\![t]\!])^n$ by defining

$$\|\tau\| \leq e^{-m} \iff \forall i : \tau_i \in (t^m)k[\![t]\!]$$

for $\tau = (\tau_1, \ldots, \tau_n)$ a tuple of power series in $\mathscr{J}_\infty(\mathbb{A}_k^n)$. It is easy to check that this defines a (non–archimedean) norm. In Table 1 the similarities between Lebesgue and motivic measure are summarized.

4.2 Motivic integration on singular varieties

In the preceding discussion we used the assumption that our spaces are non-singular in several places in an essential way. Roughly speaking we used the fact that if X is smooth then every cylinder is a stable set, thus can be endowed with a measure in a natural way. The fact that the resulting algebra of measurable sets includes the cylinders was essential for the setup since the level sets of the functions ord_Y are cylinders in a natural way.

If X is singular however, many things one might got used to from the smooth case fail. Most prominently, the truncation maps are no longer surjective and a cylinder is in general not stable. Thus one has to work somewhat harder to obtain an algebra of measurable sets which contains the cylinders. In order to be able to setup an integration theory one expects from this algebra of measurable sets the properties asserted in the following Proposition.

Proposition 4.8 *Let X be a k–variety. Then there is an algebra of measurable subsets of $\mathscr{J}_\infty(X)$ and a measure μ_X on that algebra such that*

1. *If A is stable, then A is measurable and $\mu_X(A) = [\pi_m A]\mathbb{L}^{-nm}$ for $m \gg 0$.*
2. *A cylinder C is measurable and $\mu_X(C) = \lim_m [\pi_m C]\mathbb{L}^{-nm}$.*
3. *The measure is additive on finite disjoint unions.*
4. *If $A \subseteq B$ are measurable, then $\dim \mu_X(A) \leq \dim \mu_X(B)$.*

To achieve this, one starts with the stable sets to which we know how to assign a measure. Then one proceeds just as in the smooth case using Definition 4.1, and replacing "cylinder" by "stable set" whenever necessary the same proof holds as well. The critical point now is to show that a cylinder is measurable with volume as claimed above. Even for the cylinder $\mathscr{J}_\infty(X) = \pi^{-1}(X)$ it is not clear a priori that it is

Table 6.1 *Comparison with Lebesgue measure.*

	Lebesgue	motivic
space	$\mathbb{R}^n$	$\mathscr{J}_\infty(\mathbb{A}^n)$
values of measure	$\mathbb{Z} \subseteq \mathbb{Q} \subseteq \mathbb{R}$	$K_0(\mathrm{Var}_k) \subseteq \mathscr{M}_k \subseteq \hat{\mathscr{M}}_k$
cubes around point a	$\{x \in \mathbb{R}^n \| \|x - a\| \le 1/m\}$	$\{\gamma \in \mathscr{J}_\infty(\mathbb{A}^n) \| \|\gamma - a\| \le e^{-m}\}$
measurable set	σ-algebra generated by cubes	algebra of stable/measurable sets
volume of cube	$(2/m)^n$	$(\mathbb{L}^{-m})^n$
transformation rule	$\int_A h(f)df = \int_{g^{-1}(A)} h(f(x))\,\mathrm{Jac}(f)dx$	$\int_A \mathbb{L}^{-\mathrm{ord}_D} = \int_{f_\infty^{-1}(A)} \mathbb{L}^{-\mathrm{ord}_D + K_{X'/X}}$

measurable and what its measure should be (in fact, the measure of X to be constructed leads to new birational invariants of X).

The point is that one has to partition $\mathscr{J}_\infty(X)$ according to the order of intersection along the singular locus Sing X, defined by the nth Fitting ideal of Ω_X^1. Then we can write

$$\mathscr{J}_\infty(X) = \bigsqcup_{e \geq 0} \mathscr{J}_\infty^{(e)}(X)$$

where $\mathscr{J}_\infty^{(e)}(X) = \mathrm{ord}_{\mathrm{Sing}\,X}^{-1}(e)$. As it turns out, the sets $\mathscr{J}_\infty^{(e)}(X)$ are in fact stable at level $\geq e$. The method is analogous to partitioning the cylinders of X' according to intersection with $K_{X'/X}$ (defined by the 0th Fitting ideal of $\omega_{X'/X}$) in the proof of the transformation rule. If one treats everything subordinate to this partition according to intersection with the singular locus one can construct a working theory in the singular case, see [32, 11].

5 Birational invariants via motivic integration

As an illustration of the theory we discuss some applications of geometric motivic integration to birational geometry, namely we give a description of the log canonical threshold of a pair (X, Y), where Y is a closed subscheme of the smooth scheme X, in terms of the asymptotic behavior of the dimensions of the jet schemes $\mathscr{J}_m(Y)$. These results are due to Mustaţǎ [34] and his collaborators [14, 16, 43]. The precise statement is as follows:

Theorem 5.1 *Let $Y \subseteq X$ be a subscheme of the smooth variety X. Then the log canonical threshold of the pair (X, Y) is*

$$\mathrm{lct}(X, Y) = \dim X - \sup_m \left\{ \frac{\dim \mathscr{J}_m(Y)}{m + 1} \right\}.$$

The definition of the log canonical threshold requires the introduction of some more notation from birational geometry which will be done shortly. It is an invariant which can be read of from the data of a log resolution of the pair (X, Y). The proof of the above is a very typical application of motivic integration as its strategy is to express the quantity one is interested in (say $\dim \mathscr{J}_m(Y)$), in terms of a motivic integral. Then one uses the transformation rule (Theorem 3.1) to reduce to an integral over a normal crossing divisor which can be explicitly computed, by a similar formula as the one in Proposition 2.15.

This result is in line with the earliest investigations of jet spaces by Nash [35] who conjectured an intimate correspondence between the jet spaces of a singular space and the divisors appearing in a resolution of singularities. Even though his conjecture was disproved recently by Kollar and Ishii [26] in general, there are important cases where his prediction was true, for example in the case of toric varieties.

5.1 Notation from birational geometry

Throughout this section we fix the following setup. Let X be a smooth k-variety of pure dimension n and let k be of characteristic zero. Let $Y \subseteq X$ be a closed subscheme. Let $f : X' \to X$ be a log resolution of the pair (X, Y), that is a proper birational map such that

1. X' is smooth and $F = f^{-1}Y$ is a divisor
2. writing $F \stackrel{\text{def}}{=} f^{-1}Y = \sum_{i=1}^{s} a_i D_i$ and $K \stackrel{\text{def}}{=} K_{X'/X} = \sum_{i=1}^{s} b_i D_i$ for some $a_i, b_i \in \mathbb{Q}$ and prime divisors D_i, the divisors F, K and $F + K$ have simple normal crossing support.

The existence of log resolutions is a consequence of Hironaka's resolution of singularities. Now one defines:

Definition 5.2 Let X and Y and the datum of a log resolution be as above and let $q \geq 0$ be a rational number. Then we say that

1. $(X, q \cdot Y)$ is *Kawamata log terminal* (KLT) if and only if $b_i - qa_i + 1 > 0$ for all i.
2. $(X, q \cdot Y)$ is *log canonical* (LC) if and only if $b_i - qa_i + 1 \geq 0$ for all i.

We point out (without proof) that these notions are independent of the chosen log resolution and therefore well defined, see [31] for details.

Remark 5.3 These notions can be expressed in terms of the multiplier ideal $\mathscr{I}(I_Y^q)$ of I_Y, the sheaf of ideals which cuts out Y on X, as follows [31]:

$$(X, q \cdot Y) \text{ is KLT} \iff \mathscr{I}(I_Y^q) = \mathscr{O}_X$$
$$(X, q \cdot Y) \text{ is LC} \iff \mathscr{I}(I_Y^{q'}) = \mathscr{O}_X \quad \forall q' < q.$$

To see this observe that by definition

$$\mathscr{I}(I_Y^q) = f_* \mathscr{O}_{X'}(\ulcorner K - qF \urcorner) = f_* \mathscr{O}_{X'}\left(\sum \ulcorner b_i - qa_i \urcorner D_i\right)$$

which is equal to $\mathcal{O}_X$ if and only if $\lceil b_i - qa_i \rceil \geq 0$ for all i. As the upper corners denote the round up of an integer, this is equivalent to $b_i - qa_i + 1 > 0$ for all i as required.

Now we proceed to the definition of the log canonical threshold, which is just the largest q such that the pair $(X, q \cdot Y)$ is log canonical.

Definition 5.4 The *log canonical threshold* of the pair (X, Y) is

$$\mathrm{lct}(X, Y) = \sup\{\, q \mid (X, q \cdot Y) \text{ is KLT } \}$$
$$= \sup\{\, q \mid b_i - qa_i + 1 > 0 \quad \forall i \,\}$$
$$= \min_i \left\{ \frac{b_i + 1}{a_i} \right\}.$$

Note that in the pathological case where $X = Y$ sets $\mathrm{lct}(X, X) = 0$. More generally we define $\mathrm{lct}(X, q \cdot Y) = q^{-1}\mathrm{lct}(X, Y)$. The formula for the log canonical threshold in terms of the jet schemes which we are aiming to prove in this section is an immediate consequence of the following theorem.

Theorem 5.5 *Let X and Y be as before. Then*

$$(X, q \cdot Y) \text{ is KLT} \iff \dim \mathscr{J}_m(Y) < (m+1)(n-q) \text{ for all } m$$
$$(X, q \cdot Y) \text{ is LC} \iff \dim \mathscr{J}_m(Y) \leq (m+1)(n-q) \text{ for all } m.$$

From this the proof of Theorem 5.1 follows immediately.

Proof of Theorem 5.1 By definition we have

$$\mathrm{lct}(X, Y) = \sup\{\, q \mid (X, q \cdot Y) \text{ is KLT } \}$$
$$= \sup\{\, q \mid \dim \mathscr{J}_m(Y) < (m+1)(n-q) \,\forall m \,\}$$
$$= n - \sup_m \left\{ \frac{\dim \mathscr{J}_m(Y)}{m+1} \right\}.$$

$\square$

Moreover, the proof of Theorem 5.5 will reveal that the supremum in Theorem 5.1 is actually obtained by infinitely many m, namely whenever $m + 1$ is divisible by all the coefficients a_i of D_i in $f^{-1}Y = \sum a_i D_i$ one has $\mathrm{lct}(X, Y) = \dim X - \frac{\dim \mathscr{J}_m(Y)}{m+1}$.

Before proceeding to the proof we derive some elementary properties of the log canonical threshold.

Proposition 5.6   *Let $Y \subseteq Y' \subseteq X$ closed subschemes of X. Then*

1. $\mathrm{lct}(X, Y) \geq \mathrm{lct}(X, Y')$.
2. $0 < \mathrm{lct}(X, Y) \leq \mathrm{codim}(Y, X)$ *and if Y is a hypersurface then one has equality if (X, Y) is log canonical.*
3. $\mathrm{lct}(X, Y)$ *is independent of X of fixed dimension.*
4. *Let (X', Y') be another pair, then $\mathrm{lct}(X \times X', Y \times Y') = \mathrm{lct}(X, Y) + \mathrm{lct}(X', Y')$.*

Proof   For (1) note that $Y \subseteq Y'$ implies that $\dim \mathscr{J}_m(Y) \leq \dim \mathscr{J}_m(Y')$ then apply Theorem 5.1.

For (2) recall that in any case $\mathscr{J}_m(Y)$ contains $\mathscr{J}_m(Y_{\mathrm{reg}})$ the jet scheme over the regular locus of Y. The latter has dimension $(m + 1) \dim Y$. Therefore $\dim \mathscr{J}_m(Y) \geq (m + 1) \dim Y$ and we finish by applying Theorem 5.1.

(3) is immediate since in the formula for $\mathrm{lct}(X, Y)$ the only feature of X that appears is its dimension.

The formation of jet schemes preserves products (since the functor $\mathscr{J}_m()$ has a left adjoint by construction) and hence it follows that $\dim \mathscr{J}_m(Y \times Y') = \dim \mathscr{J}_m(Y) + \dim \mathscr{J}_m(Y')$ from which (4) is implied immediately. $\square$

5.2 Proof of threshold formula

Now we present the proof of Theorem 5.5. For this we first recall the transformation rule in a slightly more general form than stated above. Let $A \subseteq \mathscr{J}_\infty(X)$ be a measurable subset (for example a stable subset) and let $g : \mathscr{J}_\infty(X) \to \mathbb{Q} \cup \{\pm\infty\}$ be a function whose level sets are measurable. Then

$$\int_A \mathbb{L}^g d\mu_X = \int_{f_\infty^{-1}(A)} \mathbb{L}^{g \circ f_\infty - \mathrm{ord}_{K_{X'/X}}} d\mu_{X'}$$

provided that one of the integrals exists, which then implies the existence of the other. The new features are minor. Clearly, it is allowed to integrate only over a measurable subset A as long as we also only integrate over its measurable image $f_\infty^{-1}(A)$ as well (that this image is measurable can be deduced from the proof of the transformation rule). In order to make the expression $\mathbb{L}^q$ for rational q defined we have to adjoin roots of $\mathbb{L}$ to the already huge ring $\mathscr{M}_k$ and define the dimension of $\mathbb{L}^q$ as q. For the following application it is in fact enough to adjoin a single root of $\mathbb{L}$ such that the new value ring of the integral is $\hat{\mathscr{M}}_k[\mathbb{L}^{1/n}]$ for a sufficiently big n.

We apply this result with $g = q \cdot \mathrm{ord}_Y = \mathrm{ord}_{qY}$ and $A = \mathrm{ord}_Y^{-1}(\geq m + 1)$ so that $f_\infty^{-1}(A) = \mathrm{ord}_F^{-1}(\geq m + 1)$ (up to measure zero) and $g \circ f_\infty - \mathrm{ord}_{K_{X'/X}} = -\mathrm{ord}_{K_{X'/X} - qF}$. Thus we get:

$$\int_{\mathrm{ord}_Y^{-1}(\geq m+1)} \mathbb{L}^{q\,\mathrm{ord}_Y}\, d\mu_X = \int_{\mathrm{ord}_F^{-1}(\geq m+1)} \mathbb{L}^{-\mathrm{ord}_{K_{X'/X} - qF}}\, d\mu_{X'} \quad (6.6)$$

Now, the left hand side of this equation contains information about the dimension of the mth jet scheme $J_m(Y)$, whereas the right hand side allows us to express this dimension in terms of the data of the log resolution. Together this will lead to a proof of Theorem 5.5.

Recall from Section 2.4 that

$$\mathrm{ord}_Y^{-1}(\geq m + 1) = \pi_m^{-1}(\mathscr{J}_m(Y))$$

and thus for the measure we have

$$\mu_X(\mathrm{ord}_Y^{-1}(\geq m + 1)) = [\mathscr{J}_m(Y)]\mathbb{L}^{-nm}$$

Now computing the left hand side of equation (6.6) one gets

$$S_m \overset{\mathrm{def}}{=} \int_{\mathrm{ord}_Y^{-1}(\geq m+1)} \mathbb{L}^{q\,\mathrm{ord}_Y}\, d\mu_X = \mu_X(\mathrm{ord}_Y^{-1}(\geq m + 1))\mathbb{L}^{q(m+1)}$$

$$= [\mathscr{J}_m(Y)]\mathbb{L}^{-nm + q(m+1)}$$

Evaluating at the dimension (see section 2.1) we get

$$\dim S_m = \dim \mathscr{J}_m(Y) - nm + q(m + 1). \quad (6.7)$$

Now we turn to computing the right hand side of equation (6.6):

$$\int_{\mathrm{ord}_F^{-1}(\geq m+1)} \mathbb{L}^{-\mathrm{ord}_K - qF}\, d\mu_{X'}$$

$$= \sum_{i=1}^{\infty} \mu_{X'}(\mathrm{ord}_F^{-1}(\geq m + 1) \cap \mathrm{ord}_{K - qF}^{-1}(i))\mathbb{L}^{-i}$$

$$= \sum_{r \in N_{n+1}} \mu_{X'}(\bigcap_i \mathrm{ord}_{D_i}^{-1}(r_i))\mathbb{L}^{-\sum (b_i - qa_i)r_i}$$

where N_{n+1} denotes the set of all tuples $r = (r_1, r_2, \ldots, r_s)$ of non-negative integers whose sum is $\geq n + 1$. This last equality relies of course on a partitioning of $\mathrm{ord}_F^{-1}(\geq m + 1)$, according to order along each component D_i of the occurring normal crossing divisors:

$$\mathrm{ord}_F^{-1}(\geq m+1) = \bigsqcup_{r \in N_{n+1}} (\bigcap_i \mathrm{ord}_{D_i}^{-1}(r_i)).$$

Clearly, this refines the partition $\mathrm{ord}_F^{-1}(\geq m+1) = \bigsqcup_i (\mathrm{ord}_F^{-1}(\geq m + 1) \cap \mathrm{ord}_{K-qF}^{-1}(i))$ and we have for $\gamma \in \cap \mathrm{ord}_{D_i}^{-1}(r_i)$ that $\mathrm{ord}_{K-qF}(\gamma) = \sum (b_i - qa_i)r_i$ which justifies the above computation. The point now is that $\mu_{X'}(\bigcap_i \mathrm{ord}_{D_i}^{-1}(r_i))$ was computed explicitly in Lemma 2.16 to be equal to $[D_{\mathrm{supp}\, r}^0](\mathbb{L}-1)^{|\,\mathrm{supp}\, r|}\mathbb{L}^{-\sum r_i}$. Hence

$$\begin{aligned} S_m &= \int_{\mathrm{ord}_F^{-1}(\geq m+1)} \mathbb{L}^{-\mathrm{ord}_{K-qF}} d\mu_{X'} \\ &= \sum_{r \in N_{n+1}} [D_{\mathrm{supp}\, r}^0](\mathbb{L}-1)^{|\,\mathrm{supp}\, r|}\mathbb{L}^{-\sum(b_i - qa_i+1)r_i} \end{aligned}$$

Note that the dimension of each summand is equal to $n - |\,\mathrm{supp}\, r| + |\,\mathrm{supp}\, r| - \sum(b_i - qa_i+1)r_i = n - \sum(b_i - qa_i+1)r_i$, and since the coefficient of the highest dimensional part of each summand has a positive sign, there is no cancelation of highest dimensional parts in the sum. Thus we get

$$\dim S_m = \max_{r \in N_{n+1}} \left\{ n - \sum(b_i - qa_i+1)r_i \right\} \qquad (6.8)$$

Together with (6.7) this yields

$$\dim \mathcal{I}_m(Y) - nm + q(m+1) = n - \min_{r \in N_{n+1}} \left\{ \sum(b_i - qa_i+1)r_i \right\} \quad (6.9)$$

which quickly leads to a proof of Theorem 5.5:

Proof of Theorem 5.5 It is enough to prove the first equivalence of Theorem 5.5, the second one being a limiting case of the first. Slightly reformulating and using the definition of KLT we have to show that

$$b_i - qa_i + 1 > 0 \quad \forall i \iff \dim \mathcal{I}_m(Y) - nm + q(m+1) < n \quad \forall m$$

From equation (6.9) the implication "$\Rightarrow$" is immediate. So let us consider "$\Leftarrow$": In fact, we only need to assume the right hand side for one m_0 such that $m_0 + 1$ is divisible by each a_i, that is, we assume that $\min_{r \in N_{m_0+1}}\{\sum(b_i - qa_i + 1)r_i\} > 0$ by equation (6.9). Since $b_i - qa_i + 1 > 0$ holds trivially for $a_i = 0$ we only need to consider i such that $a_i \neq 0$. In this case (for fixed i) define the tuple $r = (r_1, \ldots, r_s)$ by setting $r_i = \frac{m_0+1}{a_i}$ and $r_j = 0$ otherwise. Clearly $r \in N_{m_0+1}$. But in this case we get $0 > \sum_{j=1}^s (b_j - qa_j + 1)r_j = (b_i - qa_i + 1)r_i$ which implies that $(b_i - qa_i + 1) > 0$ as desired. $\square$

We include here a more elementary and more explicit version of 4.6. This is also due to Mustaţă, we only sketch the proof here and refer the reader to [33, Lemma 3.7].

Proposition 5.7 *Let X be smooth of pure dimension n and let $Y \subseteq X$ be a subvariety. Let $a = \mathrm{ord}_y Y$ be the local multiplicity of Y at the point $y \in Y$, then*

$$\dim(\pi_Y^m)^{-1}(y) \leq m \cdot \dim X - \lfloor \tfrac{m}{a} \rfloor.$$

Thus if a is now the maximum of all local multiplicities of Y one has

$$\dim \mathscr{J}_m(Y) \leq \dim Y + m \cdot \dim X - \lfloor \tfrac{m}{a} \rfloor \leq (m+1) \cdot \dim X - \lfloor \tfrac{m}{a} \rfloor.$$

It follows that if Y is nowhere dense in X, then $\mu_X(\mathscr{J}_\infty(Y)) = 0$.

Proof The second statement clearly follows from the first which in turn immediately reduces to the case that $Y \subseteq X$ is a hypersurface. Since X is smooth it is étale over $\mathbb{A}^n$ with y mapping to 0. Since $(\pi_X^\infty)^{-1}(y)$ gets thereby identified with $(\pi_{\mathbb{A}^n}^\infty)^{-1}(0)$ we may assume that $Y \subseteq \mathbb{A}^n$ is given by the vanishing of $f \in k[x] = k[x_1, \ldots, x_n]$ and the point y is the origin.

The condition that the local multiplicity of Y at 0 is equal to a means that the smallest degree monomial of f has degree a in $x_1, \ldots, x_n$. For simplicity we assume now that f is homogeneous of degree a, in general one can combine the following proof with a deformation argument to reduce to this case along the way, see [33] for this general case.

Exercise 2.8 states that $\mathscr{J}_m(Y) \subseteq \mathscr{J}_m(\mathbb{A}^n)$ is given by $m+1$ equations $f^{(0)}, \ldots, f^{(m)}$ in the coordinates of $\mathscr{J}_m(\mathbb{A}^n)$ described in Example 2.7. Concretely, $f^{(i)} \in k[x^{(0)}, \ldots, x^{(i)}]$ is given as the coefficient of t^i in the power series

$$f\Big(\sum_i x_1^{(i)} t^i, \ldots, \sum_i x_n^{(i)} t^i\Big).$$

With the notation $f_0^{(i)} \overset{\text{def}}{=} f^{(i)}(0, x^{(1)}, x^{(2)}, \ldots, x^{(i)})$ (we abbreviated the tuples $x_1^{(i)}, \ldots, x_n^{(i)}$ by $x^{(i)}$) the fiber $(\pi_Y^m)^{-1}(0)$ is given by the vanishing of $f_0^{(1)}, \ldots, f_0^{(m)}$ in the fiber $(\pi_X^m)^{-1}(0) \cong \mathrm{Spec}\, k[x^{(1)}, \ldots, x^{(m)}]$. Recall that we need to show that the dimension of $(\pi_Y^m)^{-1}(0)$ is at most $mn - \lfloor \tfrac{m}{a} \rfloor$. Thus it is enough to show that the dimension of the variety given by the vanishing of the ideal $I_p = (f_0^{(a)}, f_0^{(2a)}, \ldots, f_0^{(pa)})$ is at most $(pa)n - p$.

To show this we turn to an initial ideal of I_p with respect to the degree reverse lexicographic order on $k[x^{(1)}, \ldots, x^{(pa)}]$ where the underlying ordering of the variables $x_i^{(j)}$ is first by upper index and then by lower.

That is we order the variables according to $x_i^{(j)} \leq x_{i'}^{(j')}$ iff $j > j'$ or $j = j'$ and $i < i'$. The degree reverse lexicographic order on a polynomial ring $k[x_1, \ldots, x_n]$ with $x_1 > x_2 > \ldots > x_n$ is given by

$$A = x_1^{a_1} \cdots x_n^{a_n} > x_1^{b_1} \cdots x_n^{b_n} = B$$

if $\deg A > \deg B$ or $\deg A = \deg B$ and $a_i < b_i$ for the last index k for which $a_i \neq b_i$. Roughly speaking, a monomial in degree rev lex is big if it contains fewest of the cheap variables. For example $x_2^4 > x_1^3 x_3 > x_1^2 x_2 x_3$ (Consult [17, Chapter 15] for details). The initial term $\mathrm{in}_<(f)$ of a polynomial f is then the biggest monomial appearing in f with respect to that ordering. Now it is a matter of unraveling the definitions (of the $f^{(i)}$, of the order $\ldots$) to see that

$$\mathrm{in}_<(f_0^{(aj)}) = \mathrm{in}_<(f)(x^{(j)}) \text{ for } j = 1, \ldots, p.$$

As these are p many nontrivial equations in disjoint variables it follows that the dimension of the vanishing locus of the initial ideal of I_p is at most $(pa)n - p$. Thus the same upper bound holds for the dimension of the vanishing locus of I_p itself, and a forteriori for the dimension of $(\pi_Y^{pa})^{-1}(0)$. $\square$

5.3 Bounds for the log canonical threshold

In this section we show how the just derived description of the log canonical threshold in terms of the dimension of the jet spaces leads to some interesting bounds for $\mathrm{lct}(X, Y)$.

Proposition 5.8 *Let a be the maximal local multiplicity of a point Y. Then*

$$\frac{1}{a} \leq \mathrm{lct}(X, Y) \leq \frac{\dim X}{a}.$$

Proof If $Y = \emptyset$ then the statement is empty. If $Y = X$ then $a = \infty$ and $\mathrm{lct}(X, X) = 0$ so the statement is clear. Hence we may assume Y is a nonempty proper closed subscheme of X. Let p be a point with maximal multiplicity a. The second part of Exercise 5.9 shows that $\dim \mathscr{J}_{a-1}(Y) \geq \dim(\pi^{a-1})^{-1}(p) \geq \dim X \cdot (a - 1)$. Hence by Theorem 5.1

$$\mathrm{lct}(X, Y) \leq \dim X - \frac{\dim \mathscr{J}_{a-1}(Y)}{a} \leq \dim X - \dim X \frac{a-1}{a} = \frac{\dim X}{a}.$$

For the lower bound Proposition 5.7 gives $\dim \mathscr{J}_m(Y) \leq \dim Y + m \cdot \dim X - \lfloor \frac{m}{a} \rfloor$ and again using Theorem 5.1 this yields

$$\mathrm{lct}(X, Y) \geq \frac{\dim X - \dim Y}{m+1} + \frac{\lfloor \frac{m}{a} \rfloor}{m+1}$$

for all m. For sufficiently divisible and sufficiently big m this implies that $\mathrm{lct}(X, Y) \geq \frac{1}{a}$ as claimed. □

The next application of the arc space techniques is a bound for the log canonical threshold of a homogeneous hypersurface. This uses the following exercise as a key ingredient.

Exercise 5.9 Let $Y \subseteq \mathbb{A}^n$ be a homogeneous hypersurface of degree d. Show that one has an isomorphism

$$(\pi_Y^m)^{-1}(0) \cong \mathscr{J}_{m-d}(Y) \times \mathbb{A}^{n(d-1)}$$

for all $m \geq d - 1$, where we set $\mathscr{J}_{-1}(Y)$ to be a point.

Drop the assumption *homogeneous* and assume instead that the local multiplicity of Y at p be equal to a. Show that $(\pi_Y^{a-1})^{-1}(p) \cong \mathbb{A}^{n(a-1)}$.

Proposition 5.10 *Let $Y \subseteq \mathbb{A}^n$ be a homogeneous hypersurface of degree d. Then*

$$\mathrm{lct}(\mathbb{A}^n, Y) \geq \min\left\{\frac{n-r}{d}, 1\right\}$$

where $r = \dim \mathrm{Sing}\, Y$.

Proof One key ingredient is the observation of the previous Exercise 5.9 that for $m \geq d - 1$

$$(\pi_Y^m)^{-1}(0) \cong \mathscr{J}_{m-d}(Y) \times \mathbb{A}^{n(d-1)}$$

By semicontinuity of $\dim(\pi_Y^m)^{-1}(p)$ the inequality

$$\dim(\pi_Y^m)^{-1}(p) \leq \dim \mathscr{J}_{m-d}(Y) + n(d-1)$$

holds for all p, and in particular for the $p \in \mathrm{Sing}\, Y$. Hence all together we get the estimate

$$\dim \mathscr{J}_m(Y) \leq \max\{\dim \mathscr{J}_{m-d}(Y) + n(d-1)$$
$$+ \dim \mathrm{Sing}\, Y, (n-1)(m+1)\}$$

where $(n-1)(m+1)$ is equal to $\dim \mathscr{J}_m(Y - \mathrm{Sing}\, Y)$ which is always a lower bound for $\dim \mathscr{J}_m(Y)$. Now set $m = pd - 1$ and apply the inequality repeatedly to get

$$\dim \mathscr{J}_{pd-1}(Y) \leq \max\{(nd - n + r) \cdot p, (n-1)pd\}$$

which amounts to

$$\text{lct}(\mathbb{A}^n, Y) = n - \frac{\dim \mathscr{J}_{pd-1}(Y)}{pd} \geq \min\left\{\frac{n-r}{d}, 1\right\}$$

since the log canonical threshold is computed via the dimension of the jet spaces $\mathscr{J}_{pd-1}(Y)$ for sufficiently divisible pd. $\square$

In [14] the main achievement is to characterize the extremal case as follows: In the case that $\text{lct}(\mathbb{A}^n, Y) \neq 1$ one has $\text{lct}(\mathbb{A}^n, Y) = \frac{n-r}{d}$ if and only if $Y \cong Y' \times \mathbb{A}^r$ for some $Y' \subseteq \mathbb{A}^{n-r}$ a hypersurface.

5.4 Inversion of adjunction

One of the most celebrated applications of motivic integration to birational geometry is a much improved understanding of the Inversion of Adjunction conjecture of Shokurov [40] and Kollár [28]. The conjecture describes how certain invariants of singularities of pairs behave under restriction. The following proposition goes in this direction as it shows that under restriction to a smooth hypersurface the log canonical threshold can only decrease, that is the singularities cannot get better under restriction.

Proposition 5.11   *Let (X, Y) be a pair and H a smooth hypersurface in X. Then*

$$\text{lct}_H(X, Y) \geq \text{lct}(H, Y \cap H)$$

where $c_H(X, Y)$ is the log canonical threshold of the pair (X, Y) around H, that is the minimum $\text{lct}(U, Y \cap U)$ over all open $U \subseteq X$ with $H \subseteq U$.

Proof   The proof uses a straightforward extension of the formula for the log canonical threshold to include this more general case of $\text{lct}_H(X, Y)$. In fact the same proof as above shows that one has

$$\text{lct}_H(X, Y) = \dim X - \sup_m \left\{\frac{\dim_H \mathscr{J}_m(Y)}{m+1}\right\}$$

where the supremum is achieved for sufficiently divisible $m + 1$. Here $\dim_H \mathscr{J}_m(Y)$ is the *dimension of $\mathscr{J}_m(Y)$ along H*, that is the maximal dimension of an irreducible component T of $\mathscr{J}_m(Y)$ such that $\pi^m(T) \cap H \neq \emptyset$.

Let T be an irreducible component of $\mathscr{J}_m(Y)$ such that $\pi^m(T) \cap H \neq \emptyset$. This implies that $T \cap \mathscr{J}_m(Y \cap H)$ is also nonempty since it contains the constant jets over $\pi^m(T) \cap H$ since these are contained in *any* irreducible component (use that the irreducible components are preserved by the k^*-action given by $t \mapsto \lambda t$). Since $H \subseteq X$ is locally given by one equation the same is true for $H \cap Y \subseteq Y$. Hence $\mathscr{J}_m(H \cap Y) \subseteq \mathscr{J}_m(Y)$ is given by at most $m + 1$ equations (see Exercise 2.8). Hence $\dim_H \mathscr{J}_m(Y) \leq \dim \mathscr{J}_m(Y \cap H) + (m + 1)$ which shows

$$\sup_m \left\{ \frac{\dim_H \mathscr{J}_m(Y)}{m + 1} \right\} \leq \sup_m \left\{ \frac{\dim_H \mathscr{J}_m(Y \cap H)}{m + 1} \right\} + 1$$

which implies the claimed inequality $\mathrm{lct}_H(X, Y) \geq \mathrm{lct}(H, Y \cap H)$. □

The *Inversion of Adjunction* Conjecture of Kollár and Shokurov describes how the singularities of pairs behave under restriction to a Cartier divisor. More precisely, let (X, Y) be a pair where we allow $Y = \sum q_i Y_i$ to be any formal integer combination (rational or real combination even) of closed subschemes of X. With the notation $f : X' \to X$ of a log resolution as above (in particular $f^{-1}Y = \sum a_i E_i$ and $K_{X'/X} = \sum b_i E_i$) and a subscheme $W \subseteq X$ fixed we define the *minimal log discrepancy*

$$\mathrm{mld}(W; X, Y) \overset{\mathrm{def}}{=} \begin{cases} \min\{b_i - a_i + 1 | f(E_i) \subseteq W\} & \text{if this minimum} \\ & \text{is non-negative} \\ -\infty & \text{otherwise.} \end{cases}$$

It follows that (X, Y) is log canonical on an open subset containing W iff $\mathrm{mld}(W; X, Y) \neq -\infty$. The inversion of adjunction conjecture now states:

Conjecture 5.12 *With (X, Y) and as above with Y effective, let D be a normal effective Cartier divisor on X such that $D \not\subseteq Y$ and let $W \subset D$ a proper closed subset. Then we have*

$$\mathrm{mld}(W; X, Y + D) = \mathrm{mld}(W; D, Y|_D).$$

The inequality "$\leq$" is the *adjunction* part and is well known to follow from the adjunction formula $K_D = (K_X + D)|_D$. The reverse inequality "$\geq$" is the critical part of this conjecture.

In [16] the conjecture was proved in the case that X is smooth and Y is effective. In [15] this was established even for X a complete intersection (and Y effective). The proof of these results use a description of the minimal log discrepancies in terms of dimensions of certain cylinders of the jet spaces of X, analogous to the one for the log canonical threshold.

Proposition 5.13 *With the notation as above and for X smooth and Y effective,*

$$\mathrm{mld}(W; X, Y) \geq \tau$$

$$\Longleftrightarrow$$

$$\mathrm{codim}_{\mathscr{J}_\infty(X)}(\mathrm{Cont}_Y^\nu \cap \pi_0^{-1}(W)) \geq \sum q_i \nu_i + \tau \; \textit{for all multi-indices } \nu,$$

where $\mathrm{Cont}_Y^\nu = \cap_i \mathrm{ord}_{Y_i}^{-1}(\nu_i)$.

The proof of this is not more than a technical complication of the proof of the log canonical threshold formula we gave above. With this characterization of $\mathrm{mld}(W; X, Y)$ the proof of inversion of adjunction becomes a matter of determining the co-dimensions of the cylinders involved. In the case that X and D are both smooth this is quite easy (and could be done as an exercise). In general (X a complete intersection) the combinatorics involved can become quite intricate, *cf.* [15].

One should point out that after these results on inversion of adjunction were obtained, Takagi [41] found an alternative approach using positive characteristic methods.

5.5 Geometry of arc spaces without explicit motivic integration.

As it should have become apparent by now, the applications of motivic integration to birational geometry are by means of describing certain properties of a variety X in terms of (mostly simpler) properties of its jet spaces $\mathscr{J}_m(X)$. Motivic integration serves as the path to make this connection. However, due to some combinatorial difficulties one encounters along this path, one can ask if there is a more direct relationship. This is indeed the case and it is the content of the paper of [13] of Ein, Lazarsfeld and Mustaţă which is the source of the material in this section.

Their point is that instead of using the birational transformation rule to control the dimension of components of the jet spaces one uses the Key Theorem 3.5 and its proof to more directly get to the desired information.

We keep the notation of a subvariety $Y \subseteq X$ of a smooth variety X and define.

Definition 5.14 Let Y be a subvariety of X and $p \geq 0$ an integer define the *contact locus* to be the cylinder

$$\mathrm{Cont}_Y^p \overset{\text{def}}{=} \mathrm{ord}_Y^{-1}(p) \subseteq \mathscr{J}_\infty(X).$$

The aim is to understand the components (or at least the dimension) of the cylinders Cont_Y^p in terms of the data coming from a log resolution of the pair (X, Y). Using the notation of Section 5.1 we fix a log resolution $f : X' \to X$ of the pair (X, Y) that is an isomorphism over $X \setminus Y$ and denote $f^{-1}Y = \sum_1^k a_i E_i$ and $K_{X'/X} = \sum_1^k b_i E_i$ where the support of $\sum E_i$ is a simple normal crossing divisor.

Definition 5.15 Given $E = \sum_1^k E_i$, a simple normal crossing divisor of X', and a multi-index $\nu = (\nu_1, \ldots, \nu_k)$ define the *multi contact locus*

$$\mathrm{Cont}_E^\nu \overset{\mathrm{def}}{=} \{\, \gamma' \in \mathscr{J}_\infty(X') \mid \mathrm{ord}_{E_i}(\gamma') = \nu_i \text{ for } i = 1 \ldots k \,\}.$$

Definition 5.16 For every cylinder $C \subseteq \mathscr{J}_\infty(X)$ there is a well defined notion of codimension, namely

$$\mathrm{codim}\, C \overset{\mathrm{def}}{=} \mathrm{codim}(\mathscr{J}_m(X), \pi_m C)$$

for $m \gg 0$.

Of course one must check that this is independent of the chosen $m \gg 0$. This however is immediately clear from the definition of cylinder.

Exercise 5.17 For $C \subseteq \mathscr{J}_\infty(X)$ a cylinder, show that $\mathrm{codim}\, C = \dim X - \dim \mu_X(C)$.

The following proposition replaces in this new setup the computation of the motivic integral of a normal crossing divisor in Proposition 2.15. Note the comparative simplicity!

Proposition 5.18 For $E = \sum_1^k E_i$ a simple normal crossing divisor Cont_E^ν is a smooth cylinder of codimension

$$\mathrm{codim}\, \mathrm{Cont}_E^\nu = \sum_1^k \nu_i$$

provided Cont_E^ν *is nonempty.*

Proof This is a computation in local coordinates. Assume E is locally given by the vanishing of the first k of the coordinates $x_1 = \ldots = x_k = 0$. Then, for an arc γ, which is determined by $\gamma(x_i) = \sum \gamma_i^{(j)} t^j$ for $i = 1 \ldots n$, to have the prescribed contact order v_i with the E_i means precisely that $\gamma_i^{(j)} = 0$ for $j < \nu_i$, and $\gamma_i^{(\nu_i)} \neq 0$. Hence for $m \gg 0$ we have

$$\pi_m(\mathrm{Cont}_E^\nu) \cong (\mathbb{A}^1 - \{0\})^n \times \prod \mathbb{A}^{m-\nu_i}$$

and therefore Cont_E^ν is smooth irreducible and of codimension $\sum \nu_i$. $\square$

The central result (replacing the transformation rule) is the following
Theorem

Theorem 5.19   *With the notation as above one has for all $p > 0$ a
finite partition*

$$\mathrm{Cont}_Y^p = \bigsqcup_{\nu} f_\infty \, \mathrm{Cont}_E^\nu$$

where the disjoint union is over all multi-indices ν such that $\sum \nu_i a_i = p$.

*For every multi-index ν, the set $f_\infty \, \mathrm{Cont}_E^\nu$ is an irreducible cylinder
of codimension*

$$\sum \nu_i(b_i + 1).$$

*In particular, for each irreducible component Z of Cont_Y^p there is a
unique multi-index ν such that Cont_E^ν dominates Z.*

Proof   As in the proof of the transformation rule the key ingredient is
Theorem 3.5. With this at hand the proof is not difficult.

The condition $\sum \nu_i a_i = p$ ensures that $f_\infty \, \mathrm{Cont}_E^\nu \subseteq \mathrm{Cont}_Y^p$. The surjectivity of f_∞ (*cf.* Exercise 3.4) on the other hand implies the reverse
inclusion. The disjoint-ness of the union follows from the fact that there
is a one-to-one map between

$$\mathscr{J}_\infty(X') - \mathscr{J}_\infty(E) \xrightarrow{\ 1-1\ } \mathscr{J}_\infty(X) - \mathscr{J}_\infty(Y)$$

(which is an implication of the valuative criterion for properness as explained in Section 3) and the observation that each Cont_E^ν is contained
in the left hand side.

Since $\mathrm{Cont}_E^\nu \subseteq \mathrm{Cont}_{K_{X'/X}}^{\sum b_i \nu_i} = \mathrm{ord}_{K_{X'/X}}^{-1}(\sum b_i \nu_i)$ it follows from Theorem 3.5 (a) and Proposition 3.3 that $f_\infty \, \mathrm{Cont}_E^\nu$ is a cylinder. Part (b) of
Theorem 3.5 shows that

$$f_m : \mathrm{Cont}_E^\nu \longrightarrow f_m \, \mathrm{Cont}_E^\nu$$

is a piecewise trivial $\mathbb{A}^{\sum b_i \nu_i}$–fibration. By Exercise 5.17 the codimension
of Cont_E^ν is equal to $\sum \nu_i$. Hence the codimension of its image under f_∞
is

$$\mathrm{codim}(f_\infty \, \mathrm{Cont}_E^\nu) = \sum \nu_i + \sum \nu_i b_i = \sum \nu_i(b_i + 1)$$

as claimed.   □

As an illustration of this result we recover a very clean proof of the
log canonical threshold formula of Theorem 5.1.

Proof of Theorem 5.1   Let V_m be an irreducible component of $\mathscr{J}_m(Y)$. For some $p \geq m + 1$ the set $\mathrm{Cont}_Y^p \cap (\pi_m^X)^{-1}V_m$ is open in $(\pi_m^X)^{-1}V_m$, namely for the smallest p (automatically $\geq m+1$ since each arc in V_m has contact order $\geq m+1$ with Y) such that $\mathrm{Cont}_Y^p \cap (\pi_m^X)^{-1}V_m \neq \emptyset$. Hence there is an irreducible component W of Cont_Y^p such that the closure $\overline{W}$ contains $(\pi_m^X)^{-1}V_m$. By Theorem 5.19 there is a unique multi-index ν (necessarily $\sum \nu_i b_i = p$) such that Cont_E^ν dominates W.

By definition of the log canonical threshold we have $b_i + 1 \geq \mathrm{lct}(X, Y)a_i$ for all i such that we obtain the following inequalities:

$$\mathrm{codim}(V_m, \mathscr{J}_m(X)) \geq \mathrm{codim}\, W$$
$$= \mathrm{codim}\, f_\infty \mathrm{Cont}_E^\nu$$
$$\geq \sum \nu_i(b_i + 1)$$
$$\geq \sum \nu_i \mathrm{lct}(X, Y)a_i$$
$$\geq \mathrm{lct}(X, Y) \cdot p = \mathrm{lct}(X, Y) \cdot (m + 1).$$

As this holds for every irreducible component of $\mathscr{J}_m(Y)$ we get

$$\mathrm{lct}(X, Y) \leq \frac{\mathrm{codim}(\mathscr{J}_m(Y), \mathscr{J}_m(X))}{m + 1}.$$

To see that there is equality for some m we pick an index i such that $\mathrm{lct}(X, Y) = \frac{b_i + 1}{a_i}$ and $m + 1$ divisible by a_i. Let ν be the multi-index which is zero everywhere except at the ith spot, where it is $\frac{m+1}{a_i}$. Then $f_\infty \mathrm{Cont}_E^\nu \subseteq \mathrm{Cont}_Y^{m+1} \subseteq (\pi_m^X)^{-1}\mathscr{J}_m(Y)$ and by Theorem 5.19 the codimension $f_\infty \mathrm{Cont}_E^\nu$ is equal to $\frac{m+1}{a_i}(b_i + 1) = \mathrm{lct}(X, Y) \cdot (m+1)$. Hence in particular $\mathrm{codim}(\mathscr{J}_m(Y), \mathscr{J}_m(X)) \leq \mathrm{lct}(X, Y) \cdot (m + 1)$ for this chosen $m + 1$. This finishes the argument. $\square$

In summary, the above argument shows that the irreducible components V of $\mathscr{J}_m(Y)$ of maximal possible dimension, that is the ones that compute the log canonical threshold as $\mathrm{lct}(X, Y) = \mathrm{codim}(\mathscr{J}_m(X), V)$ are dominated by multi-contact loci Cont_E^ν with $\nu_i \neq 0$ with the property that for each $\nu_i \neq 0$ one has $\mathrm{lct}(X, Y) = \frac{b_i + 1}{a_i}$.

We want to finish these notes with Mustaţă's characterization of rational singularities for complete intersections in terms of arc spaces. This was indeed the first application of motivic integration to characterizing singularities. With the just developed viewpoint this result is not too difficult anymore.

Theorem 5.20 *Let $Y \subseteq X$ be a reduced and irreducible locally complete intersection subvariety of codimension c. Then the jet spaces $\mathscr{J}_m(Y)$ are irreducible for all m if and only if Y has rational singularities.*

Proof Let $f : X' \to X$ be a log resolution of (X, Y) which dominates the blowup of X along Y. Keeping the previous notation we may assume that E_1 is the exceptional divisor of this blowup. In [33] Theorem 2.1 it is shown that Y has at worst rational singularities if and only if $b_i \geq ca_i$ for every $i \geq 2$. Hence we must show

$$\mathscr{J}_m(Y) \text{ is irreducible for all } m \geq 1 \iff b_i \geq ca_i \text{ for } i \geq 2.$$

Assume that $\mathscr{J}_m(Y)$ is not irreducible, that is we have a component $V \subseteq \mathscr{J}_m(Y)$ other than the main component $\overline{\mathscr{J}_m(Y - \operatorname{Sing} Y)}$. As in the previous proof we have $W \subseteq \operatorname{Cont}_Y^p$ with $p \geq m+1$ whose closure contains $\pi_m^{-1}(V)$. By Theorem 5.19 this component is dominated by some multi-contact locus $\operatorname{Cont}_E^\nu$ for $\nu \neq (p+1, 0, \ldots, 0)$ since the latter is the multi-index corresponding to the multi-contact locus dominating $\pi_m^{-1}(\mathscr{J}_m(Y - \operatorname{Sing} Y))$. Since $Y \subseteq X$ is a local complete intersection of codimension c we have $\operatorname{codim}(V, \mathscr{J}_m(X)) \leq (m+1) \cdot c$. To arrive at a contradiction assume now that Y has rational singularities, that is assume that $b_i \geq ca_i$ for $i \geq 2$. Then

$$(m+1) \cdot c \geq \operatorname{codim}(W)$$
$$= \nu_1 \cdot c + \sum_{i \geq 2} \nu_i(b_i + 1)$$
$$\geq c \cdot \sum_{i \geq 1} \nu_i a_i + \sum_{i \geq 2} \nu_i \qquad (\text{since } b_i \geq ca_i)$$
$$= c \cdot p + \sum_{i \geq 2} \nu_i \geq c \cdot (m+1) + \sum_{i \geq 2} \nu_i.$$

Hence for $i \geq 2$ we must have $\nu_i = 0$, a contradiction.

Conversely, suppose $b_i < c \cdot a_i$ for some $i \geq 2$. Setting ν to be the multi-index with all entries zero except the ith equal to 1. Let $(m+1) = a_i$, then $\operatorname{Cont}_E^\nu$ maps to an irreducible subset $W \subseteq \operatorname{Cont}_Y^{m+1}$ of codimension $b_i < (m+1) \cdot c$. Hence $\pi_{m+1}(W)$ has dimension strictly bigger than that of $\overline{\mathscr{J}_m(Y \setminus \operatorname{Sing} Y)}$. Hence $\mathscr{J}_m(Y)$ cannot be irreducible. $\square$

Appendix A An elementary proof of the transformation rule

We present Looijenga's [32] elementary proof of Theorem 3.5 which then leads to a proof of the transformation formula avoiding weak factorization. For this we have to investigate more carefully the definition of the relative canonical divisor $K_{X'/X}$ and suitably interpret the contact multiplicity of an arc γ with $K_{X/X'}$.

A.1 The relative canonical divisor and differentials

Let us consider the first fundamental exact sequence for Kähler differentials, as it plays a pivotal role in all that follows.

The morphism $f : X' \to X$ induces a linear map, its derivative, $f^*\Omega_X \xrightarrow{df} \Omega_{X'}$ which is part of the first fundamental exact sequence for Kähler differentials:

$$0 \to f^*\Omega_X \xrightarrow{df} \Omega_{X'} \to \Omega_{X'/X} \to 0. \qquad (A.10)$$

Note that by our assumption of smoothness, the $\mathscr{O}_{X'}$–modules $f^*\Omega_X$ and $\Omega_{X'}$ are locally free of rank $n = \dim X$. Since, by birationality of f, $\Omega_{X'/X}$ has rank zero, the first map is injective as well. Taking the nth exterior power we obtain the map

$$0 \to f^*\Omega_X^n \xrightarrow{\wedge^n df} \Omega_{X'}^n$$

of locally free $\mathscr{O}_{X'}$ modules of rank 1. If we set $\omega = \Omega^n$ and tensor the above sequence with the invertible sheaf $\omega_{X'}^{-1}$ we obtain

$$f^*\omega_X \otimes \omega_{X'}^{-1} \subseteq \mathscr{O}_{X'}$$

thus identifying $f^*\omega_X \otimes \omega_{X'}^{-1}$ with a locally principal ideal in $\mathscr{O}_{X'}$, which we shall denote by $J_{X'/X}$ (so, by definition, $J_{X'/X}$ is the 0-th Fitting ideal of $\Omega_{X'/X}$). Now define $K_{X'/X}$ to be the Cartier divisor which is locally given by the vanishing of $J_{X'/X}$. It is important to note that $K_{X'/X}$ is defined as an effective divisor and not just as a divisor class. By choosing bases for the free $\mathscr{O}_{X'}$-modules $f^*\Omega_X$ and $\Omega_{X'}$ the map df is given by a $n \times n$ matrix with entries in $\mathscr{O}_{X'}$. Its determinant is a local defining equation for $K_{X'/X}$.

Let L be an extension field of k and let $\gamma : \operatorname{Spec} L[\![t]\!] \to X$ be a L-rational point of $\mathscr{J}_\infty(X')$, and assume that $\operatorname{ord}_{K_{X'/X}}(\gamma) = e$. By definition of contact order, this means that $(t^e) = \gamma^*(J_{X'/X}) \subseteq L[\![t]\!]$.

As $J_{X'/X}$ is locally generated by $\det df \in \mathcal{O}_{X'}$ (well defined up to unit) we obtain that $(t^e) = \det(\gamma^*(df))$. The pullback of the sequence (A.10) along γ illustrates the situation:

$$0 \longrightarrow (f \circ \gamma)^* \Omega_X \xrightarrow{\gamma^* df} \gamma^* \Omega_{X'} \longrightarrow \gamma^* \Omega_{X'/X} \longrightarrow 0. \qquad (A.11)$$

Since $L[\![t]\!]$ is a PID, we can choose bases of $(f \circ \gamma)^* \Omega_X$ and $\gamma^* \Omega_{X'}$ such that $\gamma^*(df)$, a map of free $L[\![t]\!]$ modules of rank n, is given by a diagonal matrix. With respect to this basis the exact sequence (A.10) takes the form

$$0 \longrightarrow L[\![t]\!]^n \xrightarrow{\begin{pmatrix} t^{e_1} & & 0 \\ & \ddots & \\ 0 & & t^{e_n} \end{pmatrix}} L[\![t]\!]^n \longrightarrow \oplus \frac{L[\![t]\!]}{(t^{e_i})} \longrightarrow 0. \qquad (A.12)$$

The condition that $\mathrm{ord}_{K_{X'/X}}(\gamma) = e$ translates into $\sum_{i=1}^n e_i = e$ or, equivalently, into saying that the rightmost module is torsion of length e.

A.2 Proof of Theorem 3.5

We start by recalling the statement of Theorem 3.5 we want to prove slightly reformulated in order to set up the notation that is used in its proof below.

Theorem A.1 *Let $f : X' \longrightarrow X$ be a proper birational morphism of smooth k-varieties. Let $C'_e = \mathrm{ord}^{-1}_{K_{X'/X}}(e)$ where $K_{X'/X}$ is the relative canonical divisor and let $C_e \stackrel{\mathrm{def}}{=} f_\infty C'_e$. Let $\gamma \in C'_e$ an L-point of $\mathscr{J}_\infty(X')$, that is a map $\gamma^* : \mathcal{O}_{X'} \longrightarrow L[\![t]\!]$, satisfying $\gamma^*(J_{X'/X}) = (t^e)$, with $L \supseteq k$ a field extension. Then for $m \geq 2e$ one has:*

(a'). *For all $\xi \in \mathscr{J}_\infty(X)$ such that $\pi_m^X(\xi) = f_m(\pi_m^{X'}(\gamma))$ there is $\gamma' \in \mathscr{J}_\infty(X')$ such that $f_\infty(\gamma') = \xi$ and $\pi_{m-e}^{X'}(\gamma') = \pi_{m-e}^{X'}(\gamma)$. In particular, the fiber of f_m over $f_m(\gamma_m)$ lies in the fiber of π_{m-e}^m over γ_{m-e}.*

(a). *$\pi_m(C'_e)$ is a union of fibers of f_m.*

(b). *The map $f_m : \pi_m^{X'}(C'_e) \longrightarrow C_e$ is a piecewise trivial $\mathbb{A}^e$ fibration.*

Proof To ease notation we will denote truncation by lower index, i.e. write γ_m as shorthand for $\pi_m^X(\gamma)$. We already pointed out before that (a') implies (a):

The proof of (b) can be divided into two steps. First we show that the fiber of f_m over $f_m(\gamma_m)$ can be naturally identified with $\mathrm{Der}_{\mathscr{O}_X}(\mathscr{O}_{X'}, \frac{L[\![t]\!]}{(t^{m+1})})$. Then we have to show that the latter is an affine space of dimension e. As this is easy let's do it first: Immediately preceding this proposition we noted that the cokernel of $\gamma^*(df)$ is torsion of length e as a $L[\![t]\!]$-module. This cokernel is $\gamma^*\Omega_{X'/X}$. Since $m > e$ the dual, $\mathrm{Hom}_{L[\![t]\!]}(\gamma^*\Omega_{X'/X}, \frac{L[\![t]\!]}{(t^{m+1})})$, is also torsion of length e. Using adjointness of γ^* and γ_* this Hom is just $\mathrm{Hom}_{\mathscr{O}_{X'}}(\Omega_{X'/X}, \gamma_* \frac{L[\![t]\!]}{(t^{m+1})})$, which is equal to $\mathrm{Der}_{\mathscr{O}_X}(\mathscr{O}_{X'}, \gamma_* \frac{L[\![t]\!]}{(t^{m+1})})$ essentially by definition of $\Omega_{X'/X}$. This shows that $\mathrm{Der}_{\mathscr{O}_X}(\mathscr{O}_{X'}, \gamma_* \frac{L[\![t]\!]}{(t^{m+1})})$ is isomorphic to $\mathbb{A}_L^e$. Thus we are left to show the identification $(\diamondsuit\diamondsuit\diamondsuit)$ of the following diagram the last line of which is the first exact sequence for derivations, analogous to the above exact sequence of Kähler differentials.

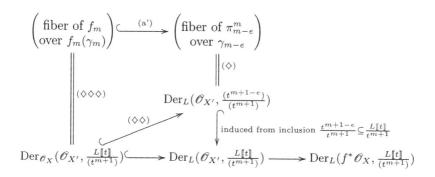

The identification $(\diamondsuit)$ is given by sending γ'_m to $\gamma'_m - \gamma_m$ which, since $m \geq 2e$, can easily be checked (see Lemma A.2 below) to define an L-derivation $\mathscr{O}_{X'} \longrightarrow \frac{(t^{m+1-e})}{(t^{m+1})}$. In this way (and using (a')) we think of $f_m^{-1}(f_m(\gamma_m))$ as a subspace of $\mathrm{Der}_L(\mathscr{O}_{X'}, \frac{(t^{m+1-e})}{(t^{m+1})})$. As this is the t^e-torsion part of $\mathrm{Der}_L(\mathscr{O}_{X'}, \frac{L[\![t]\!]}{(t^{m+1})})$ and since we just observed that $\mathrm{Der}_{\mathscr{O}_X}(\mathscr{O}_{X'}, \frac{L[\![t]\!]}{(t^{m+1})})$ is torsion of length e the inclusion $(\diamondsuit\diamondsuit)$ is also clear, and thus $(\diamondsuit\diamondsuit\diamondsuit)$ becomes a statement about subsets of $\mathrm{Der}_L(\mathscr{O}_{X'}, \frac{(t^{m+1-e})}{(t^{m+1})})$.

Let $(\gamma'_m - \gamma_m) \in \mathrm{Der}_L(\mathscr{O}_{X'}, \frac{(t^{m+1-e})}{(t^{m+1})})$. The image of $(\gamma'_m - \gamma_m)$ in $\mathrm{Der}_L(f^*\mathscr{O}_X, \frac{L[\![t]\!]}{(t^{m+1})})$ is $\gamma'_m \circ f - \gamma_m \circ f$. This is zero (i.e. $\gamma'_m \in \mathrm{Der}_{\mathscr{O}_X}(\mathscr{O}_{X'}, \frac{L[\![t]\!]}{(t^{m+1})})$) if and only if $f_m(\gamma'_m) = f_m(\gamma_m)$, that is if and only if γ'_m is in the fiber of f_m over $f_m(\gamma_m)$. This concludes the proof of (b).

In order to come by the element $\gamma' \in \mathscr{J}_\infty(X')$ as claimed in (a') we construct a sequence of arcs $\gamma^k \in \mathscr{J}_\infty(X')$ satisfying the following two properties for all $k \geq m$:

1. $\pi_k(f_\infty(\gamma^k)) = \pi_k(\xi)$ and
2. $\pi_{k-1-e}(\gamma^k) = \pi_{k-1-e}(\gamma^{k-1})$ and $\pi_{m-e}(\gamma^k) = \pi_{m-e}(\gamma)$.

Clearly, setting $\gamma^{m-1} = \gamma^m = \gamma$ these conditions hold for $k = m$. Furthermore, the second condition implies that the limit $\gamma' \overset{\text{def}}{=} \lim_k \gamma^k$ exists and that $\pi_{m-e}(\gamma') = \pi_{m-e}(\gamma)$. The first condition shows that $f_\infty(\gamma') = \xi$. Thus we are left with constructing the sequence γ^k. This is done inductively. As we already verified the solution for $k = m$ we assume to have γ^k and γ^{k-1} as claimed – now γ^{k+1} is constructed as follows:

Since $\pi_k(f(\gamma^k)) = \pi_k(\xi)$ we can view their difference as a derivation $\delta = \xi - f \circ \gamma^k \in \text{Der}_L(\mathscr{O}_X, \frac{(t^{k+1})}{(t^{k+2})})$ which we identify with $\text{Hom}_{L[t]}$ $(\gamma^{k*}f^*\Omega_X, \frac{(t^{k+1})}{(t^{k+2})})$. The latter module appears in $\text{Hom}_{L[t]}(\_, \frac{L[t]}{(t^{k+2})})$ applied to the sequence (A.11), where γ^k takes the place of γ:

$$\text{Hom}(\gamma^{k*}\Omega_{X/X'}, \tfrac{L[t]}{(t^{k+2})}) \hookrightarrow \text{Hom}(\gamma^{k*}\Omega_{X'}, \tfrac{L[t]}{(t^{k+2})}) \xrightarrow{df} \text{Hom}(\gamma^{k*}f^*\Omega_X, \tfrac{L[t]}{(t^{k+2})})$$

$$\cup \qquad\qquad\qquad\qquad \cup$$

$$\text{Hom}(\gamma^{k*}\Omega_{X'}, \tfrac{(t^{k+1-e})}{(t^{k+2})}) \ni \delta' \mapsto \delta \in \text{Hom}(\gamma^{k*}f^*\Omega_X, \tfrac{(t^{k+1})}{(t^{k+2})})$$

$$(A.13)$$

In order to understand this better we turn to the same sequence, but with respect to the basis as in sequence (A.12), where it takes this form:

$$\oplus \tfrac{(t^{k+1-e_i})}{(t^{k+2})} \hookrightarrow (\tfrac{L[t]}{(t^{k+2})})^n \xrightarrow{\begin{pmatrix} t^{e_1} & & 0 \\ & \ddots & \\ 0 & & t^{e_n} \end{pmatrix}} (\tfrac{L[t]}{(t^{k+2})})^n$$

$$\cup \qquad\qquad\qquad\qquad \cup$$

$$(\tfrac{(t^{k+1-e})}{(t^{k+2})})^n \qquad\qquad \delta \in (\tfrac{t^{k+1}}{(t^{k+2})})^n$$

Now it becomes clear that δ lies in the image of df since e and therefore all e_i are less than $m+1 \leq k+1$. Furthermore, any pre-image δ' must lie in $\text{Hom}(\gamma^{k*}\Omega_{X'}, \frac{(t^{k+1-e})}{(t^{k+2})})$ by the shape of the matrix and the fact that $e_i \leq e$ for all e. Now pick any such pre-image δ' and define $\gamma^{k+1} \overset{\text{def}}{=} \delta' + \gamma^k$. This is an arc in X' with $\pi_{k-e}(\gamma^{k+1}) = \pi_{k-e}(\gamma^k)$. Furthermore since $df(\delta') = \delta$ we get

$$\gamma^{k+1} \circ f - \gamma^k \circ f = \delta = \xi - \gamma^k \circ f \mod (t^{k+2})$$

and thus $\pi_{k+1}(\gamma^{k+1} \circ f) = \pi_{k+1}(f_\infty(\gamma^{k+1})) = \pi_{k+1}(\xi)$. $\qquad\square$

The following statement was used in the above proof. For completeness we provide a detailed argument.

Lemma A.2 *Fix an homomorphism $\gamma : R \longrightarrow S$ of k-algebras which makes S into an R-algebra. For any ideal $I \in S$ with $I^2 = 0$ one has a map*

$$\{ \gamma' \in \operatorname{Hom}_{k-alg}(R, S) \mid \operatorname{Im}(\gamma' - \gamma) \subseteq I \} \longrightarrow \operatorname{Der}_k(R, I) \qquad (A.14)$$

by sending γ' to $\gamma' - \gamma$.

Proof To check that $(\gamma' - \gamma)$ is indeed a derivation one has to make the following calculation verifying the Leibniz rule (note that the R algebra structure on S is via γ):

$$\begin{aligned}
(\gamma' &- \gamma)(xy) - ((\gamma' - \gamma)(x)\gamma(y) + \gamma(x)(\gamma' - \gamma)(y)) \\
&= \gamma'(x)\gamma'(y) - \gamma(x)\gamma(y) - \gamma'(x)\gamma(y) + \gamma(x)\gamma(y) \\
&\quad - \gamma(x)\gamma'(y) + \gamma(x)\gamma(y) \\
&= \gamma'(x)(\gamma'(y) - \gamma(y)) - \gamma(x)(\gamma'(y) - \gamma(y)) \\
&= (\gamma' - \gamma)(x) \cdot (\gamma' - \gamma)(y) = 0.
\end{aligned}$$

The last line is zero by the assumption that $\operatorname{Im}(\gamma' - \gamma) \subseteq I$ and $I^2 = 0$. The obvious inverse map sending a derivation δ to $\gamma + \delta$ shows that the two sets in (A.14) are equal. □

With this proof of Theorem 3.5 at hand a proof of the Transformation rule follows immediately as indicated in Section 3.

Appendix B Solutions to some exercises

Solution to 2.14 The answer in the first case is $[X - Y] + [Y]\frac{\mathbb{L}^c - 1}{\mathbb{L}^{c+1} - 1}$. Otherwise it can be read off from the general formula below; e.g. in the second case it is $[X - D] + \frac{[D]}{[\mathbb{P}^a]}$.

Solution to 2.17 Clearly, for an arc ϑ membership in $\cap \operatorname{ord}_{D_i}^{-1}(r_i)$ only depends on its truncation $\pi_t(\vartheta)$ as long as $t \geq \max\{ r_i \}$. With the notation of Example 2.7 we write

$$\pi_t(\vartheta(x_i)) = \sum_{j=0}^{t} \vartheta_i^{(j)} t^j$$

such that ϑ is determined by the coefficients $\vartheta_i^{(j)}$. Now we determine what the condition $\vartheta \in \cap \operatorname{ord}_{D_i}^{-1}(r_i)$ imposes on the coefficients $\vartheta_i^{(j)}$ (here it is convenient to set $r_i = 0$ for $i > s$).

1. For $j < r_i$ one has $\vartheta_i^{(j)} = 0$.
2. For $j = r_i$ one has $\vartheta_i^{(j)} \neq 0$.
3. For $j > r_i$ one has no condition $\vartheta_i^{(j)}$.

Thus the set $\pi_t(\cap_i \operatorname{ord}_{D_i}^{-1}(r_i))$ is the product made up from the factors $D_{\operatorname{supp} r}^0$ corresponding to all possible $\vartheta_i^{(0)}$, a copy of $k - \{0\}$ for each possible $\vartheta_i^{(r_i)}$ and $r_i > 0$ and a copy of k for each $r_i < j \leq k$. Putting this together we obtain the above formula.

Solution to 3.2   The valuative criterion of properness shows that any $\gamma \notin \mathscr{J}_\infty(Z)$ lies in the image (where $Z \subseteq X$ is such that f is an isomorphism from $X' - f^{-1}(Z) \to X - Z$). But for any $\gamma_m \in \mathscr{J}_m(X)$ the cylinder $\pi_m^{-1}(\gamma_m)$ cannot be contained in $\mathscr{J}_\infty(Z)$ since the latter has measure zero.

Solution to 3.4   The previous proposition shows that $f_\infty(\mathscr{J}_\infty(X'))$ is a cylinder. The surjectivity of f_m which was showed in exercise 3.2 now implies the result (in fact even the surjectivity of $f_0 = f$ is enough at this point).

Solution to 3.8   We may assume that C is irreducible. Let e be smallest such that $C \cap \operatorname{Cont}_{K_{X'/X}}^e \neq 0$. This intersection is open and dense in C, and hence we may replace C by $C \cap \operatorname{Cont}_{K_{X'/X}}^e$. Now Theorem 3.5 applies.

Solution to 3.10   Part (a) is easy. Not sure if part (c) is really feasible. Part (d) is straightforward from (c). Let me outline part (b): Recall that $K_{X'/X} = K_{X'/X''} + f'''^* K_{X''/X}$ such that

$$\int_X \mathbb{L}^{-\operatorname{ord}_D} d\mu_X = \int_{X'} \mathbb{L}^{-\operatorname{ord}_{f'^* D} - K_{X'/X}} d\mu_{X'}$$

$$= \int_{X'} \mathbb{L}^{-\operatorname{ord}_{f'^* D} - f'''^* K_{X''/X} + K_{X'/X''}} d\mu_{X'}$$

$$= \int_{X''} \mathbb{L}^{-\operatorname{ord}_{f^* D} - K_{X''/X}} d\mu_{X''}$$

where the first and last equality is the Transformation rule for f'' and f.

Solution to 5.9 This exercise is solved by explicitly writing down the equations which define $(\pi_Y^m)^{-1}(0)$ within $(\pi_{\mathbb{A}^n}^m)^{-1}(0)$. These turn out to be the same as the defining equations of the right hand side; just the variables are different.

Appendix References

[1] Dan Abramovich, Kalle Karu, Kenji Matsuki, and Jarosław Włodarczyk, *Torification and factorization of birational maps*, J. Amer. Math. Soc. **15** (2002), no. 3, 531–572 (electronic).

[2] Victor V. Batyrev, *Stringy Hodge numbers of varieties with Gorenstein canonical singularities*, Integrable systems and algebraic geometry (Kobe/Kyoto, 1997), World Sci. Publishing, River Edge, NJ, 1998, pp. 1–32.

[3] Victor V. Batyrev and Lev A. Borisov, *Mirror duality and string-theoretic Hodge numbers*, Invent. Math. **126** (1996), no. 1, 183–203.

[4] Franziska Bittner, *The universal Euler characteristic for varieties of characteristic zero*, Compos. Math. **140** (2004), no. 4, 1011–1032.

[5] Siegfried Bosch, Werner Lütkebohmert, and Michel Raynaud, *Néron models*, Ergebnisse der Mathematik und ihrer Grenzgebiete (3), vol. 21, Springer-Verlag, Berlin, 1990.

[6] Alastair Craw, *An introduction to motivic integration*, Strings and geometry, Clay Math. Proc., vol. 3, Amer. Math. Soc., Providence, RI, 2004, pp. 203–225.

[7] Pierre Deligne, *Théorie de Hodge. I*, Actes du Congrès International des Mathématiciens (Nice, 1970), Tome 1, Gauthier-Villars, Paris, 1971, pp. 425–430.

[8] ———, *Théorie de Hodge. II*, Inst. Hautes Études Sci. Publ. Math. (1971), no. 40, 5–57.

[9] ———, *Théorie de Hodge. III*, Inst. Hautes Études Sci. Publ. Math. (1974), no. 44, 5–77.

[10] Jan Denef and François Loeser, *Geometry on arc spaces of algebraic varieties*, In European Congress of Mathematics, Vol. I (Barcelona, 2000), 201:327–348. Progr. Math. Basel: Birkhäuser.

[11] ———, *Germs of arcs on singular algebraic varieties and motivic integration*, Invent. Math. **135** (1999), 201–232, arXiv:math.AG/9803039.

[12] ———, *Geometry on Arc Spaces*, Notes of a talk available at http://www.wis.kuleuven.ac.be/algebra/NotesCambridge/.

[13] Lawrence Ein, Robert Lazarsfeld, and Mircea Mustaţă, *Contact loci in arc spaces*, Compos. Math. **140** (2004), no. 5, 1229–1244.

[14] Lawrence Ein and Mircea Mustaţă, *The log canonical threshold of homogeneous affine hypersurfaces*, arXiv:math.AG/0105113.

[15] Lawrence Ein and Mircea Mustaţă, *Inversion of adjunction for local complete intersection varieties*, Amer. J. Math. **126** (2004), no. 6, 1355–1365.

[16] Lawrence Ein, Mircea Mustaţă, and Takehiko Yasuda, *Jet schemes, log discrepancies and inversion of adjunction*, Invent. Math. **153** (2003), no. 3, 519–535.

[17] David Eisenbud, *Commutative algebra*, Springer-Verlag, New York, 1995.

[18] Marvin J. Greenberg, *Schemata over local rings*, Ann. of Math. (2) **73** (1961), 624–648.

[19] _____, *Schemata over local rings. II*, Ann. of Math. (2) **78** (1963), 256–266.

[20] _____, *Rational points in Henselian discrete valuation rings*, Inst. Hautes Études Sci. Publ. Math. (1966), no. 31, 59–64.

[21] Alexandre Grothendieck, *Éléments de géométrie algébrique. I. Le langage des schémas*, Inst. Hautes Études Sci. Publ. Math. (1960), no. 4, 228.

[22] _____, *Éléments de géométrie algébrique. IV. Étude locale des schémas et des morphismes de schémas IV*, Inst. Hautes Études Sci. Publ. Math. (1967), no. 32, 361.

[23] Thomas C. Hales, *Can p-adic integrals be computed?*, Contributions to automorphic forms, geometry, and number theory, Johns Hopkins Univ. Press, Baltimore, MD, 2004, pp. 313–329.

[24] _____, *What is motivic measure?*, Bull. Amer. Math. Soc. (N.S.) **42** (2005), no. 2, 119–135 (electronic).

[25] Robin Hartshorne, *Algebraic geometry*, Springer, New York, 1973.

[26] Shihoko Ishii and János Kollár, *The Nash problem on arc families of singularities*, Duke Math. J. **120** (2003), no. 3, 601–620.

[27] Tetsushi Ito, *Stringy Hodge numbers and p-adic Hodge theory*, Compos. Math. **140** (2004), no. 6, 1499–1517.

[28] János Kollár, *Singularities of pairs*, Algebraic geometry—Santa Cruz 1995, Proc. Sympos. Pure Math., vol. 62, Amer. Math. Soc., Providence, RI, 1997, pp. 221–287.

[29] János Kollár and Shigefumi Mori, *Birational geometry of algebraic varieties*, Cambridge Tracts in Mathematics, vol. 134, Cambridge University Press, Cambridge, 1998, With the collaboration of C. H. Clemens and A. Corti, Translated from the 1998 Japanese original.

[30] Maxim Kontsevich, *Lecture at Orsay*, 1995.

[31] Robert Lazarsfeld, *Positivity in algebraic geometry. II*, Ergebnisse der Mathematik und ihrer Grenzgebiete. 3. Folge. A Series of Modern Surveys in Mathematics [Results in Mathematics and Related Areas. 3rd Series. A Series of Modern Surveys in Mathematics], vol. 49, Springer-Verlag, Berlin, 2004, Positivity for vector bundles, and multiplier ideals.

[32] Eduard Looijenga, *Motivic measures*, Astérisque (2002), no. 276, 267–297, Séminaire Bourbaki, Vol. 1999/2000.

[33] Mircea Mustaţă, *Jet Schemes of Locally Complete Intersection Canonical Singularities*, Invent. Math. **145** (2001), 397–424, arXiv:math.AG/0008002.

[34] ———, *Singularities of Pairs via Jet Schemes*, J. Amer. Math. Soc. **15** (2002), 599–615, arXiv:math.AG/0102201.

[35] John F. Nash, Jr., *Arc structure of singularities*, Duke Math. J. **81** (1995), no. 1, 31–38 (1996), A celebration of John F. Nash, Jr.

[36] Niko Naumann, *Algebraic independence in the Grothendieck ring of varieties*, Transactions of the American Mathematical Society 359, no. 4: 1653–1683 (electronic)

[37] Bjorn Poonen, *The Grothendieck ring of varieties is not a domain*, Math. Res. Lett. **9** (2002), no. 4, 493–497.

[38] Miles Reid, *La correspondance de McKay*, Astérisque (2002), no. 276, 53–72, Séminaire Bourbaki, Vol. 1999/2000.

[39] Julien Sebag, *Intégration motivique sur les schémas formels*, Bull. Soc. Math. France **132** (2004), no. 1, 1–54.

[40] Vyacheslav V. Shokurov, *Letters of a bi-rationalist. IV. Geometry of log flips*, Algebraic geometry, de Gruyter, Berlin, 2002, pp. 313–328.

[41] Shunsuke Takagi, *F-singularities of pairs and Inversion of Adjunction of arbitrary codimension*, Invent. Math. 157 (2004), no. 1, 123–146.

[42] Willem Veys, *Arc spaces, motivic integration and stringy invariants*. In: S. Izumiya et al. (eds.), *Singularity Theory and its applications*, pages 529–572. Volume 43 of *Advanced Studies in Pure Mathematics*. Mathematical Society of Japan, Tokyo, 2006.

[43] Takehiko Yasuda, *Dimensions of jet schemes of log singularities*, Amer. J. Math. **125** (2003), no. 5, 1137–1145.

7

Motivic invariants of rigid varieties, and applications to complex singularities

Johannes Nicaise and Julien Sebag

1 Introduction

Let R be a complete discrete valuation ring with quotient field K and perfect residue field k. In [47], the second author developed a theory of motivic integration on formal R-schemes $\mathfrak{X}$ that are separated and topologically of finite type. It generalizes motivic integration on algebraic varieties over a field (see Chapter 6), replacing the arc scheme by the *Greenberg scheme* $Gr(\mathfrak{X})$ of $\mathfrak{X}$. The Greenberg scheme is a k-scheme that parameterizes the étale sections on $\mathfrak{X}$. More precisely, for every finite unramified extension R' of R, there is a canonical bijection between the set $\mathfrak{X}(R')$ of R'-sections on $\mathfrak{X}$ and the set $Gr(\mathfrak{X})(k')$ of k'-points on $Gr(\mathfrak{X})$, where k' denotes the residue field of R'.

In [29], Loeser and the second author used this theory to define motivic integrals of differential forms of maximal degree on a certain class of smooth rigid K-varieties. Their construction is based on Raynaud's approach to rigid geometry in terms of formal models [44, 9] (see also Chapter 4 in this volume). Raynaud constructed a *generic fiber functor*

$$(\cdot)_\eta : \mathfrak{X} \mapsto \mathfrak{X}_\eta$$

from the category of formal R-schemes topologically of finite type to the category of rigid K-varieties. He showed that every quasi-separated quasi-compact rigid K-variety X admits a flat formal model $\mathfrak{X}$ over R, and that the generic fiber $X = \mathfrak{X}_\eta$ of $\mathfrak{X}$ does not change under *admissible blow-ups* (i.e., formal blow-ups with center in the special fiber). The

Motivic Integration and its Interactions with Model Theory and Non-Archimedean Geometry (Volume I), ed. Raf Cluckers, Johannes Nicaise, and Julien Sebag. Published by Cambridge University Press. © Cambridge University Press 2011.

model $\mathfrak{X}$ is separated iff X is separated. Moreover, for every morphism $f : X \to Y$ of quasi-separated quasi-compact rigid K-varieties, and for all flat formal R-models $\mathfrak{X}$ and $\mathfrak{Y}$ of X, resp. Y, there exists an admissible blow-up $\mathfrak{X}' \to \mathfrak{X}$ such that f extends to a morphism of formal R-schemes $\mathfrak{X}' \to \mathfrak{Y}$. These results make it possible to understand the geometry of rigid K-varieties entirely in terms of formal R-schemes. If X is a separated smooth quasi-compact rigid K-variety, and ω is a differential form of maximal degree on X, then Loeser and the second author defined the motivic integral of ω on X by choosing a formal R-model $\mathfrak{X}$ for X and considering a function $\mathrm{ord}(\omega)$ on the Greenberg scheme $Gr(\mathfrak{X})$ that measures the order of vanishing of ω at the closed points of étale sections on $\mathfrak{X}$. The motivic integral of ω on X is then defined as the integral of the motivic function $\mathbb{L}^{-\mathrm{ord}(\omega)}$ on $Gr(\mathfrak{X})$. The *change of variables formula* for motivic integrals on formal schemes guarantees that this definition does not depend on the chosen model $\mathfrak{X}$. The first author generalized the construction of the motivic integral to a larger class of formal schemes and rigid varieties in [34].

The aim of the present paper is to give an introduction to this theory of motivic integration, and to explain some applications to the theory of complex hypersurface singularities and the study of rational points on varieties over a discretely valued field that were developed by the authors. The applications to singularities are centered around the construction of the *analytic Milnor fiber*, a non-archimedean model for the classical Milnor fibration associated to the germ of a complex hypersurface singularity. It is a smooth rigid variety over the field $\mathbb{C}((t))$ of complex Laurent series. The analytic Milnor fiber yields a non-archimedean interpretation for many fundamental invariants of the singularity. In particular, it allows to interpret Denef and Loeser's *motivic zeta function* of the singularity as a "Weil generating series" that measures the set of rational points on the analytic Milnor fiber over finite extensions of the base field $\mathbb{C}((t))$. This interpretation sheds new light on the *motivic monodromy conjecture*, that relates the poles of the zeta function to eigenvalues of monodromy.

The applications to rational points on (rigid) varieties X over a discretely valued field K are based on the *motivic Serre invariant*, introduced by Loeser and the second author in [29] and further refined by the authors in [42, 34, 37]. The motivic Serre invariant is defined in terms of so-called *weak Néron models*, and provides a motivic measure for the set of rational points on X. We showed in [41, 34, 37] that, under certain tameness assumptions, the motivic Serre invariant admits a cohomological interpretation by means of a *trace formula*. When applied to the analytic Milnor fiber, it yields a cohomological interpretation of the motivic zeta function in terms of Lefschetz numbers of monodromy.

Notation

Let R be a complete discrete valuation ring with quotient field K and residue field k. We assume that the valuation v_K on K is non-trivial. We denote by p the characteristic exponent of k. Additional assumptions on k will be indicated at the beginning of each section.

We denote by π a uniformizer in R, and we fix a prime $\ell \neq p$. The choice of a value $|\pi|_K \in \,]0,1[$ defines a v_K-adic absolute value $|\cdot|_K$ on K and turns K into a non-archimedean field. We denote by K^a an algebraic closure of K, and by K^s and K^t the separable, resp. tame closure of K in K^a. We denote the respective completions by $\widehat{K^a}$, $\widehat{K^s}$ and $\widehat{K^t}$. We denote by R^{sh} the strict henselization of R in K^s, and by K^{sh} its quotient field. We denote by

$$\chi_{\text{top}} : K_0(\text{Var}_k) \to \mathbb{Z}$$

the ℓ-adic Euler characteristic (it is independent of ℓ).

We denote by $\mathbb{N}'$ the set of elements in $\mathbb{Z}_{>0}$ prime to p. For each $d \in \mathbb{Z}_{>0}$, we put $R(d) = R[t]/(t^d - \pi)$, and we denote by $K(d)$ its quotient field. If k is algebraically closed and $d \in \mathbb{N}'$, then $K(d)$ is K-isomorphic to the unique degree d extension of K in K^s, but for general k, the K-isomorphism class of $K(d)$ depends on π.

We denote by $\mathfrak{M}$ the maximal ideal of R. We endow R with its $\mathfrak{M}$-adic topology, and k with the discrete topology. For each integer $i \geq 0$ we put $R_i = R/\mathfrak{M}^{i+1}$.

An extension of R of ramification index one is a local morphism of discrete valuation rings $R \to R'$ such that $\mathfrak{M}$ generates the maximal ideal of R', and such that the residue field of R' is a separable extension of k (here we follow the terminology from [11]). We say that R' is unramified over R if, in addition, R' is finite over R. In this case, we also say that the quotient field K' of R' is an unramified extension of K.

If R has equal characteristic, then we fix a k-algebra structure on R such that $k \to R \to k$ is the identity. Recall that, if R has mixed characteristic and k is perfect, then R has a canonical stucture of finite $W(k)$-algebra, with $W(k)$ the ring of Witt vectors over k [49, II§5, Thm. 4].

For any adic topological ring A with ideal of definition I, and any tuple of variables $(x_1, \ldots, x_n)$, we denote by $A\{x_1, \ldots, x_n\}$ the A-algebra of converging power series, endowed with its I-adic topology [21, **0**.7.5]. It is defined by

$$A\{x_1, \ldots, x_n\} = \varprojlim_{m>0} ((A/I^m)[x_1, \ldots, x_n])$$

and this definition does not depend on the choice of ideal of definition I. If A is Noetherian, then so is $A\{x_1, \ldots, x_n\}$ [21, **0**.7.5.4].

For any scheme S, we denote by S_{red} its maximal reduced closed subscheme. An S-variety is a reduced separated S-scheme of finite type.

If S is a Noetherian scheme, we denote by $K_0(\mathrm{Var}_S)$ the Grothendieck ring of S-varieties, and by $\mathcal{M}_S$ its localization with respect to the class $\mathbb{L}$ of the affine line over S (see Section 3.1 of Chapter 5). We denote by $K_0^{\mathrm{mod}}(\mathrm{Var}_S)$ the *modified* Grothendieck ring of varieties, and by $\mathcal{M}_S^{\mathrm{mod}}$ its localization with respect to the class of the affine line (see Section 3.8 of Chapter 5).

Let X be a Noetherian k-scheme. If R has equal characteristic, then we put $K_0^R(\mathrm{Var}_X) = K_0(\mathrm{Var}_X)$ and $\mathcal{M}_X^R = \mathcal{M}_X$. If R has mixed characteristic, then we put $K_0^R(\mathrm{Var}_X) = K_0^{\mathrm{mod}}(\mathrm{Var}_X)$ and $\mathcal{M}_X^R = \mathcal{M}_X^{\mathrm{mod}}$. If X is the spectrum of a Noetherian k-algebra A, we write $K_0^R(\mathrm{Var}_A)$ and $\mathcal{M}_A^R$ instead of $K_0^R(\mathrm{Var}_X)$ and $\mathcal{M}_X^R$.

2 Motivic integration on formal schemes

2.1 Formal schemes and rigid varieties

In this section, we define the categories of formal schemes and rigid varieties we'll be working with. For the basic theory of formal schemes, we refer to [21, 27, 33]. For the theory of rigid varieties, see [8, 18, 33].

Definition 2.1 (Special formal schemes) Let A be a Noetherian adic topological ring. A special formal scheme over A is a separated Noetherian adic formal scheme $\mathfrak{X}$, endowed with a morphism of formal schemes $\mathfrak{X} \to \mathrm{Spf}\, A$, such that for any ideal of definition $\mathscr{J}$ on $\mathfrak{X}$, the closed subscheme $V(\mathscr{J})$ of $\mathfrak{X}$ defined by $\mathscr{J}$ is of finite type over A. The category of special formal A-schemes will be denoted by (SpF/A).

A special formal A-scheme $\mathfrak{X}$ is topologically of finite type over A iff the morphism $\mathfrak{X} \to \mathrm{Spf}\, A$ is adic. We'll call such special formal A-schemes stft formal A-schemes (*separated topologically of finite type*).

If $V(\mathscr{J})$ is of finite type over A for one ideal of definition $\mathscr{J}$ on $\mathfrak{X}$, then this holds for every ideal of definition. The terminology "special formal scheme" was introduced by Berkovich in [5, § 1]. Our definition differs slightly from his one, because we impose the additional conditions of being separated and quasi-compact. Special formal schemes appear in the literature under various names; for instance, they are called *formal schemes formally of finite type* in [43], and *formal schemes of pseudo-finite type* in [2].

An affine special formal A-scheme can be characterized in a more explicit way as follows: if I is an ideal of definition of A and $\mathfrak{X} = \operatorname{Spf} B$ is an affine formal A-scheme, then $\mathfrak{X}$ is special iff the topological A-algebra B is of the form

$$B = A\{x_1, \ldots, x_m\}[[y_1, \ldots, y_n]]/(f_1, \ldots, f_q)$$

with ideal of definition generated by I and $y_1, \ldots, y_n$. Moreover, $\mathfrak{X}$ is stft iff B is of the form

$$B = A\{x_1, \ldots, x_m\}/(g_1, \ldots, g_q)$$

with ideal of definition I (see the proof of [2, 1.7]).

Example 2.2 1. Any separated k-scheme of finite type X is an stft formal k-scheme. By composition of the structural morphism $X \to \operatorname{Spec} k$ with the morphism of formal schemes $\operatorname{Spec} k \to \operatorname{Spf} R$, one can also view X as an stft formal R-scheme.

2. Let X be a separated R-scheme of finite type, and Z a subscheme of the special fiber $X_s = X \times_R k$. The formal completion $\widehat{X/Z}$ of X along Z is a special formal R-scheme. If Z is open in X_s, then $\widehat{X/Z}$ is stft.

We'll consider the special fiber functor

$$(\cdot)_s : (\operatorname{SpF}/R) \to (\operatorname{SpF}/k) : \mathfrak{X} \to \mathfrak{X}_s = \mathfrak{X} \times_R k.$$

If A is either R or k, we also have a reduction functor

$$(\cdot)_0 : (\operatorname{SpF}/A) \to (\operatorname{sft}/k) : \mathfrak{X} \to \mathfrak{X}_0.$$

Here (sft/k) denotes the category of separated k-schemes of finite type, and $\mathfrak{X}_0$ denotes the reduction of $\mathfrak{X}$, i.e., the closed subscheme of $\mathfrak{X}$ defined by the largest ideal of definition. The closed immersion $\mathfrak{X}_0 \to \mathfrak{X}$ is a homeomorphism.

Example 2.3 1. If X is a separated k-scheme of finite type, then the largest ideal of definition on X is the ideal sheaf of nilpotent functions on X. Hence, X_0 is canonically isomorphic to X_{red}. If $\mathfrak{X}$ is an stft formal R-scheme, then $\mathfrak{X}_s$ is a k-scheme and $\mathfrak{X}_0 = (\mathfrak{X}_s)_{red}$.

2. If $\mathfrak{X}$ is the completion of a separated R-scheme of finite type X along a subscheme Z of X_s, then $\mathfrak{X}_0 = Z_{red}$ and $\mathfrak{X}_s$ is the completion of X_s along Z.

3. If $\mathfrak{X}$ is a special formal R-scheme of the form

$$\mathfrak{X} = \mathrm{Spf}\,(R\{x_1, \ldots, x_m\}[[y_1, \ldots, y_n]]/(f_1, \ldots, f_q)),$$

then we have

$$\mathfrak{X}_s = \mathrm{Spf}\,(k[x_1, \ldots, x_m][[y_1, \ldots, y_n]]/(f_1, \ldots, f_q))$$
$$\mathfrak{X}_0 = (\mathrm{Spec}\,(k[x_1, \ldots, x_m]/(f_1, \ldots, f_q)))_{red}$$

where we abused notation by writing f_i for the image of f_i in the quotient rings $k[x_1, \ldots, x_m][[y_1, \ldots, y_n]]$, resp. $k[x_1, \ldots, x_m]$, of

$$R\{x_1, \ldots, x_m\}[[y_1, \ldots, y_n]].$$

In [7, 0.2.6], Berthelot associated a generic fiber $\mathfrak{X}_\eta$ to any special formal R-scheme $\mathfrak{X}$. He constructed a generic fiber functor

$$(\cdot)_\eta : (\mathrm{SpF}/R) \to (\mathrm{Rig}/K) : \mathfrak{X} \mapsto \mathfrak{X}_\eta$$

to the category (Rig/K) of rigid varieties over K. This construction is explained in more detail in [14, § 7]. The generic fiber $\mathfrak{X}_\eta$ of a special formal R-scheme is a separated rigid K-variety. If $\mathfrak{X}$ is stft, then $\mathfrak{X}_\eta$ is quasi-compact, and the construction coincides with the one by Raynaud [44, 9]. We say that $\mathfrak{X}$ is generically smooth if its generic fiber $\mathfrak{X}_\eta$ is a smooth rigid K-variety.

Example 2.4 Let $\mathfrak{X}$ be the affine special formal R-scheme

$$\mathfrak{X} = \mathrm{Spf}\,(R\{x_1, \ldots, x_m\}[[y_1, \ldots, y_n]]/(f_1, \ldots, f_q)).$$

Denote by D_m the rigid unit polydisc

$$D_m = \mathrm{Sp}\,K\{x_1, \ldots, x_m\}$$

and by D_n^- the open rigid unit disc

$$D_n^- = \{z \in \mathrm{Sp}\,K\{y_1, \ldots, y_n\} \mid |y_i(z)| < 1 \text{ for all } i\}.$$

The elements of $R\{x_1, \ldots, x_m\}[[y_1, \ldots, y_n]]$ converge and define analytic functions on $D_m \times_K D_n^-$. The generic fiber $\mathfrak{X}_\eta$ of $\mathfrak{X}$ is the closed rigid subvariety of $D_m \times_K D_n^-$ defined by the equations

$$f_1 = \ldots = f_q = 0.$$

There is a canonical specialization morphism of locally ringed sites

$$sp_{\mathfrak{X}} : \mathfrak{X}_\eta \to \mathfrak{X}$$

(w.r.t. the strong G-topology on $\mathfrak{X}_\eta$). If Z is a locally closed subset of $\mathfrak{X}_s$, then one can show that $sp_{\mathfrak{X}}^{-1}(Z)$ is an open rigid subvariety of $\mathfrak{X}_\eta$, canonically isomorphic to the generic fiber of the formal completion $\widehat{\mathfrak{X}/Z}$ of $\mathfrak{X}$ along Z [7, 0.2.7]. If $\mathcal{N}$ is a coherent $\mathscr{O}_{\mathfrak{X}}$-module, we denote by $\mathcal{N}_{rig}$ the induced $\mathscr{O}_{\mathfrak{X}_\eta}$-module $sp_{\mathfrak{X}}^*\mathcal{N}$.

For any finite extension K' of K, with valuation ring R', applying the generic fiber functor yields a map

$$\iota : \mathrm{Hom}_{(\mathrm{SpF}/R)}(\mathrm{Spf}\, R', \mathfrak{X}) \to \mathrm{Hom}_{(\mathrm{Rig}/K)}(\mathrm{Sp}\, K', \mathfrak{X}_\eta)$$

and this map is a bijection. If x is a point on $\mathfrak{X}_\eta$ with residue field K', then $sp_{\mathfrak{X}}(x)$ is the image in $\mathfrak{X}$ of the unique point of $\mathrm{Spf}\, R'$ under the morphism $\iota^{-1}(x)$.

Proposition 2.5 *Let $\mathfrak{X}$ be a special formal R-scheme of pure relative dimension d. If x is a k-point on $\mathfrak{X}_0$ and $\mathfrak{X}$ is smooth at x, then the completed local ring $\widehat{\mathscr{O}}_{\mathfrak{X},x}$ of $\mathfrak{X}$ at x is isomorphic to $R[[x_1, \ldots, x_d]]$ as an R-algebra, and $sp_{\mathfrak{X}}^{-1}(x)$ is isomorphic to the open rigid unit disc*

$$D_d^- = \{z \in \mathrm{Sp}\, K\{y_1, \ldots, y_d\} \mid |y_i(z)| < 1 \text{ for all } i\}.$$

Proof Since the formal completion of $\mathfrak{X}$ at x is $\mathrm{Spf}\,\widehat{\mathscr{O}}_{\mathfrak{X},x}$, it suffices to show that there exists an isomorphism of R-algebras

$$\widehat{\mathscr{O}}_{\mathfrak{X},x} \cong R[[x_1, \ldots, x_d]].$$

By the infinitesimal lifting criterion for formal smoothness [23, 17.1.1], the k-point x lifts to a section in $\mathfrak{X}(R)$. Now the result follows from [11, 3.1.2]. $\qquad\square$

2.2 Greenberg schemes

In this section, we assume that k is perfect or R has equal characteristic.

The integration space of a motivic integral on an algebraic k-variety X is the arc space of X. In the theory of motivic integration on formal R-schemes [47], the arc space is replaced by the so-called Greenberg scheme [19, 20]. In this section we briefly recall its definition and basic properties, and we explain the relation with arc schemes.

For each $n \geq 0$, we define a functor

$$\mathscr{R}_n : (k - \mathrm{alg}) \to (R_n - \mathrm{alg}) : A \mapsto \mathscr{R}_n(A)$$

from the category of k-algebras to the category of R_n-algebras, as follows. If R has equal characteristic, then we put

$$\mathscr{R}_n(A) = A \otimes_k R_n.$$

If R has mixed characteristic and R is absolutely unramified, i.e., R is the ring of Witt vectors $W(k)$, then we put

$$\mathscr{R}_n(A) = W_{n+1}(A)$$

with $W_{n+1}(A)$ the ring of Witt vectors of length $n+1$ with coordinates in A.

If R has mixed characteristic and R is absolutely ramified, then we define $\mathscr{R}_n$ as the *fppf* sheaf associated to the presheaf

$$A \mapsto W(A) \otimes_{W(k)} R_n.$$

For any k-algebra A, the natural map

$$W(A) \otimes_{W(k)} R_n \to \mathscr{R}_n(A)$$

is surjective. If $A^p = A$, then it is an isomorphism [28, A.2], but this does not hold for all k-algebras A; see [42, §2] for a counterexample and further discussions.

In all of the above cases, we have $\mathscr{R}_n(k) = R_n$. One can show that $\mathscr{R}_n$ is representable by a k-scheme in R_n-algebras, whose underlying k-scheme is isomorphic to $\mathbb{A}_k^{n+1}$.

Example 2.6   1. If $R = W(k)$, then $\mathscr{R}_n$ is simply the Witt scheme $\mathscr{W}_{n+1}$ over k.

2. If R has equal characteristic, a choice of isomorphism k-algebras $R \cong k[[t]]$ yields an isomorphism

$$\mathscr{R}_n \cong \mathbb{A}_k^{n+1} = \operatorname{Spec} k[x_0, \dots, x_n].$$

The ring scheme structure on $\mathscr{R}_n$ is given by the morphisms $+$ and $\cdot$ from $\mathbb{A}_k^{n+1} \times \mathbb{A}_k^{n+1}$ to $\mathbb{A}_k^{n+1}$ given by

$$\begin{cases} (a_0, \dots, a_n) + (b_0, \dots, b_n) & = & (a_0 + b_0, \dots, a_n + b_n) \\ (a_0, \dots, a_n) \cdot (b_0, \dots, b_n) & = & (a_0 b_0, a_0 b_1 + a_1 b_0, \dots, \sum_{i+j=n} a_i b_j). \end{cases}$$

These morphisms correspond precisely to the rules for addition and multiplication for the truncated power series $a_0 + a_1 t + \cdots a_n t^n$ and $b_0 + b_1 t + \cdots b_n t^n$ modulo t^{n+1}.

Remark 2.7   In the absolutely ramified case, Greenberg [19] originally defined $\mathscr{R}_n$ as follows. There exists a unique ring morphism $W(k) \to R$ [49, II§5, Thm.4]. It makes R into a finite $W(k)$-algebra that is free as a $W(k)$-module. Its rank is equal to the absolute ramification index e of R (the valuation of p in R). More precisely, R is isomorphic to $W(k)[\pi]/(P(\pi))$, where $P(\pi)$ is an Eisenstein polynomial

$$P(\pi) = \pi^e + a_1 \pi^{e-1} + \ldots + a_e$$

over $W(k)$ for some $e > 1$. The ring R_n has characteristic p^a, with $a = \lceil (n+1)/e \rceil$ the smallest integer bigger than or equal to $(n+1)/e$. Hence, R_n is a finite algebra over $W_a(k)$. As a $W_a(k)$-module, it can be written as an internal direct sum

$$R_n = W_a(k) \oplus W_a(k) \cdot \pi \oplus \ldots \oplus W_a(k) \cdot \pi^r$$

with $r = \min\{e - 1, n\}$. The multiplication in R_n is defined by the rules $P(\pi) = 0$ and $\pi^{n+1} = 0$. Each component $W_a(k) \cdot \pi^i$ is isomorphic to $W_{n_i}(k)$, with n_i the largest integer such that $i + e(n_i - 1) < n + 1$. In particular, $n_0 = a$.

We have $n + 1 = n_1 + \ldots + n_r$. Considering the Witt coordinates on each of these rings $W_{n_i}(k)$, we obtain the affine space

$$\mathbb{A}_k^{n+1} = \mathbb{A}_k^{n_1} \times \cdots \times \mathbb{A}_k^{n_r}$$

endowed with the rules of addition and multiplication defined by the chosen presentation for R_n; this is by definition the ring scheme $\mathscr{R}_n$ (see [19, §1.Prop. 4]). The presentation of R_n depends on the choice of uniformizer π, but Greenberg showed that $\mathscr{R}_n$ is independent of this choice. We adopted the more intrinsic, but less explicit equivalent definition in the appendix of [28].

Fix an integer $n \geq 0$, and let X be a separated R_n-scheme of finite type. In [20], Greenberg proved that the functor

$$(k - \mathrm{alg}) \to (\mathrm{Sets}) : A \mapsto X(\mathscr{R}_n(A))$$

is representable by a separated k-scheme of finite type $Gr_n(X)$. A morphism of R_n-schemes $f : Y \to X$ induces a morphism of Greenberg schemes

$$Gr_n(f) : Gr_n(Y) \to Gr_n(X)$$

in the obvious way, and $Gr_n(\cdot)$ defines a functor from the category of separated R_n-schemes of finite type to the category of separated k-schemes of finite type. If R has equal characteristic, then it follows immediately from the definition of the functor $\mathscr{R}_n$ that $Gr_n(\cdot)$ is nothing but the Weil restriction functor $\prod_{R_n/k}$.

If $\mathfrak{X}$ is an stft formal R-scheme, then we put

$$Gr_n(\mathfrak{X}) = Gr_n(\mathfrak{X} \times_R R_n)$$

for any integer $n \geq 0$. We call this scheme the Greenberg scheme of $\mathfrak{X}$ of level n. For any pair of integers $m \geq n$, the truncation morphism $R_m \to R_n$ induces a morphism of k-schemes in R_m-algebras $\mathscr{R}_m \to \mathscr{R}_n$, and hence a truncation morphism of k-schemes

$$\theta_n^m : Gr_m(\mathfrak{X}) \to Gr_n(\mathfrak{X}).$$

A morphism of stft formal R-schemes $g : \mathfrak{Y} \to \mathfrak{X}$ induces a morphism of R_n-schemes $g \times_R R_n : \mathfrak{Y} \times_R R_n \to \mathfrak{X} \times_R R_n$, and hence a morphism of k-schemes $Gr_n(g) : Gr_n(\mathfrak{Y}) \to Gr_n(\mathfrak{X})$, for any integer $n \geq 0$. The morphisms $Gr_m(g)$ and $Gr_n(g)$ commute with the respective truncation morphisms θ_n^m.

Example 2.8 1. Since $R_0 = k$, $\mathscr{R}_0$ is the identity functor on the category of k-algebras, and we have a canonical isomorphism

$$Gr_0(\mathfrak{X}) \cong \mathfrak{X}_s$$

for any stft formal R-scheme $\mathfrak{X}$. The truncation morphism θ_0^n endows $Gr_n(\mathfrak{X})$ with a structure of $\mathfrak{X}_s$-scheme, for each $n \geq 0$.

2. If $\mathfrak{X} = \operatorname{Spf} R\{x_1, \ldots, x_d\}$, then there exist (non-canonical) isomorphisms

$$Gr_n(\mathfrak{X}) \cong \mathbb{A}_k^{d(n+1)}$$

such that the truncation morphisms θ_m^n are linear projections. This can be seen as follows.

If R has equal characteristic, a choice of uniformizer t in R determines an isomorphism $R \cong k[[t]]$ (recall that we already fixed a k-algebra structure on R in the introduction). For every $n \in \mathbb{N}$ and every k-algebra A, we have canonical bijections

$$Gr_n(\mathfrak{X})(A) = \mathfrak{X}(A[t]/(t^{n+1})) = (A[t]/(t^{n+1}))^d.$$

There exists a unique isomorphism

$$Gr_n(\mathfrak{X}) \to \mathbb{A}_k^{d(n+1)}$$

that maps

$$(a_{1,0} + a_{1,1}t + \ldots + a_{1,n}t^n, \ldots, a_{d,0} + a_{d,1}t + \ldots + a_{d,n}t^n) \in Gr_n(\mathfrak{X})(A)$$

to

$$(a_{1,0}, \ldots, a_{d,0}, \ldots, a_{1,n}, \ldots, a_{d,n}) \in \mathbb{A}_k^{d(n+1)}(A) = A^{d(n+1)}$$

for every k-algebra A. For every $m \leq n$ in $\mathbb{N}$, the truncation morphism θ_m^n corresponds to the projection on the first $d(m+1)$ coordinates in $\mathbb{A}^{d(n+1)}$.

If R has unequal characteristic, we choose a uniformizer π in R, and we consider for every $n \in \mathbb{N}$ the isomorphism of $W(k)$-modules

$$R_n \cong W_m(k) \oplus W_{n_1}(k) \cdot \pi \oplus \ldots \oplus W_{n_r}(k) \cdot \pi^r$$

from Remark 2.7. For every k-algebra A, we have canonical bijections

$$Gr_n(\mathfrak{X})(A) = \mathfrak{X}(\mathscr{R}_n(A)) = (\mathscr{R}_n(A))^d$$
$$= (W_{n_0}(A) \oplus W_{n_1}(A) \oplus \cdots \oplus W_{n_r}(A))^d.$$

Considering the Witt coordinates on the components $W_{n_i}(A)$ as affine coordinates, we obtain a canonical bijection

$$(W_{n_0}(A) \oplus W_{n_1}(A) \oplus \cdots \oplus W_{n_r}(A))^d = A^{d(n+1)} = \mathbb{A}_k^{d(n+1)}(A).$$

This yields an isomorphism of k-schemes

$$Gr_n(\mathfrak{X}) \to \mathbb{A}_k^{d(n+1)}.$$

For every $m \geq n$, the truncation morphism θ_m^n corresponds to the projection onto the first m_i coordinates in every component $W_{n_i}(A)$, where m_i is the largest integer such that $i + e(m_i - 1) < m$.

3. If $\mathfrak{X}$ is affine, then we can write $\mathfrak{X}$ as

$$\mathrm{Spf}\, R\{x_1, \ldots, x_d\}/(f_1, \ldots, f_\ell).$$

The equations $f_j = 0$ impose algebraic conditions on the components $a_{1,0}, \ldots, a_{d,n}$ from point 2. These algebraic conditions define $Gr_n(\mathfrak{X})$ as a closed formal subscheme of

$$Gr_n(\mathrm{Spf}\, R\{x_1, \ldots, x_d\}) \cong \mathbb{A}_k^{d(n+1)}$$

for every $n \in \mathbb{N}$.

Consider, for instance, the case where k is a field of characteristic $p > 0$, R is the ring of Witt vectors $W(k)$, and $\mathfrak{X}$ is the formal R-scheme

$$\mathfrak{X} = \mathrm{Spf}\, R\{x, y\}/(xy - p).$$

We are going to describe the Greenberg scheme $Gr_1(\mathfrak{X})$. We consider two couples (x_0, x_1) and (y_0, y_1) as Witt vectors in $W_2(k)$. By the rules for multiplication of Witt vectors, expressing that

$$(x_0, x_1) \cdot (y_0, y_1) = p = (0, 1)$$

in $W_2(k)$ yields the equations

$$\begin{cases} x_0 \cdot y_0 &= 0 \\ x_0^p y_1 + y_0^p x_1 &= 1. \end{cases}$$

These equations describe $Gr_1(\mathfrak{X})$ as a closed subscheme of

$$\operatorname{Spec} k[x_0, x_1, y_0, y_1].$$

Lemma 2.9 *If $h : \mathfrak{X} \to \mathfrak{Y}$ is an étale morphism of stft formal R-schemes, then for all integers $m \geq n \geq 0$ the square*

$$\begin{array}{ccc} Gr_m(\mathfrak{X}) & \xrightarrow{Gr_m(h)} & Gr_m(\mathfrak{Y}) \\ \theta_n^m \downarrow & & \downarrow \theta_n^m \\ Gr_n(\mathfrak{X}) & \xrightarrow[Gr_n(h)]{} & Gr_n(\mathfrak{Y}) \end{array}$$

is cartesian.

Proof This follows immediately from the infinitesimal lifting criterion for étale morphisms [23, 17.1.1], since for any k-algebra A and any integer $n \geq 0$, the truncation morphism $\mathscr{R}_{n+1}(A) \to \mathscr{R}_n(A)$ is a surjection, and its kernel I satisfies $I^2 = 0$. □

Proposition 2.10 *If $\mathfrak{X}$ is a smooth stft formal R-scheme of pure relative dimension d, then there exists a cover $\{\mathfrak{U}_1, \ldots, \mathfrak{U}_r\}$ of $\mathfrak{X}$ by open formal subschemes such that for all integers $m \geq n \geq 0$,*

$$\theta_n^m : Gr_m(\mathfrak{U}_i) \to Gr_n(\mathfrak{U}_i)$$

is a trivial fibration with fiber $\mathbb{A}_k^{d(m-n)}$ for each $i \in \{1, \ldots, r\}$.

Proof Since $\mathfrak{X}$ is smooth, we can cover $\mathfrak{X}$ by open formal subschemes $\mathfrak{U}$ which admit an étale morphism of formal R-schemes

$$\mathfrak{U} \to \operatorname{Spf} R\{x_1, \ldots, x_d\}$$

Now the result follows from Lemma 2.9 and Example 2.8(2). □

It is not hard to see that $Gr_n(\mathfrak{X})$ is affine if $\mathfrak{X}$ is affine. By Lemma 2.9, it follows that the truncation morphisms θ_n^m are affine morphisms, and that the projective limit

$$Gr(\mathfrak{X}) := \varprojlim_{n \geq 0} Gr_n(\mathfrak{X})$$

is representable by a k-scheme, which we call the Greenberg scheme of $\mathfrak{X}$. It is not Noetherian, in general. It comes with natural truncation morphisms of k-schemes $\theta_n : Gr(\mathfrak{X}) \to Gr_n(\mathfrak{X})$, for $n \geq 0$. A morphism of stft formal R-schemes $g : \mathfrak{Y} \to \mathfrak{X}$ induces a morphism of k-schemes $Gr(g) : Gr(\mathfrak{Y}) \to Gr(\mathfrak{X})$ by passage to the limit, and $Gr(g)$ has the obvious compatibility property with the truncation morphisms.

Let A be a k-algebra. If R has mixed characteristic, then we assume that $A = A^p$. We put

$$\mathscr{R}(A) = \begin{cases} A \widehat{\otimes}_k R & \text{if } R \text{ has equal characteristic} \\ W(A) \widehat{\otimes}_{W(k)} R & \text{else.} \end{cases}$$

It follows from the definition that we have a natural bijection

$$Gr(\mathfrak{X})(A) = \mathfrak{X}(\mathscr{R}(A))$$

for any stft formal R-scheme $\mathfrak{X}$. In particular, let k' be a separable field extension of k', and assume that k' is perfect if R has mixed characteristic. Then $\mathscr{R}(k')$ is a complete discrete valuation ring of ramification index one over R, with residue field k'. So we see that the Greenberg scheme $Gr(\mathfrak{X})$ parameterizes étale sections on $\mathfrak{X}$.

Example 2.11   1. We continue Example 2.8(3). We consider two infinite tuples $x = (x_0, x_1, \ldots)$ and $y = (y_0, y_1, \ldots)$ as Witt vectors in $W(k)$. For every $m \in \mathbb{N}$, there exists a polynomial P_m with integer coefficients in $2(m+1)$ variables such that $P_m(x_0, y_0, \ldots, x_m, y_m)$ is the $(m+1)$-th component of the Witt vector $x \cdot y$. For instance, we have $P_0(x_0, y_0) = x_0 y_0$ and

$$P_1(x_0, y_0, x_1, y_1) = x_0^p y_1 + y_0^p x_1.$$

Looking at the equation

$$x \cdot y = p = (0, 1, 0, \ldots)$$

we see that the Greenberg scheme $Gr(\mathfrak{X})$ is the closed subscheme of

$$\operatorname{Spec} k[x_0, y_0, x_1, y_1, \ldots]$$

defined by the infinite system of equations

$$\begin{cases} P_1(x_0, x_1, y_0, y_1) & = & 1 \\ P_m(x_0, y_0, \ldots, x_m, y_m) & = & 0 \text{ for all } m \in \mathbb{N} \setminus \{1\}. \end{cases}$$

2. Assume that R has equal characteristic. Let X be a separated k-scheme of finite type, and put $\mathfrak{X} = X \times_{\operatorname{Spec} k} \operatorname{Spf} R$, i.e., $\mathfrak{X}$ is the formal $\mathfrak{M}$-adic completion of $X \times_k R$. We denote by $\mathscr{J}_\infty(X)$ and $\mathscr{J}_n(X)$ the *arc scheme*, resp. n-th *jet scheme* of X (see Chapter 6). Then it follows

immediately from the definitions that we have canonical isomorphisms of k-schemes

$$Gr_n(\mathfrak{X}) \cong \mathscr{J}_n(X)$$
$$Gr(\mathfrak{X}) \cong \mathscr{J}_\infty(X)$$

which are compatible with the respective truncation morphisms.

2.3 Motivic integration on smooth formal schemes

In this section, we assume that k is perfect.

Definition 2.12 Let $\mathfrak{X}$ be a smooth stft formal R-scheme, and n an element of $\mathbb{N}$. A cylinder C in $Gr(\mathfrak{X})$ of level n is a subset of the form

$$C = (\theta_n)^{-1}(C_n)$$

with C_n a constructible subset of $Gr_n(\mathfrak{X})$. A cylinder in $Gr(\mathfrak{X})$ is a cylinder of level n for some $n \geq 0$.

If $\mathfrak{X}$ has pure relative dimension d, then the motivic measure of a cylinder C of level n is defined by

$$\mu_{\mathfrak{X}}(C) = \mathbb{L}^{-nd}[C_n] \quad \in \mathscr{M}^R_{\mathfrak{X}_0}.$$

In general, we put

$$\mu_{\mathfrak{X}}(C) = \sum_{\mathfrak{U} \in \pi_0(\mathfrak{X})} \mu_{\mathfrak{U}}(C \cap Gr(\mathfrak{U})) \quad \in \mathscr{M}^R_{\mathfrak{X}_0}$$

where $\pi_0(\mathfrak{X})$ is the set of connected components of $\mathfrak{X}$.

It follows immediately from Proposition 2.10 that this definition does not depend on the choice of n. Recall that $Gr_n(\mathfrak{X})$ carries a natural structure of $\mathfrak{X}_0$-scheme via the truncation morphism

$$\theta_0^n : Gr_n(\mathfrak{X}) \to Gr_0(\mathfrak{X}) = \mathfrak{X}_s = \mathfrak{X}_0.$$

The reason for working with $\mathscr{M}^R_{\mathfrak{X}_0}$ instead of $\mathscr{M}_{\mathfrak{X}_0}$ will be explained in Section 2.4 below; this only makes a difference when R has mixed characteristic.

Example 2.13 Let $\mathfrak{X}$ be a smooth stft formal R-scheme. The Greenberg scheme $Gr(\mathfrak{X})$ is a cylinder, with motivic measure $[\mathfrak{X}_0] \in \mathscr{M}^R_{\mathfrak{X}_0}$.

Remark 2.14 In the literature (e.g. [29, 34, 41, 47]) one often adds a normalizing factor $\mathbb{L}^{-d}$ in the definition of the motivic measure, in analogy with the p-adic case, where the closed unit disc $\mathbb{Z}_p$ conventionally gets Haar measure 1, rather than the cardinality p of the residue field. For geometric applications, it is more natural to omit it, as we do in this paper.

Definition 2.15 Let $\mathfrak{X}$ be a smooth stft formal R-scheme. A function

$$\alpha : Gr(\mathfrak{X}) \to \mathbb{Z} \cup \{\infty\}$$

is integrable if it takes only finitely many values, and if $\alpha^{-1}(i)$ is a cylinder for each $i \in \mathbb{Z}$. We define the motivic integral of α by

$$\int_{Gr(\mathfrak{X})} \mathbb{L}^\alpha = \sum_{i \in \mathbb{Z}} \mu(\alpha^{-1}(i))\mathbb{L}^i \quad \in \mathcal{M}_{\mathfrak{X}_0}^R.$$

For our purposes, the natural integrable functions to consider arise in the following way.

Definition 2.16 Let $\mathfrak{X}$ be a smooth stft formal R-scheme. A gauge form ω on $\mathfrak{X}_\eta$ is the datum of a nowhere vanishing differential form of maximal degree on each connected component of $\mathfrak{X}_\eta$.

Assume that $\mathfrak{X}$ has pure relative dimension d. Let ω be a gauge form on $\mathfrak{X}_\eta$. We have a natural isomorphism of $\mathscr{O}_{\mathfrak{X}}$-modules

$$sp_* \Omega^d_{\mathfrak{X}_\eta} \cong \Omega^d_{\mathfrak{X}/R} \otimes_R K.$$

Since $\mathfrak{X}$ is quasi-compact, the natural map

$$\Omega^d_{\mathfrak{X}/R}(\mathfrak{X}) \otimes_R K \to (\Omega^d_{\mathfrak{X}/R} \otimes_R K)(\mathfrak{X}) = \Omega^d_{\mathfrak{X}_\eta/K}(\mathfrak{X}_\eta)$$

is bijective. We choose a value $a \in \mathbb{N}$ such that $\pi^a \omega$ belongs to the image of the injection

$$i : \Omega^d_{\mathfrak{X}/R}(\mathfrak{X}) \to \Omega^d_{\mathfrak{X}/R}(\mathfrak{X}) \otimes_R K = \Omega^d_{\mathfrak{X}_\eta/K}(\mathfrak{X}_\eta)$$

and we denote the inverse image $i^{-1}(\pi^a \omega)$ by ω'.

Let C be a connected component of $\mathfrak{X}_s$, and denote by ξ its generic point. The local ring $\mathscr{O}_{\mathfrak{X},\xi}$ is a discrete valuation ring with uniformizer π.

Definition 2.17 We define the order of ω along C by

$$\mathrm{ord}_C\omega = \mathrm{length}_{\mathscr{O}_{\mathfrak{X},\xi}} \left((\Omega^d_{\mathfrak{X}/R})_\xi / (\mathscr{O}_{\mathfrak{X},\xi} \cdot \omega') \right) - a.$$

For any extension R' of R of ramification index one and any element ψ of $\mathfrak{X}(R')$, we define the order of ω at ψ by

$$\mathrm{ord}(\omega)(\psi) = \mathrm{length}_{R'} \left(\psi^* \Omega^d_{\mathfrak{X}/R} / (R' \cdot \omega') \right) - a.$$

where we abused notation by writing ω' for the image of ω' in the free rank one R'-module $\psi^* \Omega^d_{\mathfrak{X}/R}$.

It is easily seen that these definitions do not depend on the choice of a or on the choice of uniformizer π. Note that $\mathrm{ord}_C \omega$ and $\mathrm{ord}(\omega)(\psi)$ are finite, since ω is a gauge form.

Let x be any point on $Gr(\mathfrak{X})$ and let F be the perfect closure of the residue field $k(x)$. By definition of the Greenberg scheme, the point x defines an element ψ_x of $\mathfrak{X}(\mathscr{R}(F))$, and $\mathscr{R}(F)$ is an extension of R of ramification index one.

Definition 2.18 Let $\mathfrak{X}$ be a smooth stft formal R-scheme, and let ω be a gauge form on $\mathfrak{X}_\eta$. If $\mathfrak{X}$ has pure relative dimension, we define a function

$$\mathrm{ord}(\omega) : Gr(\mathfrak{X}) \to \mathbb{Z}$$

by putting

$$\mathrm{ord}(\omega)(x) = \mathrm{ord}(\omega)(\psi_x)$$

for each point x of $Gr(\mathfrak{X})$. In the general case, we define $\mathrm{ord}(\omega)$ as the unique function $Gr(\mathfrak{X}) \to \mathbb{Z}$ whose restriction to $Gr(\mathfrak{U})$ is $\mathrm{ord}(\omega|_{\mathfrak{U}_\eta})$, for each connected component $\mathfrak{U}$ of $\mathfrak{X}$.

Proposition 2.19 *Let $\mathfrak{X}$ be a smooth stft formal R-scheme. For any gauge form ω on $\mathfrak{X}_\eta$, the function*

$$\mathrm{ord}(\omega) : Gr(\mathfrak{X}) \to \mathbb{Z}$$

is integrable. In fact, for each connected component $\mathfrak{U}$ of $\mathfrak{X}$, the function $\mathrm{ord}(\omega)$ is constant on $Gr(\mathfrak{U})$, with value $\mathrm{ord}_{\mathfrak{U}_0}(\omega)$.

Proof The statement is clearly local on $\mathfrak{X}$, so we may assume that $\mathfrak{X}$ is affine and connected. Denote by d the relative dimension of $\mathfrak{X}$ over R. We may also assume that $\mathfrak{X}$ admits a relative gauge form ϕ, i.e., an element of $\Omega^d_{\mathfrak{X}/R}(\mathfrak{X})$ that generates the stalk $(\Omega^d_{\mathfrak{X}/R})_x$ at every point x of $\mathfrak{X}$. Multiplying ω with a power of π, we may moreover assume that ω extends to an element ω' of $\Omega^d_{\mathfrak{X}/R}(\mathfrak{X})$. Then we have $\omega' = f \cdot \phi$ for a unique regular function f on $\mathfrak{X}$.

It is clear from the definitions that $\mathrm{ord}_{\mathfrak{X}_0} \omega$ equals the multiplicity of f along $\mathfrak{X}_0$ (i.e., the π-adic valuation of f in $\mathscr{O}_{\mathfrak{X},\xi}$, with ξ the generic point of $\mathfrak{X}_s$) and that for each extension R' of R of ramification index one and each element ψ of $\mathfrak{X}(R')$, the value $\mathrm{ord}(\omega)(\psi)$ equals the π-adic valuation of $\psi^* f \in R'$.

Since ω is a gauge form on $\mathfrak{X}_\eta$, the function f is a unit on $\mathfrak{X}_\eta$. The ring of regular functions on $\mathfrak{X}_\eta$ is precisely $A \otimes_R K$, so f is a unit in $A \otimes_R K$. Since π is a prime in A by smoothness of $\mathfrak{X}$, the element f equals π^b times a unit in A, with $b \in \mathbb{N}$. It follows that

$$\mathrm{ord}_{\mathfrak{X}_0}(\omega) = \mathrm{ord}(\omega)(\psi) = b.$$

$\square$

Definition 2.20 If $\mathfrak{X}$ is a smooth stft formal R-scheme, and ω is a gauge form on $\mathfrak{X}_\eta$, we define the motivic integral of ω on $\mathfrak{X}$ by

$$\int_{\mathfrak{X}} |\omega| := \int_{Gr(\mathfrak{X})} \mathbb{L}^{-\mathrm{ord}(\omega)} \quad \in \mathcal{M}^R_{\mathfrak{X}_0}.$$

Proposition 2.21 *For any smooth* stft *formal R-scheme $\mathfrak{X}$, and any gauge form ω on $\mathfrak{X}_\eta$, we have*

$$\int_{\mathfrak{X}} |\omega| = \sum_{U \in \pi_0(\mathfrak{X}_0)} [U]\mathbb{L}^{-\mathrm{ord}_U(\omega)} \quad \in \mathcal{M}^R_{\mathfrak{X}_0}$$

where we denote by $\pi_0(\mathfrak{X}_0)$ the set of connected components of $\mathfrak{X}_0$.

Proof This follows immediately from Proposition 2.19. $\square$

In [47, 8.0.5], the second author proved a change of variables formula for motivic integrals on formal schemes. For motivic integrals of gauge forms, this formula takes the following particularily elegant form.

Theorem 2.22 *Let $h : \mathfrak{Y} \to \mathfrak{X}$ be a morphism of smooth* stft *formal R-schemes. We assume that $h_\eta : \mathfrak{Y}_\eta \to \mathfrak{X}_\eta$ is an open immersion, and that the map*

$$h_\eta(K') : \mathfrak{Y}_\eta(K') \to \mathfrak{X}_\eta(K')$$

is a bijection for every finite unramified extension K' of K. If ω is a gauge form on $\mathfrak{X}_\eta$, then

$$\int_{\mathfrak{X}} |\omega| = \int_{\mathfrak{Y}} |h_\eta^* \omega| \quad \in \mathcal{M}^R_{\mathfrak{X}_0}$$

where we view the right hand side as an element of $\mathcal{M}^R_{\mathfrak{X}_0}$ via the forgetful morphism $\mathcal{M}^R_{\mathfrak{Y}_0} \to \mathcal{M}^R_{\mathfrak{X}_0}$ (see Sections 3.1 and 3.8 of Chapter 5).

Proof See [29, 4.1.2], [40, 4.19] and Section 2.4 below. $\square$

With slight abuse of notation, we'll often write $\int_{\mathfrak{Y}} |\omega|$ instead of $\int_{\mathfrak{Y}} |h_\eta^* \omega|$, identifying $\mathfrak{Y}_\eta$ with an open rigid subvariety of $\mathfrak{X}_\eta$ via the open immersion h_η, and considering the restriction of ω to $\mathfrak{Y}_\eta$.

Remark 2.23 If R has equal characteristic, the condition that k is perfect is not necessary for the construction of the motivic measure, but it is essential in Theorem 2.22. Put $\mathfrak{X} = \operatorname{Spf} R\{x\}$ and $\omega = dx$, and let y be a closed point of $\mathfrak{X}_s$ whose residue field is inseparable over k. If $\mathfrak{Y} = \mathfrak{X} \setminus \{y\}$ then the open immersion $\mathfrak{Y} \to \mathfrak{X}$ satisfies the conditions of Theorem 2.22, but $\int_{\mathfrak{X}} |\omega| = \mathbb{L}$ while $\int_{\mathfrak{Y}} |\omega| = \mathbb{L} - [y] \neq \mathbb{L}$ in $\mathscr{M}_k^R$.

Remark 2.24 Replacing $\mathscr{M}_{\mathfrak{X}_0}^R$ by its dimensional completion $\widehat{\mathscr{M}}_{\mathfrak{X}_0}^R$, the second author defined in [47] a much larger class of measurable sets and integrable functions. In this way, one can define motivic integrals on singular formal R-schemes, and integrate arbitrary differential forms of maximal degree on smooth rigid varieties. See [29].

2.4 The change of variables formula

The definition of the motivic measure (Definition 2.12) is still valid if we replace $\mathscr{M}_{\mathfrak{X}_0}^R$ by the finer ring $\mathscr{M}_{\mathfrak{X}_0}$, and this is the definition that was originally stated in the literature [47]. However, the proof of the change of variables theorem for motivic integrals on formal schemes in mixed characteristic [47, 8.0.5] is not correct. As we will explain below, in order to correct it, it is necessary to replace $\mathscr{M}_{\mathfrak{X}_0}$ by its quotient $\mathscr{M}_{\mathfrak{X}_0}^R$ (this solution was suggested to us by A. Chambert-Loir; we thank him heartily for his advice). We emphasize that this correction only affects the theory of motivic integration in *mixed* characteristic; in equal characteristic $p \geq 0$, the results in [47] are valid as stated. We also recall that it is not known whether the projection $\mathscr{M}_{\mathfrak{X}_0} \to \mathscr{M}_{\mathfrak{X}_0}^R$ is an isomorphism. Passing to the value ring $\mathscr{M}_{\mathfrak{X}_0}^R$ is harmless for all known applications, because the standard realization morphisms on $\mathscr{M}_{\mathfrak{X}_0}$ all factor through $\mathscr{M}_{\mathfrak{X}_0}^R$ (see Proposition 4.13 in Chapter 5).

The idea behind the change of variables formula. We adopt the notations and hypotheses of Theorem 2.22. A first important observation is the following.

Lemma 2.25 *The morphism*

$$Gr(h) : Gr(\mathfrak{Y}) \to Gr(\mathfrak{X})$$

is bijective.

Proof For every finite unramified extension K' of K, with ring of integers R' and residue field k', we have canonical bijections

$$\mathfrak{X}_\eta(K') = \mathfrak{X}(R') = Gr(\mathfrak{X})(k')$$

and the analogous ones for $\mathfrak{Y}$. Hence, our assumptions on h immediately imply that the map

$$Gr(\mathfrak{Y})(k') \to Gr(\mathfrak{X})(k')$$

is bijective for every finite extension k' of k.

We'll see in Lemma 2.28 below that the image of $Gr(h)$ is a cylinder. It follows that $Gr(h)$ must be surjective. Injectivity follows from the fact that, for *any* complete extension R' of R of ramification index one, with quotient field K', the morphism

$$\mathfrak{Y}_\eta \times_K K' \to \mathfrak{X}_\eta \times_K K'$$

is still an open immersion [8, 3.9.6.1].   □

Lemma 2.25 makes it plausible that there exists a change of variables formula in this setting, since $Gr(h)$ defines a bijection between our integration spaces. The motivic measure, however, is *not* preserved under this morphism, because the morphisms $Gr_n(h)$ are not isomorphisms (they are always surjective, but they are not injective, in general). To compare the motivic measures on $Gr(\mathfrak{Y})$ and $Gr(\mathfrak{X})$, we introduce the *Jacobian* of h.

Let y be a point of $Gr(\mathfrak{Y})$, and let F be the perfect closure of its residue field. By definition of the Greenberg scheme, the point y defines an element ψ_y of $\mathfrak{Y}(\mathscr{R}(F))$, and $\mathscr{R}(F)$ is an extension of R of ramification index one. We denote by d the relative dimension over R of the connected component of $\mathfrak{Y}$ containing ψ_y. From the canonical morphism of line bundles on $\mathfrak{Y}$

$$h^* \Omega^d_{\mathfrak{X}/R} \to \Omega^d_{\mathfrak{Y}/R}$$

we deduce a morphism of free rank one $\mathscr{R}(F)$-modules

$$\psi_y^* h^* \Omega^d_{\mathfrak{X}/R} \to \psi_y^* \Omega^d_{\mathfrak{Y}/R}.$$

We denote the length of its cokernel by

$$\operatorname{ord}(\operatorname{Jac}_h)(y) \quad \in \mathbb{N},$$

and we call this value the order of the Jacobian of h at y. We obtain in this way a map

$$\operatorname{ord}(\operatorname{Jac}_h) : Gr(\mathfrak{Y}) \to \mathbb{N}.$$

We'll see in Lemma 2.28 below that this map measures precisely the "contraction" of the Greenberg schemes $Gr_n(\mathfrak{Y})$ under $Gr_n(\mathfrak{X})$, and, hence, the difference between the motivic measures on $Gr(\mathfrak{Y})$ and $Gr(\mathfrak{X})$.

The relation with gauge forms is the following.

Proposition 2.26 (Chain rule) *Let $h : \mathfrak{Y} \to \mathfrak{X}$ be a morphism of smooth stft formal R-schemes of pure relative dimension d. Let ω be a gauge form on $\mathfrak{X}_\eta$, and y a point on $Gr(\mathfrak{Y})$. If we denote by x the image $Gr(h)(y)$ of y in $Gr(\mathfrak{X})$, then*

$$\mathrm{ord}(h_\eta^*\omega)(y) = \mathrm{ord}(\omega)(x) + \mathrm{ord}(\mathrm{Jac}_h)(y).$$

Proof This follows immediately from the definitions. □

In particular, combining this result with Proposition 2.19, we see that the function $\mathrm{ord}(\mathrm{Jac}_h)$ is constant on $Gr(\mathfrak{U})$, for every connected component $\mathfrak{U}$ of $\mathfrak{Y}$.

Proof of the change of variables formula. We consider, more generally, a morphism of formal R-schemes $h : \mathfrak{Y} \to \mathfrak{X}$, where $\mathfrak{X}$ and $\mathfrak{Y}$ are flat stft formal R-schemes of pure relative dimension d, and where $\mathfrak{Y}$ is smooth over R.

The proof of the change of variables theorem [47, 8.0.5] is based on the following lemma [47, 7.1.3]. For the definition of $Gr^{(e')}(\mathfrak{X})$ and $c_\mathfrak{X} = 0$, we refer to [47]. For our purposes, we only need the case where $\mathfrak{X}$ is smooth over R. Then $Gr^{(e')}(\mathfrak{X}) = Gr(\mathfrak{X})$ and $c_\mathfrak{X} = 0$.

Unfortunately, if R has mixed characteristic, the proof of the lemma in [47] is not correct, and the statement might be false. Before explaining the problem and presenting a corrected version, we state the original formulation from [47].

Lemma 2.27 ([47], Lemme 7.1.3 - possibly erroneous form) *Let $B \subset Gr(\mathfrak{Y})$ be a cylinder of level m, for some $m \in \mathbb{N}$, and put $A = h(B)$. Assume that e and e' are elements of $\mathbb{N}$ such that $\mathrm{ord}(\mathrm{Jac}_h)(\varphi) = e$ for all $\varphi \in B$, and $A \subset Gr^{(e')}(\mathfrak{X})$. Then A is a cylinder.*

Moreover, if the restriction of h to B is injective, then, for all integers $n \geq \max(2e + c_\mathfrak{X}, m + e)$, the following properties hold.

1. *If φ and φ' belong to B and $\theta_n(h(\varphi)) = \theta_n(h(\varphi'))$, then $\theta_{n-e}(\varphi) = \theta_{n-e}(\varphi')$,*
2. *We have $[\theta_n(B)] = [\theta_n(A)]\mathbb{L}^e$ in $\mathcal{M}_k$.*

If R has mixed characteristic, the proof of point (2) in [47, 7.1.3] only computes the fibers

$$Gr_n(Y) \times_{Gr_n(\mathfrak{X})} \mathrm{Spec}\, F$$

when F is a *perfect* field containing k and $\operatorname{Spec} F \to Gr_n(\mathfrak{X})$ a morphism of k-schemes. Indeed, only when F is perfect, we can identify the set $Gr(X)(F)$ with the set $X(R_F)$, where

$$R_F = R\widehat{\otimes}_{W(k)}W(F)$$

is a complete discrete valuation ring of ramification index one over R. Then the proof of [47, 7.1.3] shows that there exists an isomorphism of F-schemes

$$Gr_n(Y) \times_{Gr_n(X)} \operatorname{Spec} F \cong \mathbb{A}_F^e$$

so that we can deduce from Proposition 3.14 in Chapter 5 that

$$[\theta_n(B)] = [\theta_n(A)]\mathbb{L}^e$$

in $\mathscr{M}_{\mathfrak{X}_0}^R$ (and hence also in $\mathscr{M}_k^R$). However, we cannot conclude that it holds in $\mathscr{M}_k$.

Therefore, we have to replace Lemma 2.27 by the following statement. We express it in a slightly stronger form, replacing the equality in the Grothendieck ring over k by an equality in the Grothendieck ring over $\mathfrak{X}_0$.

Lemma 2.28 ([47], Lemme 7.1.3 - Corrected form) *Let $B \subset Gr(\mathfrak{Y})$ be a cylinder of level m, for some $m \in \mathbb{N}$, and put $A = h(B)$. Assume that e and e' are elements of $\mathbb{N}$ such that $\operatorname{ord}(\operatorname{Jac}_h)(\varphi) = e$ for all $\varphi \in B$, and $A \subset Gr^{(e')}(\mathfrak{X})$. Then A is a cylinder.*

Moreover, if the restriction of h to B is injective, then, for all integers $n \geq \max(2e + c_{\mathfrak{X}}, m + e)$, the following properties hold.

1. *If φ and φ' belong to B and $\theta_n(h(\varphi)) = \theta_n(h(\varphi'))$, then $\theta_{n-e}(\varphi) = \theta_{n-e}(\varphi')$,*
2. *We have $[\theta_n(B)] = [\theta_n(A)]\mathbb{L}^e$ in $\mathscr{M}_{\mathfrak{X}_0}^R$.*

The statements in [47, 8.0.3] and [47, 8.0.5] (change of variables theorem) have to be modified accordingly, replacing the completed Grothendieck ring $\widehat{\mathscr{M}_k}$ by the completed modified Grothendieck ring $\widehat{\mathscr{M}_k^{mod}}$.

Example 2.29 Assume that k has characteristic $p > 0$ and that $R = W(k)$. Consider the morphism of smooth stft formal R-schemes

$$h : \mathfrak{Y} = \operatorname{Spf} R\{y\} \to \mathfrak{X} = \operatorname{Spf} R\{x\}$$

defined by $x \mapsto py$. Then the function $\mathrm{ord}(\mathrm{Jac}_h)(\varphi)$ is constant on $Gr(\mathfrak{Y})$ with value $e = 1$. We consider the cylinder $B = Gr(\mathfrak{Y})$ of level $m = 0$. Then the morphism

$$Gr_2(h) : Gr_2(\mathfrak{Y}) = \mathrm{Spec}\, k[y_0, y_1, y_2] \to Gr_2(\mathfrak{X}) = \mathrm{Spec}\, k[x_0, x_1, x_2]$$

can be described as follows: we interpret the coordinates y_j and x_j as coordinates of Witt vectors in $W_2(k)$, and we express that

$$(x_0, x_1, x_2) = p \cdot (y_0, y_1, y_2).$$

By the definition of multiplication of Witt vectors, we have

$$p \cdot (y_0, y_1, y_2) = (0, (y_0)^p, (y_1)^p)$$

so that the morphism $Gr_2(h)$ is given by

$$(x_0, x_1, x_2) \mapsto (0, (y_0)^p, (y_1)^p).$$

We see that the image of $Gr(h)$ is the level zero cylinder

$$A = (\theta_0)^{-1}(O) \subset Gr(\mathfrak{X})$$

where O denotes the origin of the special fiber $\mathfrak{X}_s = \mathbb{A}_k^1$. However, the morphism

$$\theta_2(B) \to \theta_2(A)$$

induced by h is *not* a piecewise trivial fibration, since it is not separable. Note that, in this example, we nevertheless still have

$$[\theta_2(B)] = [\theta_2(A)]\mathbb{L}$$

in $\mathcal{M}_{\mathfrak{X}_0}$.

With Lemma 2.28 at our disposal, it is not difficult to prove the change of variables formula in Theorem 2.22.

Proof of the change of variables formula in Theorem 2.22 We may assume that $\mathfrak{X}$ is connected, since we can write the motivic integral of ω on $\mathfrak{X}$ as a sum of integrals on the connected components. We denote by $\mathfrak{Y}_1, \ldots, \mathfrak{Y}_r$ the connected components of $\mathfrak{Y}$. For each i, we put $A_i = Gr(h)(Gr(\mathfrak{Y}_i))$ and we denote by e_i the value of $\mathrm{ord}(\mathrm{Jac}_h)$ on $Gr(\mathfrak{Y}_i)$.

By the assumptions on h, $\mathfrak{Y}$ has pure relative dimension. Lemma 2.25 implies that

$$Gr(\mathfrak{X}) = \bigsqcup_{i=1}^{r} A_i.$$

By Lemma 2.28, A_i is a cylinder in $Gr(\mathfrak{X})$ with motivic measure

$$\mu_{\mathfrak{X}}(A_i) = [(\mathfrak{Y}_i)_s]\mathbb{L}^{-e_i} \quad \in \mathcal{M}_{\mathfrak{X}_0}^R.$$

By Proposition 2.26, we find

$$\int_{\mathfrak{X}} |\omega| = [\mathfrak{X}_s]\mathbb{L}^{-\mathrm{ord}_{\mathfrak{X}_s}\omega}$$

$$= \sum_{i=1}^{r} [(\mathfrak{Y}_i)_s]\mathbb{L}^{-e_i - \mathrm{ord}_{\mathfrak{X}_s}\omega}$$

$$= \sum_{i=1}^{r} [(\mathfrak{Y}_i)_s]\mathbb{L}^{-\mathrm{ord}_{(\mathfrak{Y}_i)_s}h_\eta^*\omega}$$

$$= \int_{\mathfrak{Y}} |h_\eta^*\omega|$$

in $\mathcal{M}_{\mathfrak{X}_0}^R$.   □

Further corrections to the literature. The modification of the change of variables theorem in [47, 8.0.5] also affects the theory of motivic integration on rigid K-varieties developed in [29] (*only* in the mixed characteristic case; in equal characteristic $p \geq 0$, all the results in [29] are valid as stated). In the appropriate places, $\mathcal{M}_k$ and its dimensional completion $\widehat{\mathcal{M}_k}$ should be replaced by $\mathcal{M}_k^R$ and $\widehat{\mathcal{M}_k^R}$ (this applies in particular to Theorem-Definition 4.1.2 and Theorems 4.5.1 and 4.5.3 in [29]). For the theory of motivic integration of gauge forms on smooth rigid varieties, and of the motivic Serre invariant, the correct statements can be found in the following sections.

The theory of motivic integration on formal schemes and rigid varieties in mixed characteristic has been applied in several other articles. All of them can easily be corrected, replacing the Grothendieck ring of varieties by the modified Grothendieck ring of varieties. As we said before, this is harmless for the applications of the theory, since all the realization morphisms that are used factor through the modified Grothendieck ring (see Proposition 4.13 of Chapter 5). Let us indicate some of the changes that should be made (these changes *only* affect the mixed characteristic case; in equal characteristic $p \geq 0$, all the results are valid as stated).

In [40, 4.18], the last two lines of the statement should be replaced by: "... $[\pi_n(B)] = [\pi_n(A)]\mathbb{L}_{X_s}^e$ in $K_0^R(\mathrm{Var}_{X_s})$." In [40, 4.19], the ring $\widehat{\mathcal{M}_{X_s}}$ should be replaced by $\widehat{\mathcal{M}_{X_s}^R}$, and in [40, 4.20(2)], the ring $\mathcal{M}_{X_s}$ should be replaced by $\mathcal{M}_{X_s}^R$. Likewise, in [40, §6], the Grothendieck rings have to be replaced by their modified analogues. In particular, the motivic Serre invariant of a generically smooth stft formal R-scheme X_∞ of pure relative dimension [40, 6.2] is well-defined in $K_0^R(\mathrm{Var}_{X_s})/(\mathbb{L} - [X_s])$.

In [41, § 3.2], the various Grothendieck rings should be replaced by the modified Grothendieck rings. The trace formula in [41, 5.4] remains valid, because the ℓ-adic Euler characteristic factors through the modified Grothendieck ring.

In [42, § 5], the various Grothendieck rings should be replaced by the modified Grothendieck rings.

In Sections 4 and 5.3 of [34], the various Grothendieck rings should be replaced by the modified Grothendieck rings. The trace formula in [34, 6.4] remains valid, because the ℓ-adic Euler characteristic factors through the modified Grothendieck ring.

In [37, § 5], in particular in Theorem 5.4 and Definition 5.5, the motivic Serre invariant should take its values in $K_0^R(\mathrm{Var}_k)/(\mathbb{L} - 1)$. The results in Sections 6 and 7 of [37] remain valid, because the Poincaré polynomial and the ℓ-adic Euler characteristic factor through the modified Grothendieck ring.

2.5 Néron smoothening and dilatation

There is a convenient tool which allows to reduce certain constructions for special formal R-schemes to the case of stft formal R-schemes: the notion of dilatation [34, 2.20].

Definition 2.30 Let $\mathfrak{X}$ be a flat special formal R-scheme. Denote by $\mathscr{J}$ its largest ideal of definition, and by $\mathfrak{X}'' \to \mathfrak{X}$ the formal blow-up of $\mathfrak{X}$ at $\mathfrak{X}_0$ [34, 2.15]. The dilatation $d : \mathfrak{X}' \to \mathfrak{X}$ of $\mathfrak{X}$ is the largest open formal $\mathfrak{X}$-scheme of $\mathfrak{X}''$ where $\mathscr{J}\mathcal{O}_{\mathfrak{X}''}$ is generated by π.

If $\mathfrak{X}$ is any special formal R-scheme, we denote by $\mathfrak{X}^{flat}$ the maximal R-flat closed formal subscheme of $\mathfrak{X}$ (obtained by killing π-torsion) and we define the dilatation $\mathfrak{X}' \to \mathfrak{X}$ of $\mathfrak{X}$ as the composition of the closed immersion $\mathfrak{X}^{flat} \to \mathfrak{X}$ and the dilatation $\mathfrak{X}' \to \mathfrak{X}^{flat}$ of $\mathfrak{X}^{flat}$.

Example 2.31 If $\mathfrak{X} = \mathrm{Spf}\, R[[x]]$, then the dilatation of $\mathfrak{X}$ is the morphism

$$\mathfrak{X}' = \mathrm{Spf}\, R\{y\} \to \mathfrak{X}$$

defined by $x \mapsto \pi y$.

The dilatation of $\mathfrak{X}$ was called the dilatation of $\mathfrak{X}$ with center $\mathfrak{X}_0$ in [34]. By [34, 2.22], the dilatation is characterized by the following universal property.

Proposition 2.32 *Let $\mathfrak{X}$ be a flat special formal R-scheme, and denote by $d : \mathfrak{X}' \to \mathfrak{X}$ the dilatation of $\mathfrak{X}$. Then $\mathfrak{X}'$ is a flat stft formal R-scheme, and the morphism $d_s : \mathfrak{X}'_s \to \mathfrak{X}_s$ factors through the closed subscheme $\mathfrak{X}_0$ of $\mathfrak{X}_s$.*

Moreover, for any morphism of formal R-schemes $h : \mathfrak{V} \to \mathfrak{X}$ such that $\mathfrak{V}$ is a flat stft formal R-scheme and h_s factors through $\mathfrak{X}_0$, there exists a unique morphism of formal R-schemes $g : \mathfrak{V} \to \mathfrak{X}'$ such that $h = d \circ g$.

Definition 2.33 A Néron smoothening of a special formal R-scheme $\mathfrak{X}$ is a morphism of formal R-schemes $h : \mathfrak{Y} \to \mathfrak{X}$ such that

1. $\mathfrak{Y}$ is a smooth stft formal R-scheme
2. $h_\eta : \mathfrak{Y}_\eta \to \mathfrak{X}_\eta$ is an open immersion of rigid varieties
3. for each finite unramified extension K' of K, the map

$$h_\eta(K') : \mathfrak{Y}_\eta(K') \to \mathfrak{X}_\eta(K')$$

is a bijection.

Proposition 2.34 *Let $\mathfrak{X}$ be a special formal R-scheme, and denote by $d : \mathfrak{X}' \to \mathfrak{X}$ the dilatation of $\mathfrak{X}$. The morphism $d_\eta : \mathfrak{X}'_\eta \to \mathfrak{X}_\eta$ is an open immersion. For each finite unramified extension K' of K, the map*

$$d_\eta(K') : \mathfrak{X}'_\eta(K') \to \mathfrak{X}_\eta(K')$$

is a bijection.

If $\mathfrak{Y}$ is an stft formal R-scheme such that $\mathfrak{Y}_s$ is reduced, and $h : \mathfrak{Y} \to \mathfrak{X}$ is a morphism of formal R-schemes, then there exists a unique morphism $g : \mathfrak{Y} \to \mathfrak{X}'$ such that $h = d \circ g$. Moreover, g is a Néron smoothening iff h is a Néron smoothening.

Proof Since d is the composition of an open immersion and a formal blow-up of an ideal that contains π, the morphism d_η is an open immersion [34, 2.19]. We denote by R' the valuation ring of K', and by k' its residue field. If ψ is an element of $\mathfrak{X}(R')$, then $\psi_s : \operatorname{Spec} k' \to \mathfrak{X}_s$ factors though $\mathfrak{X}_0$ (note that $(\operatorname{Spf} R')_s = \operatorname{Spec} k'$ because R' is unramified over R). Hence, ψ lifts uniquely to an element of $\mathfrak{X}'(R')$ by Proposition 2.32. It follows that the map

$$d_\eta(K') : \mathfrak{X}'_\eta(K') \to \mathfrak{X}_\eta(K')$$

is a bijection.

The existence and uniqueness of the morphism g also follow from Proposition 2.32, because $\mathfrak{Y}_s = \mathfrak{Y}_0$. The property that g is a Néron smoothening iff h is a Néron smoothening follows immediately from the first part of the proof. $\qquad\square$

Theorem 2.35 *Any generically smooth special formal R-scheme $\mathfrak{X}$ admits a Néron smoothening $\mathfrak{Y} \to \mathfrak{X}$. If $\mathfrak{X}$ is stft, then there exists an admissible blow-up $\mathfrak{Z} \to \mathfrak{X}$ such that $Sm(\mathfrak{Z}) \to \mathfrak{X}$ is a Néron smoothening, where we denote by $Sm(\mathfrak{Z})$ the R-smooth locus of $\mathfrak{Z}$.*

Proof Taking the dilatation $d : \mathfrak{X}' \to \mathfrak{X}$, we may assume that $\mathfrak{X}$ is stft, by Proposition 2.34. This case was proved by Bosch and Schlöter in [12, 3.1]. $\qquad\square$

Proposition 2.36 *Let R' be a complete extension of R of ramification index one, and let $\mathfrak{X}$ be a special formal R-scheme. If $h : \mathfrak{Y} \to \mathfrak{X}$ is a Néron smoothening, then the morphism $h' : \mathfrak{Y} \times_R R' \to \mathfrak{X} \times_R R'$ obtained from h by base change is again a Néron smoothening.*

Proof It follows from the universal property in Proposition 2.32 that dilatation commutes with any base change of ramification index one (here we use that $A \times_k k'$ is reduced if A is a reduced k-algebra and k'/k a separable field extension). Hence, we may assume that $\mathfrak{X}$ is stft, by Proposition 2.34. Let L be a finite unramified extension of the quotient field K' of R', and denote by S the normalization of R' in L. Since $\mathfrak{Y} \times_R R'$ is smooth and stft, and h'_η is an open immersion, it suffices to show that the map

$$h'_\eta(L) : (\mathfrak{Y} \times_R R')_\eta(L) \to (\mathfrak{X} \times_R R')_\eta(L)$$

is surjective. Let ψ be an element of the set

$$(\mathfrak{X} \times_R R')_\eta(L) = \mathfrak{X}(S).$$

It is enough to prove that ψ lifts to an element of $\mathfrak{Y}(S)$.

Since h_η is an open immersion, there exists a commutative diagram

$$
\begin{array}{ccc}
\mathfrak{Y}' & \xrightarrow{\ g\ } & \mathfrak{X}' \\
{\scriptstyle f_1}\downarrow & & \downarrow{\scriptstyle f_2} \\
\mathfrak{Y} & \xrightarrow[\ h\ ]{} & \mathfrak{X}
\end{array}
$$

such that $(f_i)_\eta$ is an isomorphism for $i = 1, 2$, and such that g is an open immersion [10, 5.4(b)]. Moreover, taking an admissible blow-up of $\mathfrak{X}'$, we may assume that $Sm(\mathfrak{X}') \to \mathfrak{X}$ is a Néron smoothening, by

Theorem 2.35. The result in [11, 3.6.3] immediately carries over to formal schemes, so that, blowing up closed subschemes of $\mathfrak{X}'_s$ that are disjoint from the smooth locus, we also may assume that the unique lifting ψ' of ψ to $\mathfrak{X}'(S)$ factors through $Sm(\mathfrak{X}')$. It remains to show that ψ' belongs to $\mathfrak{Y}'(S)$.

Denote by ξ the image of ψ_s in $\mathfrak{X}'_s$, and by Z the Zariski closure of ξ in $Sm(\mathfrak{X}')_s$. Since the residue field of S is a separable extension of k, we know that Z is generically smooth over k, so that there exists a closed point x of Z whose residue field $k(x)$ is separable over k. Denote by S' a finite unramified extension of R with residue field $k(x)$. By the infinitesimal lifting criterion for formal smoothness, the point x extends to a section φ in $Sm(\mathfrak{X}')(S')$, and since h is a Néron smoothening, φ belongs to $\mathfrak{Y}'(S')$. Since x belongs to the Zariski-closure of ξ, this implies that ξ belongs to $\mathfrak{Y}'$, so that ψ' belongs to $\mathfrak{Y}'(S')$. $\qquad\square$

Proposition 2.37 *Let $\mathfrak{X}$ be a regular stft formal R-scheme. If we denote by $Sm(\mathfrak{X})$ the R-smooth locus of $\mathfrak{X}$, then the open immersion*

$$Sm(\mathfrak{X}) \to \mathfrak{X}$$

is a Néron smoothening.

Proof Let R' be a finite unramified extension of R, and let ψ be a section of $\mathfrak{X}(R')$. It suffices to show that ψ factors trough $Sm(\mathfrak{X})$. Since $Sm(\mathfrak{X}) \times_R R' = Sm(\mathfrak{X} \times_R R')$ by faithfully flat descent, we may assume that $R = R'$. Denote by x the image of ψ_s in $\mathfrak{X}_s$. By [11, 3.1.2], the completed local ring $\widehat{\mathscr{O}}_{\mathfrak{X},x}$ is isomorphic as a R-algebra to $R[[x_1, \ldots, x_n]]$. Hence, $\mathfrak{X}$ is smooth at x. $\qquad\square$

Definition 2.38 Let X be a rigid K-variety. A weak Néron model for X is a smooth stft formal R-scheme $\mathfrak{U}$, endowed with an open immersion $i : \mathfrak{U}_\eta \to X$ such that the map $i(K') : \mathfrak{U}_\eta(K') \to X(K')$ is a bijection for every finite unramified extension K' of K.

A weak Néron model of X can be seen as a smooth stft model for the unramified points of X.

Lemma 2.39 *A morphism of special formal R-schemes $f : \mathfrak{Y} \to \mathfrak{X}$ is a Néron smoothening iff $(\mathfrak{Y}, f_\eta)$ is a weak Néron model of $\mathfrak{X}_\eta$.*

Proof Immediate from the definitions. $\qquad\square$

Proposition 2.40 *If $\mathfrak{X}$ is a generically smooth special formal R-scheme, then $\mathfrak{X}_\eta$ admits a weak Néron model.*

Proof Immediate from Theorem 2.35 and Lemma 2.39. $\qquad\square$

Proposition 2.41 *A rigid K-variety X admits a weak Néron model iff there exists a separated smooth quasi-compact open rigid subvariety U of X such that $U(K') = X(K')$ for every finite unramified extension K' of K.*

Proof The condition is obviously necessary, so let us prove that it is also sufficient. If we take any stft formal R-model $\mathfrak{U}$ of U, then $\mathfrak{U}$ admits a Néron smoothening $h : \mathfrak{V} \to \mathfrak{U}$ by Theorem 2.35. The smooth stft formal R-scheme $\mathfrak{V}$, endowed with the open immersion

$$h_\eta : \mathfrak{V}_\eta \to \mathfrak{U}_\eta \cong U \subset X,$$

is a weak Néron model for X. $\qquad\square$

The following proposition is crucial for the definition of the motivic integrals and the motivic Serre invariant in Sections 2.6 and 3.1.

Proposition 2.42
1. *Let $\mathfrak{X}$ be a special formal R-scheme, and let $f_i : \mathfrak{X}_i \to \mathfrak{X}$ be a Néron smoothening, for $i = 1, 2$. Then there exists a commutative diagram*

$$
\begin{array}{ccc}
\mathfrak{X}_3 & \xrightarrow{\;g_2\;} & \mathfrak{X}_2 \\
{\scriptstyle g_1}\big\downarrow & & \big\downarrow{\scriptstyle f_2} \\
\mathfrak{X}_1 & \xrightarrow[\;f_1\;]{} & \mathfrak{X}
\end{array}
$$

such that g_1, g_2 and $f_1 \circ g_1 = f_2 \circ g_2$ are Néron smoothenings.
2. *Let X be a separated rigid K-variety, and let $(\mathfrak{U}, i)$ and $(\mathfrak{V}, j)$ be weak Néron models for X. There exist a weak Néron model $(\mathfrak{W}, \kappa)$ of X and Néron smoothenings $u : \mathfrak{W} \to \mathfrak{U}$ and $v : \mathfrak{W} \to \mathfrak{V}$ such that $\kappa = i \circ u_\eta = j \circ v_\eta$.*

Proof First, we show that (2) implies (1). Putting $X = \mathfrak{X}_\eta$, $(\mathfrak{U}, i) = (\mathfrak{X}_1, (f_1)_\eta)$ and $\mathfrak{V} = (\mathfrak{X}_2, (f_2)_\eta)$, it follows from Lemma 2.39 that the triple $(\mathfrak{W}, u, v)$ provided by (2) satisfies the conditions on $(\mathfrak{X}_3, g_1, g_2)$ (note that $f_1 \circ u = f_2 \circ v$ since this holds on the generic fibers and $\mathfrak{W}$ is R-flat).

So let us prove (2). Denote by Y the intersection of $i(\mathfrak{U}_\eta)$ and $j(\mathfrak{V}_\eta)$ in X. Since X is separated, Y is separated and quasi-compact, so that Y

admits an stft formal R-model $\mathfrak{Y}$. By [9, 4.1] there exists an admissible blow-up $\mathfrak{Y}' \to \mathfrak{Y}$ such that the open immersions $\mathfrak{Y}'_\eta \cong Y \to \mathfrak{U}_\eta$ and $\mathfrak{Y}'_\eta \cong Y \to \mathfrak{V}_\eta$ extend to morphisms of formal R-schemes $\mathfrak{Y}' \to \mathfrak{U}$, resp. $\mathfrak{Y}' \to \mathfrak{V}$. If we take a Néron smoothening $\mathfrak{W} \to \mathfrak{Y}'$, then the composed morphisms $f : \mathfrak{W} \to \mathfrak{U}$ and $g : \mathfrak{W} \to \mathfrak{V}$ are Néron smoothenings. Putting $\kappa := i \circ f_\eta = j \circ g_\eta$, it follows that $(\mathfrak{W}, \kappa)$ is a weak Néron model of X. $\square$

2.6 Motivic integration on special formal schemes and rigid varieties

In this section, we assume that k is perfect.

Theorem-Definition 2.43 *Let $\mathfrak{X}$ be a generically smooth special formal R-scheme. Let $h : \mathfrak{Y} \to \mathfrak{X}$ be a Néron smoothening, and suppose that $\mathfrak{X}_\eta$ admits a gauge form ω. Let X be a separated smooth rigid K-variety, and assume that X admits a weak Néron model $\mathfrak{U}$. Let ϕ be a gauge form on X.*

1. The motivic integral

$$\int_{\mathfrak{X}} |\omega| := \int_{\mathfrak{Y}} |h_\eta^* \omega| = \sum_{C \in \pi_0(\mathfrak{Y}_0)} [C] \mathbb{L}^{-\mathrm{ord}_C(h_\eta^* \omega)} \quad \in \mathscr{M}_{\mathfrak{X}_0}^R$$

only depends on $(\mathfrak{X}, \omega)$, and not on the Néron smoothening.

2. Its image under the forgetful morphism $\mathscr{M}_{\mathfrak{X}_0}^R \to \mathscr{M}_k^R$ only depends on $(\mathfrak{X}_\eta, \omega)$ and is denoted by

$$\int_{\mathfrak{X}_\eta} |\omega|.$$

3. Let R' be a complete extension of R of ramification index one, with quotient field K' and perfect residue field k'. If we put $\mathfrak{X}' = \mathfrak{X} \times_R R'$ and we denote by ω' the pull-back of ω to $\mathfrak{X}'_\eta$, then $\int_{\mathfrak{X}'} |\omega'|$ is the image of $\int_{\mathfrak{X}} |\omega|$ under the base change morphism $\mathscr{M}_{\mathfrak{X}_0}^R \to \mathscr{M}_{\mathfrak{X}'_0}^R$. Likewise, $\int_{\mathfrak{X}'_\eta} |\omega'|$ is the image of $\int_{\mathfrak{X}_\eta} |\omega|$ under the base change morphism $\mathscr{M}_k^R \to \mathscr{M}_{k'}^R$.

4. If $\{\mathfrak{X}_i \,|\, i \in I\}$ is a finite cover of $\mathfrak{X}$ by open formal subschemes, then

$$\int_{\mathfrak{X}} |\omega| = \sum_{\emptyset \neq J \subset I} (-1)^{|J|-1} \int_{\cap_{j \in J} \mathfrak{X}_j} |\omega_J|$$

in $\mathscr{M}_{\mathfrak{X}_0}^R$, where ω_J denotes the restriction of ω to $\cap_{j \in J}(\mathfrak{X}_j)_\eta$.

5. We define the motivic integral

$$\int_X |\omega| \quad \in \mathscr{M}_k^R$$

as the image of

$$\int_{\mathfrak{U}} |\omega|$$

under the forgetful morphism $\mathscr{M}_{\mathfrak{U}_0}^R \to \mathscr{M}_k^R$. It only depends on (X, ω), and not on $\mathfrak{U}$. If X is the generic fiber of a special formal R-scheme, then this definition coincides with the one in (2).

Proof (1) By Proposition 2.42(1), we may assume that $\mathfrak{X}$ is smooth and stft. This case follows immediately from the change of variables formula in Theorem 2.22.

(2) We'll show by a similar argument that

$$\int_{\mathfrak{X}_\eta} |\omega| \quad \in \mathscr{M}_k^R$$

does not depend on the formal model $\mathfrak{X}$. Let $\mathfrak{Y}$ be another formal R-model for $\mathfrak{X}_\eta$, and let $f : \mathfrak{X}' \to \mathfrak{X}$ and $g : \mathfrak{Y}' \to \mathfrak{Y}$ be Néron smoothenings. Then $(\mathfrak{X}', f_\eta)$ and $(\mathfrak{Y}', g_\eta)$ are weak Néron models of $\mathfrak{X}_\eta = \mathfrak{Y}_\eta$, by Lemma 2.39. Now it follows from Proposition 2.42(2) and the change of variables formula in Theorem 2.22 that

$$\int_{\mathfrak{Y}} |\omega| := \int_{\mathfrak{Y}'} |g_\eta^* \omega| = \int_{\mathfrak{X}'} |f_\eta^* \omega| =: \int_{\mathfrak{X}} |\omega|$$

in $\mathscr{M}_k^R$.

(3) It suffices to prove the first assertion. If we put $\mathfrak{Y}' = \mathfrak{Y} \times_R R'$, then the morphism $h' : \mathfrak{Y}' \to \mathfrak{X}'$ obtained from h by base change is a Néron smoothening, by Proposition 2.36. Hence, we may assume that $\mathfrak{X}$ is smooth and stft. It is clear from the definition that for any connected component C of $\mathfrak{X}_s$ and any connected component C' of $C \times_k k'$ we have

$$\mathrm{ord}_C(\omega) = \mathrm{ord}_{C'}(\omega')$$

so the result follows from Proposition 2.21.

(4) Since $h : h^{-1}(\cap_{j \in J} \mathfrak{X}_j) \to \cap_{j \in J} \mathfrak{X}_j$ is a Néron smoothening for each non-empty subset J of I, we may assume that $\mathfrak{X}$ is smooth and stft. Now the result follows easily from the definition of the motivic integral.

(5) Immediate from Proposition 2.42, Theorem 2.22 and Lemma 2.39. $\square$

Definition 2.44 Let $\mathfrak{X}$ be a generically smooth special formal R-scheme, and let ω be a gauge form on $\mathfrak{X}_\eta$. For each $d \in \mathbb{N}'$, we denote by $\omega(d)$ the pull-back of ω to $\mathfrak{X}_\eta \times_K K(d)$. We define the motivic generating series associated to $(\mathfrak{X}, \omega)$ by

$$S(\mathfrak{X}, \omega; T) = \sum_{d \in \mathbb{N}'} \left(\int_{\mathfrak{X} \times_R R(d)} |\omega(d)| \right) T^d \quad \in \mathcal{M}^R_{\mathfrak{X}_0}[[T]].$$

Likewise, we define the motivic generating series associated to $(\mathfrak{X}_\eta, \omega)$ by

$$S(\mathfrak{X}_\eta, \omega; T) = \sum_{d \in \mathbb{N}'} \left(\int_{\mathfrak{X}_\eta \times_K K(d)} |\omega(d)| \right) T^d \quad \in \mathcal{M}^R_k[[T]].$$

It is the image of $S(\mathfrak{X}, \omega; T)$ under the forgetful morphism $\mathcal{M}^R_{\mathfrak{X}_0}[[T]] \to \mathcal{M}^R_k[[T]]$.

Note that $S(\mathfrak{X}, \omega; T)$ depends on the choice of π, since the extensions $K(d)$ of K depend on this choice. However, if k is algebraically closed, then for each $d \in \mathbb{N}'$, $K(d)$ is the unique extension of degree d of K, up to K-isomorphism. Hence, $S(\mathfrak{X}, \omega; T)$ is independent of π. In general, Theorem-Definition 2.43 immediately implies the following property.

Proposition 2.45 *Let $\mathfrak{X}$ be a generically smooth special formal R-scheme, and let ω be a gauge form on $\mathfrak{X}$. Let R' be a complete extension of R of ramification index one, with perfect residue field. Put $\mathfrak{X}' = \mathfrak{X} \times_R R'$, and denote by ω' the pullback of ω to $\mathfrak{X}'_\eta$. If we use the uniformizer π of R' to define the series $S(\mathfrak{X}', \omega'; T)$, then $S(\mathfrak{X}', \omega'; T)$ is the image of $S(\mathfrak{X}, \omega; T)$ under the base change morphism*

$$\mathcal{M}^R_{\mathfrak{X}_0}[[T]] \to \mathcal{M}^R_{\mathfrak{X}'_0}[[T]].$$

The analogous property holds for $S(\mathfrak{X}'_\eta, \omega'; T)$.

To formulate the main result on the motivic generating series, we need a technical definition. We say that the gauge form ω on $\mathfrak{X}_\eta$ is $\mathfrak{X}$-bounded if it presents at worst poles along the special fiber $\mathfrak{X}_s$, but no essential singularities. A precise definition is given in [34, 2.11]. This property is automatic if $\mathfrak{X}$ is stft. The following theorem was proven in [34, 7.14].

Theorem 2.46 *Assume that k has characteristic zero. Let $\mathfrak{X}$ be a generically smooth special formal R-scheme, and let ω be an $\mathfrak{X}$-bounded gauge form on $\mathfrak{X}_\eta$. The motivic generating series $S(\mathfrak{X}, \omega; T)$ is rational* over $\mathcal{M}_{\mathfrak{X}_0}$. More precisely, it belongs to the subring

$$\mathcal{M}_{\mathfrak{X}_0}\left[\frac{\mathbb{L}^a T^b}{1 - \mathbb{L}^a T^b}\right]_{(a,b)\in\mathbb{Z}\times\mathbb{Z}_{>0}}$$

of $\mathcal{M}_{\mathfrak{X}_0}[[T]]$.

The series $S(\mathfrak{X}, \omega; T)$ can be computed explicitly on an embedded resolution of singularities for the pair $(\mathfrak{X}, \mathfrak{X}_s)$ [34, 7.13] (we refer to Chapter 15 for results on resolution of singularities for formal schemes). Beware that our definition of $S(\mathfrak{X}, \omega; T)$ differs slightly from the definition given in [34], because of our choice of normalization in Definition 2.12 (both series coincide modulo multiplication with a power of $\mathbb{L}$).

3 The motivic Serre invariant and the trace formula

Throughout this section, we assume that k is perfect.

3.1 The motivic Serre invariant for rigid varieties

Let $f : V \to U$ be a separated morphism of finite type of Noetherian schemes, and consider the forgetful morphism of groups

$$f_! : K_0^R(\mathrm{Var}_V) \to K_0^R(\mathrm{Var}_U)$$

(see Sections 3.1 and 3.8 of Chapter 5). For every separated V-scheme of finite type X and every $i \in \mathbb{Z}$, we have

$$f_!([X]\mathbb{L}^i) = f_!([X])\mathbb{L}^i$$

(recall that we abuse notation by writing $\mathbb{L}$ for the class of $\mathbb{A}_V^1$ in $K_0^R(\mathrm{Var}_V)$ as well as for the class of $[\mathbb{A}_U^1]$ in $K_0^R(\mathrm{Var}_U)$). It follows that $f_!$ induces a morphism of groups

$$K_0^R(\mathrm{Var}_V)/(\mathbb{L} - [V]) \to K_0^R(\mathrm{Var}_U)/(\mathbb{L} - [U]).$$

Theorem-Definition 3.1 *Let $\mathfrak{X}$ be a generically smooth special formal R-scheme, and let $h : \mathfrak{Y} \to \mathfrak{X}$ be a Néron smoothening of $\mathfrak{X}$. Let X be a separated rigid K-variety, and assume that X admits a weak Néron model $\mathfrak{U}$.*

1. *The class $[\mathfrak{Y}_0]$ of $\mathfrak{Y}_0$ in $K_0^R(\mathrm{Var}_{\mathfrak{X}_0})/(\mathbb{L} - [\mathfrak{X}_0])$ depends only on $\mathfrak{X}$, and not on h. We denote it by $S(\mathfrak{X})$, and we call it the motivic Serre invariant of $\mathfrak{X}$.*

2. The class $[\mathfrak{U}_0]$ of $\mathfrak{U}_0$ in $K_0^R(\mathrm{Var}_k)/(\mathbb{L}-1)$ depends only on X, and not on $\mathfrak{U}$. We denote it by $S(X)$, and we call it the motivic Serre invariant of X.

3. The motivic Serre invariant $S(\mathfrak{X}_\eta)$ is the image of $S(\mathfrak{X})$ under the forgetful morphism of groups
$$K_0^R(\mathrm{Var}_{\mathfrak{X}_0})/(\mathbb{L}-[\mathfrak{X}_0]) \to K_0^R(\mathrm{Var}_k)/(\mathbb{L}-1).$$

4. If $\mathfrak{X}_\eta$ admits a gauge form ω, then $S(\mathfrak{X})$ is the image of $\int_{\mathfrak{X}} |\omega|$ under the projection
$$\mathscr{M}_{\mathfrak{X}_0}^R \to \mathscr{M}_{\mathfrak{X}_0}^R/(\mathbb{L}-[\mathfrak{X}_0]) \cong K_0^R(\mathrm{Var}_{\mathfrak{X}_0})/(\mathbb{L}-[\mathfrak{X}_0]).$$

If X is smooth and X admits a gauge form ω', then $S(X)$ is the image of $\int_X |\omega'|$ under the projection
$$\mathscr{M}_k^R \to \mathscr{M}_k^R/(\mathbb{L}-1) \cong K_0^R(\mathrm{Var}_k)/(\mathbb{L}-1).$$

5. If $X(K') = \emptyset$ for each finite unramified extension K' of K, then $S(X) = 0$.

6. Any quasi-compact open subset Y of X admits a weak Néron model. If X is quasi-compact and $\{X_i \,|\, i \in I\}$ a finite cover of X by quasi-compact open subvarieties, then
$$S(X) = \sum_{\emptyset \neq J \subset I} (-1)^{|J|+1} S(\cap_{j \in J} X_j)$$
in $K_0^R(\mathrm{Var}_k)/(\mathbb{L}-1)$.

7. If R' is a complete extension of R of ramification index one, with quotient field K' and perfect residue field k', then $S(\mathfrak{X}_\eta \times_K K')$ is the image of $S(\mathfrak{X}_\eta)$ under the base change morphism
$$K_0^R(\mathrm{Var}_k)/(\mathbb{L}-1) \to K_0^R(\mathrm{Var}_{k'})/(\mathbb{L}-1).$$

Proof Point (3) follows immediately from the definition, since $\mathfrak{Y}$ is a weak Néron model for $\mathfrak{X}_\eta$. Point (5) follows from the fact that the empty formal scheme is a weak Néron model for X. Point (7) follows immediately from Proposition 2.36. So let us prove the remaining statements.

First, assume that $\mathfrak{X}_\eta$ admits a gauge form ω. The formula for $\int_{\mathfrak{X}} |\omega|$ in Theorem-Definition 2.43 shows that
$$\int_{\mathfrak{X}} |\omega| = \sum_{C \in \pi_0(\mathfrak{Y}_0)} [C] = [\mathfrak{Y}_0]$$

in $K_0^R(\mathrm{Var}_{\mathfrak{X}_0})/(\mathbb{L}-[\mathfrak{X}_0])$. In particular, the right hand side of the equality only depends on $\mathfrak{X}$. Likewise, if X is smooth and X admits a gauge form ω', then we have

$$\int_X |\omega'| = \int_{\mathfrak{U}} |\omega'| = [\mathfrak{U}_0]$$

in $K_0^R(\mathrm{Var}_k)/(\mathbb{L}-1)$. This proves (1), (2) and (4), under the assumption that $\mathfrak{X}_\eta$ and X admit gauge forms.

Now let us consider the general case. We start with point (1). By Proposition 2.42 we may assume that $\mathfrak{X}$ is smooth and stft. Let $g : \mathfrak{Z} \to \mathfrak{X}$ be a Néron smoothening. It suffices to show that $[\mathfrak{Z}_0] = [\mathfrak{X}_0]$ in $K_0^R(\mathrm{Var}_{\mathfrak{X}_0})/(\mathbb{L}-[\mathfrak{X}_0])$.

By smoothness of $\mathfrak{X}$, there exists a finite cover

$$\{\mathfrak{U}_i \,|\, i \in I\}$$

of $\mathfrak{X}$ by open formal subschemes such that $\mathfrak{U}_i$ admits a gauge form for each i. For each non-empty subset J of I, we put $\mathfrak{U}_J = \cap_{i \in J}\mathfrak{U}_i$. Then we have

$$\sum_{\emptyset \neq J \subset I} (-1)^{|J|+1}[(\mathfrak{U}_J)_0] = [\mathfrak{X}_0]$$

in $K_0^R(\mathrm{Var}_{\mathfrak{X}_0})/(\mathbb{L}-[\mathfrak{X}_0])$. If we denote by $\mathfrak{V}_J$ the inverse image of $\mathfrak{U}_J$ in $\mathfrak{Y}$, for each non-empty subset J of I, then $\mathfrak{V}_J \to \mathfrak{U}_J$ is a Néron smoothening. Since $(\mathfrak{U}_J)_\eta$ admits a gauge form, we know that (1) holds for $\mathfrak{U}_J$, so that $[(\mathfrak{V}_J)_0] = [(\mathfrak{U}_J)_0]$ in $K_0^R(\mathrm{Var}_{\mathfrak{X}_0})/(\mathbb{L}-[\mathfrak{X}_0])$, and

$$\sum_{\emptyset \neq J \subset I} (-1)^{|J|+1}[(\mathfrak{U}_J)_0] = \sum_{\emptyset \neq J \subset I} (-1)^{|J|+1}[(\mathfrak{V}_J)_0] = [\mathfrak{Y}_0].$$

This concludes the proof of (1). Point (2) immediately follows from (1) and Proposition 2.42.

Finally, we prove (6). It is clear that Y admits a weak Néron model, by Proposition 2.41. To prove the additivity property of the motivic Serre invariant, it suffices to observe that, blowing up $\mathfrak{U}$, we may assume that there exists an open covering $\{\mathfrak{U}_i \,|\, i \in I\}$ of $\mathfrak{U}$ such that $\mathfrak{U}_i$ is a weak Néron model of X_i, by [9, 4.4]. $\qquad\square$

The following example explains why killing $\mathbb{L}-1$ is necessary in the definition of the motivic Serre invariant.

Example 3.2 The formal R-scheme $\mathfrak{X} = \mathrm{Spf}\, R\{x\}$ is smooth and stft, so it is a weak Néron model for its generic fiber $\mathfrak{X}_\eta$. Let z be a k-point of the special fiber $\mathfrak{X}_s$, and consider the formal blow-up $\mathfrak{X}' \to \mathfrak{X}$ with center z. The special fiber $\mathfrak{X}'_s$ consists of a copy of $\mathbb{A}^1_k$ and a copy of $\mathbb{P}^1_k$, intersecting in a point y. We denote by $\mathfrak{Y}$ the complement in $\mathfrak{X}'$ of the unique singular point y of $\mathfrak{X}'_s$. By Proposition 2.37, $\mathfrak{Y}$ is again a weak Néron model for $\mathfrak{X}_\eta$. We have $[\mathfrak{Y}_0] = 2\mathbb{L} - 1$ and $[\mathfrak{X}_0] = \mathbb{L}$ in $K_0^R(\mathrm{Var}_k)$, so that indeed $[\mathfrak{Y}_0] = [\mathfrak{X}_0]$ in $K_0^R(\mathrm{Var}_k)/(\mathbb{L} - 1)$.

Remark 3.3 The general assumption that k is perfect is crucial in Theorem-Definition 3.1. Consider, for instance, the formal R-scheme $\mathfrak{X} = \mathrm{Spf}\, R\{x\}$. If z is a closed point of $\mathfrak{X}_s$ whose residue field is not separable over k, then $\mathfrak{Y} = \mathfrak{X} \setminus \{z\}$ is still a weak Néron model of $\mathfrak{X}_\eta$, but $[\mathfrak{Y}_s] \neq [\mathfrak{X}_s]$ in $K_0^R(\mathrm{Var}_k)/(\mathbb{L} - 1)$.

3.2 The motivic Serre invariant for algebraic varieties

The above definitions can easily be adapted to algebraic K-varieties. We recall the definition of a weak Néron model from [11].

Definition 3.4 Let X be a separated smooth K-scheme of finite type. A weak Néron model of X is a separated smooth R-scheme U of finite type, endowed with an isomorphism of K-schemes

$$i : U \times_R K \to X$$

such that the natural map

$$U(R^{sh}) \to (U \times_R K)(K^{sh})$$

is a bijection.

If X is a K-scheme of finite type, then we denote by X^{rig} the rigid K-variety associated to X by non-archimedean GAGA.

Proposition 3.5 *Let X be a separated smooth K-scheme of finite type. The following properties are equivalent:*

1. *X admits a weak Néron model*
2. *$X(K^{sh})$ is bounded in X in the sense of [11, 1.1.2]*
3. *X^{rig} admits a weak Néron model*
4. *there exists a quasi-compact open rigid subvariety Y of X^{rig} such that $Y(K') = X^{rig}(K')$ for each finite unramified extension K' of K.*

If U is a weak Néron model for X, then its formal $\mathfrak{M}$-adic completion $\hat{U}$ is a weak Néron model for X^{rig}.

Proof The equivalence of (1) and (2) follows from [11, 3.4.2] and [11, 3.5.7]. The equivalence of (3) and (4) is Proposition 2.41. The equivalence of (2) and (3) is easily shown; see [37, 4.3]. The fact that $\widehat{U}$ is a weak Néron model for X^{rig} follows from the fact that $\widehat{U}_\eta$ is an open rigid subvariety of X^{rig} such that, for any finite unramified extension R' of R, with quotient field K', we have canonical bijections

$$X^{rig}(K') = X(K') = U(R') = \widehat{U}(R') = \widehat{U}_\eta(K').\qquad\square$$

Proposition-Definition 3.6 *Let X be a separated smooth K-scheme of finite type, and assume that X admits a weak Néron model U. We define the motivic Serre invariant of X by*

$$S(X) = S(X^{rig}) \in K_0^R(\mathrm{Var}_k)/(\mathbb{L} - 1).$$

We have

$$S(X) = [U_s] \in K_0^R(\mathrm{Var}_k)/(\mathbb{L} - 1).$$

In particular, the class $[U_s]$ in $K_0^R(\mathrm{Var}_k)/(\mathbb{L} - 1)$ only depends on X, and not on U. If $X(K^{sh}) = \emptyset$, then $S(X) = 0$.

Proof This follows immediately from Proposition 3.5. $\qquad\square$

The motivic Serre invariant $S(X)$ is defined, in particular, if X is smooth and proper over k, since in this case X^{rig} is separated, smooth and quasi-compact. The following example is taken from [38, 2.4].

Example 3.7 Assume that k is algebraically closed, and let A be an abelian K-variety. Denote by $\mathscr{A}$ its Néron model, and by ϕ_A the number of connected components of $\mathscr{A}_s$. Consider the Chevalley decomposition

$$0 \longrightarrow L = (U \times_k T) \longrightarrow \mathscr{A}_s^o \overset{f}{\longrightarrow} B \longrightarrow 0$$

of the identity component $\mathscr{A}_s^o$, with U unipotent, T a torus, and B an abelian variety [11, 9.2.1 and 9.2.2]. Denote by t the dimension of T. Then we have the following equalities in $K_0^R(\mathrm{Var}_k)/(\mathbb{L} - 1)$:

$$S(A) = \begin{cases} 0 & \text{iff } t > 0, \\ \phi_A \cdot [B] & \text{iff } t = 0. \end{cases}$$

To see this, note that $\mathscr{A}$ is a weak Néron model for A, so that $S(A) = [\mathscr{A}_s]$. Since all connected components of $\mathscr{A}_s$ are isomorphic, we find $[\mathscr{A}_s] = \phi_A[\mathscr{A}_s^o]$. The morphism f is a Zariski-locally trivial fibration: any $fppf$ L-torsor is also a Zariski L-torsor, because $L \cong U \times_k T$ where

T is a split torus and U is a successive extension of $\mathbb{G}_{a,k}$ [30, III.3.7 and III.4.9]. This yields $[\mathscr{A}_s^o] = [L][B]$ in $K_0(Var_k)$. But T is isomorphic to $\mathbb{G}_{m,k}^t$, and as a k-variety, U is isomorphic to $\mathbb{A}_k^u$ for some $u \geq 0$ [48, VII n° 6]. Hence,

$$[\mathscr{A}_s^o] = \mathbb{L}^u(\mathbb{L} - 1)^t[B]$$

in $K_0(Var_k)$.

It only remains to show that $\phi_A \cdot [B] \neq 0$ in $K_0^R(Var_k)/(\mathbb{L} - 1)$. This can be seen by applying the Poincaré polynomial; see [38, 2.4].

Proposition 3.8 *Let R' be a complete extension of R of ramification index one, with quotient field K' and perfect residue field k'. Let X be a separated smooth K-scheme of finite type, and assume that X admits a weak Néron model U. Then $U \times_R R'$ is a weak Néron model for X, and $S(X \times_K K')$ is the image of $S(X)$ under the base change morphism*

$$K_0^R(Var_k)/(\mathbb{L} - 1) \to K_0(Var_{k'})/(\mathbb{L} - 1).$$

Proof It suffices to prove the first assertion, which follows from [11, 3.6.7]. □

If K has characteristic zero, then the first author showed in [37, 5.4] how one can associate a motivic Serre invariant to any K-scheme of finite type. More precisely, there exists a unique ring morphism

$$S : K_0(Var_K) \to K_0^R(Var_k)/(\mathbb{L} - 1)$$

such that $S([X]) = S(X)$ for any smooth and proper K-variety X. If Y is a K-scheme of finite type such that Y^{rig} admits a weak Néron model, then $S([Y]) = S(Y^{rig})$. Hence, we can define $S(Y) := S([Y])$ for any K-scheme of finite type Y. If $Y(K^{sh}) = \emptyset$, then $S(Y) = S(Y^{rig}) = 0$.

3.3 The trace formula

If X is a separated rigid K-variety that admits a weak Néron model $\mathfrak{U}$, then we can consider the motivic Serre invariant

$$S(X) = [\mathfrak{U}_0] \in K_0^R(Var_k)/(\mathbb{L} - 1)$$

as a measure for the set of étale points on X. Indeed, the generic fiber $\mathfrak{U}_\eta$ is an open rigid subvariety of X containing all K'-points, for each finite unramified extension K' of K, and since $\mathfrak{U}$ is smooth, its special fiber is a good measure for the set of unramified points on the generic fiber. For any k-rational point x of $\mathfrak{U}_0$, its inverse image $sp_{\mathfrak{U}}^{-1}(x)$ under the specialization morphism

$$sp_{\mathfrak{U}} : \mathfrak{U}_\eta \to \mathfrak{U}$$

is an open unit ball, by Proposition 2.5. Hence, we can view the set of étale points on X as a family of open unit balls parameterized by $\mathfrak{U}_0$.

It is natural to ask if this measure admits a cohomological interpretation, analogous to the Grothendieck-Lefschetz-Verdier trace formula for varieties over finite fields. This is indeed the case, under an appropriate tameness condition on X.

Definition 3.9 Let X be a separated smooth rigid K-variety. A tame model for X is a regular special formal R-scheme $\mathfrak{X}$, endowed with an isomorphism $\mathfrak{X}_\eta \to X$, such that $\mathfrak{X}_s$ is a strict normal crossings divisor $\sum_{i \in I} N_i \mathfrak{E}_i$ on $\mathfrak{X}$ with multiplicities N_i prime to p.

If $\mathfrak{X}$ is not stft, then the notion of strict normal crossings divisor on $\mathfrak{X}$ is a little delicate; see [34, 2.5]. If k has characteristic zero and $\mathfrak{Y}$ is a generically smooth affine special formal R-scheme, then one can deduce from resolution of singularities for quasi-excellent $\mathbb{Q}$-schemes [52] that $\mathfrak{Y}_\eta$ admits a tame model [34, 2.43].

The following trace formula was proven in [34, 6.4]. It generalizes the one in [41, 5.4].

Theorem 3.10 *Let X be a separated smooth rigid K-variety that admits a tame model. Let φ be a topological generator of the tame geometric monodromy group $G(K^t/K^{sh})$. For each integer $d > 0$ prime to p, we have*

$$\chi_{\text{top}}(S(X \times_K K(d))) = \sum_{i \geq 0} (-1)^i \text{Trace}(\varphi^d \mid H^i_{\text{ét}}(X \times_K \widehat{K^t}, \mathbb{Q}_\ell)).$$

Here $H^i_{\text{ét}}(\cdot)$ is Berkovich's ℓ-adic cohomology for analytic spaces [4]. Intuitively, the trace formula says that the "number" of $K(d)^{sh}$-points on X equals the trace of the d-th power of the monodromy on the tame ℓ-adic cohomology of X. Note that the left hand side is well-defined since $\chi_{\text{top}}(\mathbb{L}) = 1$.

The proof of Theorem 3.10 is based on explicit computations of both sides of the equality on a tame model $\mathfrak{X}$ for X, by computing the ℓ-adic nearby cycles of $\mathfrak{X}$ and constructing a Néron smoothening of $\mathfrak{X} \times_R R(d)$ for each d. It would be quite interesting to find a more conceptual proof of the result, that does not rely on resolution of singularities. This seems to be a challenging problem.

Corollary 3.11   *Assume that k has characteristic zero, and let φ be a topological generator of the geometric monodromy group $G(K^s/K^{sh})$. If $\mathfrak{X}$ is a generically smooth special formal R-scheme, then we have*

$$\chi_{\text{top}}(S(\mathfrak{X}_\eta \times_K K(d))) = \sum_{i \geq 0}(-1)^i \text{Trace}(\varphi^d \mid H^i_{\text{ét}}(\mathfrak{X}_\eta \times_K \widehat{K^s}, \mathbb{Q}_\ell))$$

for each integer $d > 0$. In particular, we have

$$\chi_{\text{top}}(S(X \times_K K(d))) = \sum_{i \geq 0}(-1)^i \text{Trace}(\varphi^d \mid H^i_{\text{ét}}(X \times_K \widehat{K^s}, \mathbb{Q}_\ell))$$

for each $d > 0$ and each separated smooth quasi-compact rigid K-variety X.

Proof   If $\mathfrak{X}$ is affine, then $\mathfrak{X}_\eta$ admits a tame model, as we noted above. The general case follows by showing that both sides of the equality are additive w.r.t. open covers of $\mathfrak{X}$ [34, 6.5].   □

Definition 3.12   If Y is a smooth and proper K-variety, then a tame model for Y is a regular proper flat R-scheme $\mathscr{Y}$, endowed with an isomorphism $\mathscr{Y} \times_R K \to Y$, such that $\mathscr{Y}_s$ is a strict normal crossings divisor $\sum_{i \in I} N_i E_i$ on $\mathscr{Y}$ with multiplicities N_i prime to p.

If k has characteristic zero, then any smooth and proper K-variety admits a tame model, by resolution of singularities. In the algebraic setting, the trace formula takes the following form.

Theorem 3.13   *Let Y be a smooth and proper K-variety that admits a tame model. Let φ be a topological generator of the geometric tame monodromy group $G(K^t/K^{sh})$. For each integer $d > 0$ prime to p, we have*

$$\chi_{\text{top}}(S(Y \times_K K(d))) = \sum_{i \geq 0}(-1)^i \text{Trace}(\varphi^d \mid H^i_{\text{ét}}(Y \times_K K^t, \mathbb{Q}_\ell)).$$

Proof   This follows from Theorem 3.10, by Berkovich's comparison theorem for étale cohomology of algebraic and rigid varieties [4, 7.5.4].   □

If $p > 1$, the condition that Y has a tame model is certainly too strong. The most natural condition to impose is that Y is *cohomologically tame*, i.e., the wild inertia $P \subset G(K^s/K^{sh})$ acts trivially on the ℓ-adic cohomology $H^i_{\text{ét}}(Y \times_K K^s, \mathbb{Q}_\ell)$ of Y. If Y has a tame model then Y is cohomologically tame [37, 6.2] but the converse implication does not hold.

If Y is a curve, then by [37, § 7], cohomological tameness indeed implies the validity of the trace formula, unless Y has genus one, purely additive reduction, and no K-point (then Y has no K^t-point either; see [37, § 7]). If E is a tamely ramified elliptic K-curve with purely additive reduction, then the trace formula fails for any non-trivial E-torsor Y over K. Indeed, since E is isomorphic to the Jacobian of Y, the cohomological side of the trace formula is the same for Y and E; on the other hand, $S(Y) = 0$ since $Y(K) \neq \emptyset$, while $\chi_{\text{top}}(S(E)) = \phi_E$ by Example 3.7.

In any dimension, it seems plausible that the following question has an affirmative answer. This question was raised by the first author in [38, § 1].

Question 3.14 Does the trace formula hold for any geometrically connected, smooth, proper, cohomologically tame K-variety X with a K^t-rational point?

The case where X is a cohomologically tame abelian K-variety is proven in [38]. By Example 3.7, $\chi_{\text{top}}(S(X))$ vanishes unless X has purely additive reduction, in which case $\chi_{\text{top}}(S(X))$ is equal to the number of connected components of the Néron model of X. The trace formula gives a cohomological interpretation for this number.

In [39, 4.4], the first author gave an explicit formula for the error term in the trace formula for an arbitrary smooth and proper K-variety X, in terms of a regular model for X with strict normal crossings. Using this expression, the trace formula for curves can be deduced from Saito's geometric criterion for cohomological tameness [46, 3.11]. A positive answer to Question 3.14 would imply a partial generalization of Saito's criterion to arbitrary dimension.

We refer to [37, 6.4] for a trace formula for singular K-varieties. One can prove, in particular, the following result [37, 6.5 and 6.6].

Theorem 3.15 *Assume that k has characteristic zero, and let φ be a topological generator of the geometric monodromy group $G(K^s/K^{sh})$. For any separated K-scheme of finite type Y, we have*

$$\chi_{\text{top}}(S(Y)) = \sum_{i \geq 0} (-1)^i \text{Trace}(\varphi \mid H_c^i(Y \times_K K^s, \mathbb{Q}_\ell)).$$

In particular, Y has a K^{sh}-point iff there exists a subvariety U of Y such that

$$\sum_{i \geq 0} (-1)^i \text{Trace}(\varphi \mid H_c^i(U \times_K K^s, \mathbb{Q}_\ell)) \neq 0.$$

4 Comparison with the p-adic theory

Throughout this section, we assume that K is a p-adic field, i.e., a finite extension of the field $\mathbb{Q}_p$ of p-adic numbers, for some prime p. We denote by q the cardinality of the residue field k of K. We normalize the absolute value on K by putting $|\pi|_K = q^{-1}$, and we endow K with the metric topology defined by $|\cdot|_K$. Then K is locally compact, so that it carries a Haar measure μ_{Haar}, which we normalize by putting $\mu_{Haar}(R) = 1$.

4.1 Analytic manifolds and p-adic integration

Definition 4.1 Fix an integer $d \geq 0$, and let U be an open subset of K^d. A K-analytic function on U is a function $f : U \to K$ such that for each point $u = (u_1, \ldots, u_d)$ of U, there exists an open neighbourhood V of u in U and a formal power series

$$c(x_1, \ldots, x_d) = \sum_{i \in \mathbb{N}^d} c_i x_1^{i_1} \ldots x_n^{i_d} \quad \in K[[x_1, \ldots, x_d]]$$

such that the infinite sum $c(v_1 - u_1, \ldots, v_d - u_d)$ converges to $f(v)$ in K for each point $v = (v_1, \ldots, v_d)$ in V.

It is clear that for any open subset Y of K^d, the analytic functions form a sheaf of K-algebras on Y, which we denote by $\mathscr{O}_Y$.

Definition 4.2 Fix an integer $d \geq 0$. A d-dimensional K-analytic manifold is a locally ringed space in K-algebras $(X, \mathscr{O}_X)$ such that X is Hausdorff and such that any point x of X has an open neighbourhood U in X with $(U, \mathscr{O}_X|_U)$ isomorphic to $(Y, \mathscr{O}_Y)$ for some open subset Y of K^d. A morphism of K-analytic manifolds is a morphism of locally ringed spaces in K-algebras.

For a definition in terms of charts and atlases, we refer to [26, 2.4]. The notion of differential form on a K-analytic manifold is defined in the usual way. A gauge form is a nowhere vanishing differential form of maximal degree.

If X is a d-dimensional K-analytic manifold, then a differential form ω of degree d on X defines a measure $|\omega|$ on the set of compact open subsets of X [26, 7.4]. We briefly recall its definition. Note that the compact open subsets of X form a basis for the topology on X, since this holds for K^d. Let U be an open compact subset of X. Breaking up U into a finite disjoint union of compact open subsets, we may assume that U admits an isomorphism of K-analytic manifolds $\phi : W \to U$ with W an open subset of K^d. Taking the standard gauge form $dx = dx_1 \wedge \ldots \wedge dx_d$ on K^d, we can write the pull-back $\phi^*\omega$ as $f \cdot dx$, with f an analytic function on W. The measure $|\omega|(U)$ is now defined by

$$|\omega|(U) = \int_W |f|_K$$

where the right hand side is integration with respect to the Haar measure on W. The change of variables formula for p-adic integrals [26, 7.4.1] guarantees that this definition does not depend on the choice of ϕ.

4.2 The p-adic Serre invariant

The motivic Serre invariant was defined in [29] as a generalization of Serre's p-adic invariant [50]. We will see in Proposition 4.4 that the motivic Serre invariant specializes to the p-adic one. This was originally proven in [29, 4.6.1].

Serre's p-adic invariant classifies the compact K-analytic manifolds. More precisely, Serre proved the following result (see also [26, 7.5]).

Theorem-Definition 4.3 *Fix an integer $d \geq 0$, and let X be a d-dimensional non-empty compact K-analytic manifold. There exists a unique value r in $\{1, \ldots, q-1\}$ such that for each $m \in \mathbb{N}$, X is iso-morphic to a disjoint union of $r + mq$ copies of the unit disc $R^d \subset K^d$. The image of r in $\mathbb{Z}/(q-1)\mathbb{Z}$ is called the p-adic Serre invariant $S_p(X)$ of X. The manifold X admits a gauge form, and for each gauge form ω on X, the value $|\omega|(X) = \int_X |\omega|$ belongs to $\mathbb{Z}[1/q]$. Moreover, we have*

$$\int_X |\omega| = S_p(X)$$

in $\mathbb{Z}/(q-1)\mathbb{Z}$.

Proof For each point x of K^d, the set $\{x + \pi^n \cdot R^d\}_{n \geq 0}$ is a funda-mental system of compact open neighborhoods of x in K^d which are all isomorphic to R^d as K-analytic manifolds. It easily follows that X can be written as a disjoint union of open compact subsets

$$X \cong \sqcup_{i=1}^r U_i$$

such that each U_i is isomorphic to R^d as a K-analytic manifold. We can always replace r by any other element in $\mathbb{Z}_{>0} \cap (r + q\mathbb{Z})$: if we choose for each $c \in k$ an element c' in the fiber of c under the projection $R \to k$, then we have isomorphisms of K-analytic manifolds

$$R^d \cong \bigsqcup_{c \in k} ((c' + \pi R) \times R^{d-1}) \cong \bigsqcup_{c \in k} R^d.$$

We fix an isomorphism $\phi_i : R^d \to U_i$, for each i.

Denote by $\alpha = dx_1 \wedge \ldots \wedge dx_d$ the standard gauge form on R^d. Then X carries a unique gauge form ω_0 whose restriction to U_i is $(\phi_i^{-1})^* \alpha$, for each $i \in \{1, \ldots, r\}$. Assume that ω is another gauge form on X. For each $i \in \{1, \ldots, r\}$, the differential form $\phi_i^* \omega$ can be written as $f_i \alpha$, with f_i a nowhere vanishing analytic function on R^d. Since the value group $|K^*|_K$ is discrete, there exists an integer $a > 0$ with the following property: for each $i \in \{1, \ldots, r\}$ and each point x of R^d, the function f_i has constant absolute value on $x + \pi^a R^d$. In other words, for each element y of $(R/\pi^a R)^d$, the function f_i has constant absolute value $q^{n_{i,y}}$ on the fiber of y under the projection morphism $R^d \to (R/\pi^a)^d$, with $n_{i,y} \in \mathbb{Z}$.

Let us compute the measures $|\omega_0|(X)$ and $|\omega|(X)$. Since the Haar measure of $x + \pi^a R^d$ equals q^{-ad}, for each $x \in R^d$, we find

$$|\omega_0|(X) = r$$

$$|\omega|(X) = \sum_{i=1}^r \sum_{y \in (R/\pi^a)^d} q^{n_{i,y} - ad}$$

in $\mathbb{Z}[1/q]$. The cardinality of $(R/\pi^a)^d$ equals q^{ad}, so that

$$|\omega|(X) = |\omega_0|(X) = r$$

in $\mathbb{Z}/(q-1)\mathbb{Z}$. Hence, the value $|\omega|(X) \in \mathbb{Z}/(q-1)\mathbb{Z}$ only depends on X, and not on ω, and also the value $r \in \mathbb{Z}/(q-1)\mathbb{Z}$ only depends on X. This concludes the proof. $\square$

We see, in particular, that there are very few isomorphism classes of compact K-analytic manifolds. This suggests that our naïve definition of K-analytic manifold is not the right one, and that we admitted too many K-analytic functions. In particular, any locally constant function $K^d \to K$ is analytic, and there are many of these, because K is totally disconnected. This observation motivates the introduction of more refined theories of non-archimedean geometry, and inspired Tate to introduce his rigid spaces.

If X is a separated and smooth rigid K-variety of pure dimension d, then the set $X(K)$ carries a natural structure of d-dimensional K-analytic manifold, and we denote this manifold by X^{an}. This construction can be extended to a functor $(\cdot)^{an}$ from the category of seperated smooth rigid K-varieties of pure dimension to the category of K-analytic manifolds. The functor $(\cdot)^{an}$ respects open immersions. If X is quasi-compact, then X^{an} is compact. The following proposition compares the motivic Serre invariant $S(X)$ and the p-adic Serre invariant $S_p(X^{an})$. We denote by

$$\sharp : K_0^R(\mathrm{Var}_k) \to \mathbb{Z}$$

the unique ring morphism which sends the class $[Y]$ of a k-variety to the number $|Y(k)|$ of rational points on Y. Since $|\mathbb{A}^1(k)| = q$, the morphism $\sharp$ factors to a ring morphism

$$\sharp : K_0^R(\mathrm{Var}_k)/(\mathbb{L}-1) \to \mathbb{Z}/(q-1)\mathbb{Z}.$$

Proposition 4.4 *If X is a separated, smooth and quasi-compact rigid K-variety, then $S_p(X^{an})$ is the image of $S(X)$ under the morphism*

$$\sharp : K_0^R(\mathrm{Var}_k)/(\mathbb{L}-1) \to \mathbb{Z}/(q-1)\mathbb{Z}.$$

Proof Since we did not give a precise definition of the functor $(\cdot)^{an}$, we'll only give a sketch of the proof. Let $\mathfrak{U}$ be a weak Néron model for X. Then $\sharp S(X) = \mathfrak{U}_s(k)$ in $\mathbb{Z}/(q-1)\mathbb{Z}$ by definition of the motivic Serre invariant. If x is any point of $\mathfrak{U}_\eta(K) = X(K)$, then its image under the specialization morphism $sp_{\mathfrak{U}} : \mathfrak{U}_\eta \to \mathfrak{U}$ is a k-point y of $\mathfrak{U}_s$. Denote by $\mathfrak{U}_y$ the formal completion of $\mathfrak{U}$ at y. By Proposition 2.5,

$$sp_{\mathfrak{U}}^{-1}(y) \cong (\mathfrak{U}_y)_\eta$$

is isomorphic to the open rigid unit disc of dimension d. One can deduce from this fact that $(\mathfrak{U}_y)_\eta^{an}$ is isomorphic to $(\pi R)^d \cong R^d$, and that $\mathfrak{U}_\eta^{an} \cong X^{an}$ is isomorphic to the disjoint union of $|\mathfrak{U}_s(k)|$ copies of R^d. This concludes the proof. $\qquad \square$

More generally, one can associate to any gauge form ω on X a gauge form ω^{an} on X^{an}, and if X has pure dimension d, the respective integrals are related by the equality

$$\sharp\left(\int_X |\omega|\right) = q^d \int_{X^{an}} |\omega^{an}| \quad \in \mathbb{Z}[1/q].$$

See [29, 4.6.1] for details. The factor q^d in the right hand side is due to our choice of normalization in Definition 2.12.

5 The motivic monodromy conjecture

5.1 The Milnor fibration

Let X be a complex manifold, and let

$$g : X \to \mathbb{C}$$

be an analytic map. We denote by X_s the special fiber of g (i.e., the analytic space defined by $g = 0$), and we fix a point $x \in X_s$.

What does X_s look like in a neighbourhood of x? If g is smooth at x this is easy: X_s is locally a complex submanifold of X. If g is not smooth at x, then the topology of X_s near x can be studied by means of the Milnor fibration [31, 17].

Working locally, we may assume that $X = \mathbb{C}^d$. Let $B = B(x, \varepsilon)$ be an open ball around x in $\mathbb{C}^d$ with radius ε, let $D = D(0, \eta)$ be an open disc around the origin 0 in $\mathbb{C}$ with radius η, and put $D^* = D \setminus \{0\}$. For $0 < \eta \ll \varepsilon \ll 1$ the map

$$g_x : g^{-1}(D^*) \cap B \to D^*$$

is a locally trivial fibration, called the Milnor fibration of g at x.

We consider the universal covering space

$$\widetilde{D^*} = \{z \in \mathbb{C} \mid \Im(z) > -\log \eta\} \to D^* : z \mapsto \exp(iz)$$

of D^*, and we put

$$F_x = (g^{-1}(D^*) \cap B) \times_{D^*} \widetilde{D^*}.$$

This is the universal fiber of the fibration g_x, and it is called the Milnor fiber of g at x. Since g_x is a locally trivial fibration and $\widetilde{D^*}$ is contractible, F_x is homotopy-equivalent to the fiber of g_x over any point of D^*.

If g is smooth at x, then the fibration g_x is trivial. In general, the defect of triviality is measured by the monodromy action on the singular cohomology of F_x, i.e., the action of the group $\pi_1(D^*)$ of covering transformations of $\widetilde{D^*}$ over D^* on $\oplus_{i \geq 0} H^i_{\mathrm{sing}}(F_x, \mathbb{Z})$. The action of the canonical generator $z \mapsto z + 2\pi$ is called the monodromy transformation and denoted by M_x. We say that a complex number γ is a monodromy eigenvalue of g at x if γ is an eigenvalue of the monodromy transformation M_x on $H^i_{\mathrm{sing}}(F_x, \mathbb{Z})$ for some $i \geq 0$. The Monodromy Theorem states that these monodromy eigenvalues are roots of unity.

We denote by $D^b_c(X_s, \mathbb{Z})$ the bounded derived category of constructible abelian sheaves on $X_s(\mathbb{C})$ (for the complex topology). The cohomology of the Milnor fiber of g at x can be computed by means of the complex of *nearby cycles* $R\psi_g(\mathbb{Z}) \in D^b_c(X_s, \mathbb{Z})$ of g in the following way: for each $i \in \mathbb{Z}$, there exists a canonical isomorphism

$$H^i(F_x, \mathbb{Z}) \cong R^i\psi_g(\mathbb{Z})_x$$

identifying the monodromy actions on both sides [17, 4.2.2].

5.2 The motivic zeta function

Let X be a smooth irreducible k-variety of dimension m, and consider a dominant morphism

$$f : X \to \mathbb{A}^1_k = \operatorname{Spec} k[x].$$

We denote by X_s the fiber of f over the origin.

Let d be an element of $\mathbb{N}$, and let ψ be a point of $\mathscr{L}_d(X)$. The morphism $\mathscr{L}_d(f)$ maps ψ to a point ψ' of $\mathscr{L}_d(\mathbb{A}^1_k)$. We denote by F its residue field, and by $f(\psi)$ the element of

$$\mathscr{L}_d(\mathbb{A}^1_k)(F) = F[t]/(t^{d+1})$$

defined by ψ'. If ψ is an arc on X, i.e., a point of $\mathscr{L}(X)$, then $f(\psi) \in F[[t]]$ is defined analogously.

For any integer $d > 0$, we consider the following closed subset of the jet scheme $\mathscr{L}_d(X)$:

$$\mathscr{X}_{d,1} := \{\psi \in \mathscr{L}_d(X) \,|\, f(\psi) = t^d \mod t^{d+1}\}.$$

We endow $\mathscr{X}_{d,1}$ with its reduced closed subscheme structure. The truncation map $\theta^d_0 : \mathscr{L}_d(X) \to X$ induces a morphism $\mathscr{X}_{d,1} \to X$. Since $d > 0$, this morphism factors through X_s, and $\mathscr{X}_{d,1}$ gets a structure of separated X_s-scheme of finite type. For any point x of X_s, we put

$$\mathscr{X}_{d,1,x} = \mathscr{X}_{d,1} \times_{X_s} x.$$

Definition 5.1 (Denef-Loeser)   The motivic zeta function associated to f is the generating series

$$Z_f(T) := \sum_{d>0} [\mathscr{X}_{d,1}]\mathbb{L}^{-dm} T^d \in \mathscr{M}_{X_s}[[T]].$$

If x is a point of X_s, then the motivic zeta function of f at x is defined by

$$Z_{f,x}(T) := \sum_{d>0} [\mathscr{X}_{d,1,x}]\mathbb{L}^{-dm} T^d \in \mathscr{M}_x[[T]].$$

It is the image of $Z_f(T)$ under the base change morphism

$$\mathscr{M}_{X_s}[[T]] \to \mathscr{M}_x[[T]].$$

Actually, Denef and Loeser defined the motivic zeta function over a more refined, equivariant Grothendieck ring, using the fact that $\mathscr{X}_{d,1}$ carries a natural action of the group k-scheme of d-th roots of unity μ_d. For any k-algebra A, the action is given by

$$\mu_d(A) \times \mathscr{X}_{d,1}(A) \to \mathscr{X}_{d,1}(A) : (\xi, \psi) \mapsto \psi \circ \Phi_\xi$$

with Φ_ξ the A-automorphism of $\operatorname{Spec} A[t]/(t^{d+1})$ defined by $t \mapsto \xi \cdot t$.

The following result is due to Denef and Loeser [15, 3.3.1].

Theorem 5.2 *Assume that k has characteristic zero. The series $Z_f(T)$ is rational. More precisely, $Z_f(T)$ belongs to the subring*

$$\mathscr{M}_{X_s} \left[\frac{\mathbb{L}^a T^b}{1 - \mathbb{L}^a T^b} \right]_{(a,b) \in \mathbb{Z}_{>0} \times \mathbb{Z}_{>0}}$$

of $\mathscr{M}_{X_s}[[T]]$.

In fact, one can say much more: the zeta function $Z_f(T)$ can be computed explicitly in terms of a log-resolution $h : X' \to X$ of the ideal sheaf (f) on X. The terms that appear in the expression are classes in the Grothendieck ring of certain strata of the total transform of the zero locus of f, and the so-called *numerical data* of the log-resolution (the multiplicities of the prime divisors of $(f \circ h)$ on X', and the multiplicities of the prime divisors of the Jacobian ideal of h on X'). We refer to [15] for more information.

Denef and Loeser formulated an intriguing conjecture, the motivic monodromy conjecture, relating the poles of $Z_f(T)$ to monodromy eigenvalues of f.

Conjecture 5.3 *Assume that k is a subfield of $\mathbb{C}$. There exists a finite subset $\mathscr{S}$ of $\mathbb{Z}_{>0} \times \mathbb{Z}_{>0}$ such that $Z_f(T)$ belongs to the subring*

$$\mathscr{M}_{X_s} \left[T, \frac{1}{1 - \mathbb{L}^{-a} T^b} \right]_{(a,b) \in \mathscr{S}}$$

of $\mathscr{M}_{X_s}[[T]]$ and such that, for each $(a,b) \in \mathscr{S}$, the value $\exp(2\pi i a / b)$ is a monodromy eigenvalue of $f^{an} : (X \times_k \mathbb{C})^{an} \to \mathbb{C}$ at some point of the fiber of f^{an} over the origin.

Here we wrote $(X \times_k \mathbb{C})^{an}$ for the complex manifold associated to $X \times_k \mathbb{C}$ by GAGA, and f^{an} for the analytification of the morphism $X \times_k \mathbb{C} \to \mathbb{A}^1_{\mathbb{C}}$ obtained from f by base change.

Denef and Loeser's motivic monodromy conjecture generalizes a conjecture of Igusa's, which predicts a similar property for the p-adic zeta function of f when f is defined over a number field. In particular, for a polynomial g in $\mathbb{Z}[x_1, \ldots, x_m]$, Igusa's conjecture predicts a relation between the asymptotic behaviour of the number of solutions of the congruence $g \equiv 0 \mod p^d$ as $d \to \infty$, and the singularities of the complex hypersurface defined by g. We refer to [36] for a detailed account, and for a survey of cases where the conjecture has been solved. The only complete answer so far has been obtained in dimension $m = 2$.

6 The analytic Milnor fiber

In this section, we put $R = k[[\pi]]$ and $K = k((\pi))$. Let X be an irreducible k-variety, and let

$$f : X \to \mathbb{A}^1_k = \operatorname{Spec} k[\pi]$$

be a dominant morphism. We denote by X_s the fiber of f over the origin, and we fix a closed point x on X_s. We denote by

$$\widehat{f} : \widehat{X} \to \operatorname{Spf} R$$

the formal π-adic completion of f. It is an stft formal R-scheme. Moreover, we denote by

$$\widehat{f}_x : \mathfrak{X} = \operatorname{Spf} \widehat{\mathscr{O}}_{X,x} \to \operatorname{Spf} R$$

the formal completion of f at x. It is a special formal R-scheme.

6.1 Construction of the analytic Milnor fiber

Definition 6.1 The analytic Milnor fiber $\mathscr{F}_x$ of f at x is the generic fiber $\mathfrak{X}_\eta$ of the special formal R-scheme $\widehat{f}_x : \mathfrak{X} \to \operatorname{Spf} R$.

The analytic Milnor fiber is a separated rigid K-variety. By [7, 0.2.7] there exists a canonical isomorphism $\mathscr{F}_x \cong sp_{\widehat{X}}^{-1}(x)$. If the generic fiber of f is smooth (e.g. when k has characteristic zero and $X \setminus X_s$ is smooth over k), then $\mathscr{F}_x$ is smooth as well.

Example 6.2   Assume that x is k-rational and that X is smooth over k at the point x. Then we can find an isomorphism of k-algebras

$$\widehat{\mathscr{O}}_{X,x} \cong k[[x_1,\ldots,x_m]]$$

and the morphism f defines an element in this algebra of power series. The analytic Milnor fiber is isomorphic to the closed rigid subvariety of the open unit disc

$$D_m^- = \{z \in \mathrm{Sp}\, K\{x_1,\ldots,x_m\} \mid |x_i(z)| < 1 \text{ for all } i\}$$

defined by the equation $f = \pi$.

We constructed the analytic Milnor fiber in [41, 9.1] as a non-archimedean model for the topological Milnor fibration from Section 5.1. The topological intuition behind the construction is the following: the formal neighborhood $\mathrm{Spf}\, R$ of the origin in $\mathbb{A}^1_k = \mathrm{Spec}\, k[\pi]$ corresponds to an infinitesimally small disc D around the origin in $\mathbb{C}$. The inverse image of D under f is realized as the π-adic completion $\widehat{X}$ of the morphism f; the formal scheme $\widehat{X}$ should be seen as a tubular neighborhood of X_s in X. One can think of the formal germ $\mathfrak{X} = \mathrm{Spf}\, \widehat{\mathscr{O}}_{X,x}$ as an infinitesimally small ball B around x in X, so that the structural morphism

$$\mathscr{F}_x \to \mathrm{Sp}\, K$$

corresponds to the Milnor fibration $B \cap f^{-1}(D^*) \to D^*$. By the dictionary between finite extensions of K and finite coverings of D^*, the non-archimedean analogue of the Milnor fiber F_x is the rigid $\widehat{K^a}$-variety $\mathscr{F}_x \times_K \widehat{K^a}$.

The analytic Milnor fiber contains a wealth of information on the singularity of f at x. In fact, we have the following result [34, 8.8].

Proposition 6.3   *If X is normal at x, then $\mathscr{F}_x$ is a complete invariant of the formal germ $\widehat{f}_x$ of f at x. The R-algebra $\widehat{\mathscr{O}}_{X,x}$ is isomorphic to the R-algebra of bounded analytic functions*

$$\{f \in \mathscr{O}_{\mathscr{F}_x}(\mathscr{F}_x) \mid |f(z)| \leq 1 \text{ for all } z \in \mathscr{F}_x\}.$$

Here the R-algebra structure on $\widehat{\mathscr{O}}_{X,x}$ is given by the morphism $\widehat{f}_x$.

6.2 Étale cohomology of the analytic Milnor fiber

Since the analytic Milnor fiber is constructed as a non-archimedean analogue of the Milnor fibration, one would expect that its cohomology is related to the ℓ-adic nearby cycles $R\psi_f(\mathbb{Q}_\ell)_x$ of f at x. This is indeed the case, by the following result [5, 3.5][41, 9.2].

Theorem 6.4 *Assume that k is algebraically closed. For each $i \geq 0$ there exists a canonical $G(K^s/K)$-equivariant isomorphism*

$$H^i(\mathscr{F}_x \times_K \widehat{K^s}, \mathbb{Q}_\ell) \cong R^i\psi_f(\mathbb{Q}_\ell)_x.$$

In particular, when $k = \mathbb{C}$ and X is smooth at x, there exists for each $i \geq 0$ a canonical isomorphism

$$H^i(\mathscr{F}_x \times_K \widehat{K^s}, \mathbb{Q}_\ell) \cong H^i(F_x, \mathbb{Q}_\ell)$$

that identifies the action of the canonical generator of $G(K^a/K) \cong \widehat{\mathbb{Z}}(1)(\mathbb{C})$ on the left hand side with the monodromy transformation on the right hand side. Here F_x denotes the (topological) Milnor fiber of f at x.

In the above theorem, the spaces $H^*(\mathscr{F}_x \times_K \widehat{K^s}, \mathbb{Q}_\ell)$ are the ℓ-adic cohomology spaces obtained from Berkovich's étale cohomology for non-archimedean spaces [4].

6.3 Points of the analytic Milnor fiber and arc spaces

The points on the analytic Milnor fiber $\mathscr{F}_x$ can be described in terms of arc spaces. To simplify notation, we'll assume that k is algebraically closed.

For each $d > 0$, we choose of a d-th root of π in K^a, denoted by $\sqrt[d]{\pi}$. This choice defines an isomorphism of k-algebras $R(d) \cong k[[\sqrt[d]{\pi}]]$. It also induces an isomorphism of R-algebras

$$\varphi_d : R(d) \to R(d)' : \sum_{i\geq 0} a_i(\sqrt[d]{\pi})^i \mapsto \sum_{i\geq 0} a_i\pi^i$$

where $R(d)'$ is the ring R with R-algebra structure given by

$$R \to R : \sum_{i\geq 0} b_i\pi^i \mapsto \sum_{i\geq 0} b_i\pi^{id}.$$

For any integer $d > 0$, we denote by $\mathscr{X}(d)$ the subset of $\mathscr{L}(X)(k)$ defined by

$$\mathscr{X}(d) = \{\psi \in \mathscr{L}(X)(k) \mid f(\psi) = t^d\}.$$

We put

$$\mathscr{X}(d)_x = \{\psi \in \mathscr{X}(d) \mid \theta_0(\psi) = x\}$$

where $\theta_0 : \mathscr{L}(X) \to X$ is the truncation morphism.

Proposition 6.5 *For each integer $d > 0$, there exists a canonical bijection*

$$\beta_d : \widehat{X}_\eta(K(d)) \to \mathscr{X}(d)$$

such that the square

$$\begin{array}{ccc}
\widehat{X}_\eta(K(d)) & \xrightarrow{\ \beta_d\ } & \mathscr{X}(d) \\
{\scriptstyle sp_{\widehat{X}}} \downarrow & & \downarrow {\scriptstyle \theta_0} \\
X_0(k) & \xrightarrow{\quad = \quad} & X_0(k)
\end{array}$$

commutes. The map β_d sends

$$\mathscr{F}_x(K(d)) = sp_{\widehat{X}}^{-1}(x)(K(d)) \subset \widehat{X}_\eta(K(d))$$

bijectively onto $\mathscr{X}(d)_x$.

Proof It suffices to prove the first part of the statement. The map

$$\gamma_d : \widehat{X}(R(d)) \to \widehat{X}_\eta(K(d))$$

obtained by passing to the generic fiber is a bijection, and for any $z \in \widehat{X}_\eta(K(d))$, the point $sp_{\widehat{X}}(z)$ is the image in $X_s(k)$ of the closed point of $\operatorname{Spec} R(d)$ under the morphism $\gamma_d^{-1}(z)$ [7, 0.2.2].

We put $X_R = X \times_{\mathbb{A}_k^1} R$. By Grothendieck's Existence Theorem [22, 5.4.1], the completion functor induces a bijection

$$X_R(R(d)) \to \widehat{X}(R(d)).$$

The R-isomorphism $\varphi_d : R(d) \to R(d)'$ induces a bijection

$$X_R(R(d)) \to X_R(R(d)').$$

But a section in $X_R(R(d)')$ is nothing but a morphism $\psi : \operatorname{Spec} R \to X$ such that $f(\psi) = \pi^d$, i.e., an arc in $\mathscr{X}(d)$, and the image of ψ_s is precisely $\theta_0(\psi)$. □

In other words, if we take an arc $\psi : \operatorname{Spec} R \to X$ with $f(\psi) = \pi^d$, then the morphism $\widehat{\psi}_\eta$ yields a $K(d)$-point on X_η, and this correspondence defines a bijection between $\mathscr{X}(d)$ and $\widehat{X}_\eta(K(d))$. Moreover, the image of ψ under the truncation map $\theta_0 : \mathscr{L}(X) \to X$ is nothing but the image of the corresponding element of $\widehat{X}_\eta(K(d))$ under the specialization morphism $sp_{\widehat{X}} : \widehat{X}_\eta \to \widehat{X}$.

The Galois group $G(K(d)/K) = \mu_d(k)$ acts on $\widehat{X}_\eta(K(d))$, and its action on the level of arcs is easy to describe: if ψ is an arc $\operatorname{Spec} R \to X$ with $f(\psi) = \pi^d$, and ξ is an element of $\mu_d(k)$, then $\xi \cdot \psi = \psi \circ \Phi_\xi$ with Φ_ξ the k-automorphism of $\operatorname{Spec} R$ defined by $\pi \mapsto \xi \cdot \pi$.

The spaces $\mathscr{X}(d)$, with their $\mu_d(k)$-action, are quite close to the arc spaces $\mathscr{X}_{d,1}$ appearing in the definition of the motivic zeta function associated to f. In fact, the motivic zeta function of f at x can be realized in terms of the motivic integral of a so-called *Gelfand-Leray form* on $\mathscr{F}_x$, as we will see in Section 6.5.

6.4 Motivic Weil generating series

In this section, we assume that k has characteristic zero. Let $\mathfrak{Y}$ be a special formal R-scheme, of pure relative dimension m. Forgetting the R-structure, we can also consider $\mathfrak{Y}$ as a special formal scheme over $\operatorname{Spec} k$. The sheaves of continuous differential forms $\Omega^j_{\mathfrak{Y}/k}$ are coherent, by [2, 3.3]. We say that a section ω in $\Omega^{m+1}_{\mathfrak{Y}/k}(\mathfrak{Y})$ is a gauge form on $\mathfrak{Y}$ over k if ω generates the stalk $(\Omega^{m+1}_{\mathfrak{Y}/k})_y$ at every point y of $\mathfrak{Y}$.

Consider the morphism of coherent $\mathscr{O}_\mathfrak{Y}$-modules

$$i : \Omega^m_{\mathfrak{Y}/k} \xrightarrow{d\pi\wedge} \Omega^{m+1}_{\mathfrak{Y}/k} : \omega \mapsto d\pi \wedge \omega$$

Since $d\pi \wedge \Omega^{m-1}_{\mathfrak{Y}/k}$ is contained in its kernel, and

$$\Omega^m_{\mathfrak{Y}/k} / \left(d\pi \wedge \Omega^{m-1}_{\mathfrak{Y}/k} \right) \cong \Omega^m_{\mathfrak{Y}/R}$$

by [2, 3.10], i descends to a morphism of coherent $\mathscr{O}_\mathfrak{Y}$-modules

$$i : \Omega^m_{\mathfrak{Y}/R} \xrightarrow{d\pi\wedge} \Omega^{m+1}_{\mathfrak{Y}/k}.$$

Passing to the generic fiber, we obtain a morphism of coherent $\mathscr{O}_{\mathfrak{Y}_\eta}$-modules

$$i : \Omega^m_{\mathfrak{Y}_\eta/K} \xrightarrow{d\pi\wedge} (\Omega^{m+1}_{\mathfrak{Y}/k})_{rig}$$

since $(\Omega^m_{\mathfrak{Y}/R})_{rig}$ is canonically isomorphic to $\Omega^m_{\mathfrak{Y}_\eta/K}$ by [14, 7.1.12]. In [34, 7.19 and 7.23] the first author proved the following result.

Proposition-Definition 6.6 *If $\mathfrak{Y}$ is a generically smooth special formal R-scheme of pure relative dimension m, then*

$$i : \Omega^m_{\mathfrak{Y}_\eta/K} \xrightarrow{d\pi\wedge} (\Omega^{m+1}_{\mathfrak{Y}/k})_{rig}$$

is an isomorphism. If ω is an element of $(\Omega^{m+1}_{\mathfrak{Y}/k})_{rig}(\mathfrak{Y}_\eta)$, we denote by $\omega/d\pi$ its inverse image in $\Omega^m_{\mathfrak{Y}_\eta/K}(\mathfrak{Y}_\eta)$, and we call it the Gelfand-Leray form associated to ω. If ϕ is an element of $\Omega^{m+1}_{\mathfrak{Y}/k}(\mathfrak{Y})$ we put $\phi/d\pi = \phi'/d\pi$, with ϕ' the image of ϕ in $(\Omega^{m+1}_{\mathfrak{Y}/k})_{rig}(\mathfrak{Y}_\eta)$. The form $\phi/d\pi$ is always $\mathfrak{Y}$-bounded (see Section 2.6). If ϕ is a gauge form on $\mathfrak{Y}$ over k, then $\phi/d\pi$ is a gauge form on $\mathfrak{Y}_\eta$.

This result allows us to state the following definition. If $\mathfrak{Y}$ is regular, then $\Omega^{m+1}_{\mathfrak{Y}/k}$ is a locally free $\mathscr{O}_{\mathfrak{Y}}$-module of rank one [34, 7.26], so that $\mathfrak{Y}$ locally admits a gauge form over k.

Definition 6.7 Let $\mathfrak{Y}$ be a regular special formal R-scheme. If $\mathfrak{Y}$ is of pure relative dimension, and $\mathfrak{Y}$ admits a gauge form ϕ over k, then we define the motivic Weil generating series of $\mathfrak{Y}$ by

$$S(\mathfrak{Y};T) = S(\mathfrak{Y},\phi/d\pi;T) \quad \in \mathscr{M}_{\mathfrak{Y}_0}[[T]].$$

In the general case, we choose a finite cover $\{\mathfrak{U}_i \mid i \in I\}$ of $\mathfrak{Y}$ by open formal subschemes of pure relative dimension, such that $\mathfrak{U}_i$ admits a gauge form over k for each $i \in I$, and we put

$$S(\mathfrak{Y};T) = \sum_{\emptyset \neq J \subset I} (-1)^{|J|+1} S(\cap_{j \in J} \mathfrak{U}_j;T) \quad \in \mathscr{M}_{\mathfrak{Y}_0}[[T]].$$

Since a gauge form ϕ on $\mathfrak{Y}$ over k is unique up to multiplication with a unit on $\mathfrak{Y}$, it is clear that $S(\mathfrak{Y};T)$ only depends on $\mathfrak{Y}$, and not on the choice of gauge form. It is also independent of the choice of the cover $\{\mathfrak{U}_i \mid i \in I\}$, by Theorem-Definition 2.43. Theorem 2.46 immediately implies the following result.

Theorem 6.8 *If $\mathfrak{Y}$ is a regular special formal R-scheme, then its motivic Weil generating series $S(\mathfrak{Y};T)$ belongs to*

$$\mathscr{M}_{\mathfrak{X}_0}\left[\frac{\mathbb{L}^a T^b}{1 - \mathbb{L}^a T^b}\right]_{(a,b)\in\mathbb{Z}\times\mathbb{Z}_{>0}}.$$

The name "motivic Weil generating series" is justified by the fact that, by Definition 2.44, we can consider the coefficient of T^d in the series $S(\mathfrak{Y};T)$ as a measure for the set of étale points on $\mathfrak{Y}_\eta \times_K K(d)$. Taking the images of the coefficients of $S(\mathfrak{Y};T)$ in

$$\mathscr{M}_k/(\mathbb{L}-1) \cong K_0(\mathrm{Var}_k)/(\mathbb{L}-1)$$

we find the series

$$\sum_{d>0} S(\mathfrak{Y}_\eta \times_K K(d))T^d$$

by Theorem-Definition 3.1.

The trace formula in Corollary 3.11 yields the following cohomological interpretation for the motivic Weil generating series. We denote by $\chi_{\text{top}}(S(\mathfrak{Y}; T))$ the element of $\mathbb{Z}[[T]]$ obtained by applying the forgetful morphism $\mathscr{M}_{\mathfrak{Y}_0} \to \mathscr{M}_k$ and the ℓ-adic Euler characteristic $\chi_{\text{top}} : \mathscr{M}_k \to \mathbb{Z}$ to the coefficients of $S(\mathfrak{Y}; T)$.

Theorem 6.9 *Let φ be a topological generator of the monodromy group $G(K^s/K^{sh})$. If $\mathfrak{Y}$ is a regular special formal R-scheme, then*

$$\chi_{\text{top}}(S(\mathfrak{Y}; T)) = \sum_{d>0}\sum_{i\geq 0}(-1)^i \text{Trace}(\varphi^d \mid H^i_{\text{ét}}(\mathfrak{Y}_\eta \times_K \widehat{K^s}, \mathbb{Q}_\ell))T^d.$$

6.5 The motivic zeta function as a Weil generating series

We return to the set-up and notation of Section 6.1. We assume that k has characteristic zero, and that X is smooth. The following theorem interprets the motivic zeta functions $Z_f(T)$ and $Z_{f,x}(T)$ as motivic Weil generating series.

Theorem 6.10 *If we denote by m the dimension of X, then we have the following equalities:*

$$Z_f(\mathbb{L}T) = S(\widehat{X}; T) \quad \in \mathscr{M}_{X_s}[[T]]$$
$$Z_{f,x}(\mathbb{L}T) = S(\mathfrak{X}; T) \quad \in \mathscr{M}_x[[T]].$$

Moreover, if we denote by k' the residue field of x and we consider $\mathscr{F}_x$ as a rigid $k'((\pi))$-variety, then

$$Z_{f,x}(\mathbb{L}T) = \sum_{d>0}\left(\int_{\mathscr{F}_x \times_K K(d)} |(\phi/d\pi)(d)|\right)T^d \quad \in \mathscr{M}_x[[T]]$$

for any gauge form ϕ on $\mathfrak{X}$ over k, where we denote by $(\phi/d\pi)(d)$ the pull-back of $\phi/d\pi$ to $\mathscr{F}_x \times_K K(d)$.

Note that $\mathfrak{X}$ admits a gauge form ϕ over k since $\mathfrak{X}$ is the formal spectrum of a regular complete local k-algebra. Theorem 6.10 was proven in [41, 9.10] and [34, 9.6 and 9.7], by computing both sides of the equality on a log-resolution for the ideal sheaf (f) on X. Note that our comparison result takes a slightly different shape, because of the choice of normalization we made in Section 2.3.

Together with Theorem 6.10, the trace formula in Theorem 6.9 yields a cohomological interpretation of $Z_{f,x}(T)$.

Theorem 6.11 *If φ is a topological generator of the geometric monodromy group $G(K^s/K^{sh})$, then*

$$\chi_{\text{top}}(Z_{f,x}(T)) = \sum_{d>0}\sum_{i\geq 0}(-1)^i\text{Trace}(\varphi^d \mid H^i_{\text{ét}}(\mathscr{F}_x \times_K \widehat{K^s}, \mathbb{Q}_\ell))T^d.$$

If we combine this cohomological interpretation with the comparison result in Theorem 6.4, we recover the following important result by Denef and Loeser [16, 1.1].

Corollary 6.12 *Assume that $k = \mathbb{C}$. For each $d > 0$, we have*

$$\chi_{\text{top}}(\mathscr{X}_{d,1,x}) = \sum_{i\geq 0}(-1)^i\text{Trace}(M_x^d \mid H^i_{\text{sing}}(F_x, \mathbb{Q}))$$

where F_x is the topological Milnor fiber of f at x, and M_x the monodromy transformation.

6.6 Non-archimedean geometry and complex singularities

The preceding results show how the analytic Milnor fiber forms a bridge between arc spaces and motivic zeta functions on the one hand, and the monodromy transformation on the other hand. Our interpretation of the motivic zeta function as a "Weil generating series" of a rigid variety makes it possible to apply ideas and techniques from arithmetic and non-archimedean geometry to the study of motivic zeta functions and complex singularities. The following proposition gives an elementary illustration of the interplay between arithmetic geometry and complex singularity theory.

Proposition 6.13 *Assume that k is algebraically closed. Let X be a smooth irreducible k-variety, let $f : X \to \text{Spec}\,k[t]$ be a dominant morphism, and let x be a closed point of X_s. The following are equivalent:*

1. *the morphism f is smooth at x,*
2. *the analytic Milnor fiber $\mathscr{F}_x$ of f at x contains a $k((t))$-rational point,*
3. *the analytic Milnor fiber $\mathscr{F}_x$ of f at x satisfies $S(\mathscr{F}_x) \neq 0$.*

If k has characteristic zero, and if we denote by φ a topological generator of the monodromy group $G(K^s/K)$, then each of the above statements is also equivalent to

4. the analytic Milnor fiber $\mathscr{F}_x$ of f at x satisfies

$$\sum_{i\geq 0}(-1)^i\mathrm{Trace}(\varphi \mid H^i_{\text{ét}}(\mathscr{F}_x \times_K \widehat{K^s}, \mathbb{Q}_\ell)) \neq 0.$$

If $k = \mathbb{C}$, and if we denote by F_x the classical topological Milnor fiber of f at x and by M_x the monodromy transformation, then each of the above statements is also equivalent to

5. we have

$$\sum_{i\geq 0}(-1)^i\mathrm{Trace}(M_x \mid H^i_{\text{sing}}(F_x, \mathbb{C})) \neq 0.$$

Proof The implication (1) $\Rightarrow$ (2) follows from Proposition 2.5. The implication (2) $\Rightarrow$ (1) follows from Proposition 2.37. The implication (4) $\Rightarrow$ (3) follows from the trace formula in Theorem 3.11, (3) $\Rightarrow$ (2) is obvious, and (1) $\Rightarrow$ (4) follows from [5, 3.5] and the triviality of the ℓ-adic nearby cycles of f at x. Finally, the equivalence (4) $\Leftrightarrow$ (5) follows from Theorem 6.4. □

The equivalence of (1) and (5) (for $k = \mathbb{C}$) is a classical result by A'Campo [1].

6.7 Singular cohomology of the analytic Milnor fiber

As another application of non-archimedean geometry to the theory of complex hypersurface singularities, we state the main result from [35]. We keep the notations of Section 6.1. We assume that $k = \mathbb{C}$, and that the generic fiber of f is smooth. Considering the rigid K-variety $\mathscr{F}_x$ as a Berkovich K-analytic space [3], we can take its singular cohomology $H^*_{\text{sing}}(\mathscr{F}_x \times_K \widehat{K^a}, \mathbb{Q})$. The following theorem relates these cohomology spaces to the natural mixed Hodge structure on the nearby cohomology $R^*\psi_f(\mathbb{Q})_x$ of f at x. This mixed Hodge structure was defined by Steenbrink in the case where X is smooth and f has an isolated singularity at x [51], and by Navarro Aznar in the general case [32].

Theorem 6.14 ([35], Theorem 5.7) *For each $i \geq 0$ there exists a canonical isomorphism*

$$H^i_{\text{sing}}(\mathscr{F}_x \times_K \widehat{K^a}, \mathbb{Q}) \cong W_0 R^i\psi_f(\mathbb{Q})_x$$

where the right hand side is the weight zero part of the mixed Hodge structure on $R^i\psi_f(\mathbb{Q})_x$.

This result is a local analogue of a result by Berkovich on the limit mixed Hodge structure of a proper family over $\mathbb{A}^1_{\mathbb{C}}$ [6].

7 Further results

7.1 The motivic volume

Assume that k has characteristic zero. By formally taking a limit $T \to \infty$ of the motivic Weil generating series, one can associate a so-called motivic volume to generically smooth special formal R-schemes and to a certain class of smooth rigid K-varieties. If k is algebraically closed, then the motivic volume of a rigid K-variety X should be considered as the motivic measure of $X \times_K \widehat{K^a}$. The motivic volume of the analytic Milnor fiber coincides, modulo normalization, with Denef and Loeser's motivic Milnor fiber, and a similar interpretation exists for the motivic nearby cycles. We refer to [34, § 7.4] for precise definitions, and to [35, § 6] for a computation in terms of semi-stable models.

7.2 The motivic zeta function of an abelian variety

Assume that k is algebraically closed. If A is an abelian K-variety of dimension g, then one can associate a motivic zeta function $Z_A(T)$ to A as follows. Denote by $\mathscr{A}$ the Néron model of A, and let ω be a translation-invariant gauge form on A such that $\mathrm{ord}_{\mathscr{A}_s \circ}\omega = 0$. Such a gauge form always exists, and it is unique up to multiplication with a unit in R.

We denote, for each $d \in \mathbb{N}'$, by $\omega(d)$ the pull-back of ω to $A \times_K K(d)$ and by $\mathscr{A}(d)$ the Néron model of $A \times_K K(d)$. The motivic zeta function $Z_A(T)$ of A is defined as

$$Z_A(T) = \sum_{d \in \mathbb{N}'} [\mathscr{A}(d)_s] \mathbb{L}^{-\mathrm{ord}_{\mathscr{A}(d)_s^o}\omega(d)} \in \mathscr{M}_k[[T]].$$

This series only depends on A, and not on ω. It measures the behaviour of the Néron model of A under tame base change. Its image in $\mathscr{M}_k^R[[T]]$ equals

$$Z_A(T) = S(A, \omega; T) \in \mathscr{M}_k^R[[T]].$$

We see that $Z_A(T)$ can be expressed in terms of motivic integrals of the "motivic Haar measure" on A defined by ω.

The *base change conductor* $c(A)$ of A was defined by Chai in [13]. It is a rational number that measures the defect of semi-abelian reduction of A. In particular, $c(A) = 0$ iff A has semi-abelian reduction. L. Halvard Halle and the first author proved in [24, 25] the following global version of the motivic monodromy conjecture.

Theorem 7.1 *Assume that A is cohomologically tame. The motivic zeta function $Z_A(T)$ belongs to*

$$\mathscr{M}_k \left[T, \frac{1}{1 - \mathbb{L}^a T^b} \right]_{(a,b)\in\mathbb{Z}\times\mathbb{Z}_{>0},\ a/b=c(A)}.$$

The zeta function $Z_A(\mathbb{L}^{-s})$ has a unique pole at $s = c(A)$, and the order of this pole equals one plus the potential toric rank $t_{pot}(A)$ of A. Moreover, for any embedding of $\mathbb{Q}_\ell$ in $\mathbb{C}$, the value $\exp(2\pi i c(A))$ is an eigenvalue of φ on $H^g(A \times_K K^t, \mathbb{Q}_\ell)$, for every topological generator φ on $G(K^t/K)$, and the size of the largest Jordan block for this eigenvalue is $1 + t_{pot}(A)$.

Recall that A is cohomologically tame iff the wild inertia acts trivially on the ℓ-adic Tate module of A, or, equivalently, iff A acquires semi-abelian reduction after base change to a finite tame extension of K. We refer to [24] for details and further results.

References

[1] N. A'Campo. Le nombre de Lefschetz d'une monodromie. *Indag. Math.*, 35:113–118, 1973.

[2] L. Alonso Tarrío, A. Jeremías López, and M. Pérez Rodríguez. Infinitesimal lifting and Jacobi criterion for smoothness on formal schemes. *Commun. Algebra*, 35(4):1341–1367, 2007.

[3] V. G. Berkovich. *Spectral theory and analytic geometry over non-archimedean fields*, volume 33 of *Mathematical Surveys and Monographs*. AMS, 1990.

[4] V. G. Berkovich. Étale cohomology for non-Archimedean analytic spaces. *Publ. Math., Inst. Hautes Étud. Sci.*, 78:5–171, 1993.

[5] V. G. Berkovich. Vanishing cycles for formal schemes, II. *Invent. Math.*, 125(2):367–390, 1996.

[6] V. G. Berkovich. A non-Archimedean interpretation of the weight zero subspaces of limit mixed Hodge structures. In *Algebra, Arithmetic and Geometry - Manin Festschrift (to appear)*. Boston: Birkhäuser.

[7] P. Berthelot. Cohomologie rigide et cohomologie rigide à supports propres. *Prepublication, Inst. Math. de Rennes*, 1996.

[8] S. Bosch, U. Güntzer, and R. Remmert. *Non-Archimedean analysis. A systematic approach to rigid analytic geometry*, volume 261 of *Grundlehren der Mathematischen Wissenschaften*. Springer Verlag, 1984.

[9] S. Bosch and W. Lütkebohmert. Formal and rigid geometry. I: Rigid spaces. *Math. Ann.*, 295(2):291–317, 1993.

[10] S. Bosch and W. Lütkebohmert. Formal and rigid geometry. II: Flattening techniques. *Math. Ann.*, 296(3):403–429, 1993.

[11] S. Bosch, W. Lütkebohmert, and M. Raynaud. *Néron models*, volume 21. Ergebnisse der Mathematik und ihrer Grenzgebiete, 1990.

[12] S. Bosch and K. Schlöter. Néron models in the setting of formal and rigid geometry. *Math. Ann.*, 301(2):339–362, 1995.

[13] C.L. Chai. Néron models for semiabelian varieties: congruence and change of base field. *Asian J. Math.*, 4(4):715–736, 2000.

[14] A. J. de Jong. Crystalline Dieudonné module theory via formal and rigid geometry. *Publ. Math., Inst. Hautes Étud. Sci.*, 82:5–96, 1995.

[15] J. Denef and F. Loeser. Geometry on arc spaces of algebraic varieties. *Progr. Math.*, 201:327–348, 2001.

[16] J. Denef and F. Loeser. Lefschetz numbers of iterates of the monodromy and truncated arcs. *Topology*, 41(5):1031–1040, 2002.

[17] A. Dimca. *Singularities and Topology of Hypersurfaces*. Springer-Verlag, 1992.

[18] J. Fresnel and M. van der Put. *Rigid analytic geometry and its applications*, volume 218 of *Progress in Mathematics*. Boston, MA: Birkhäuser, 2004.

[19] M.J. Greenberg. Schemata over local rings. *Ann. Math.*, 73(2):624–648, 1961.

[20] M.J. Greenberg. Schemata over local rings II. *Ann. Math.*, 78(2):256–266, 1963.

[21] A. Grothendieck and J. Dieudonné. Eléments de Géométrie Algébrique, I. *Publ. Math., Inst. Hautes Étud. Sci.*, 4:5–228, 1960.

[22] A. Grothendieck and J. Dieudonné. Eléments de Géométrie Algébrique, III. *Publ. Math., Inst. Hautes Étud. Sci.*, 11:5–167, 1961.

[23] A. Grothendieck and J. Dieudonné. Eléments de Géométrie Algébrique, IV. *Publ. Math., Inst. Hautes Étud. Sci.*, 32:5–361, 1967.

[24] L. Halvard Halle and J. Nicaise. Motivic zeta functions for abelian varieties, and the monodromy conjecture. *Adv. Math.*, 227:610–653, 2011.

[25] L. Halvard Halle and J. Nicaise. Jumps and monodromy of abelian varieties. preprint, arXiv:1009.3777.

[26] J. Igusa. *An introduction to the theory of local zeta functions*. Studies in Advanced Mathematics. AMS, 2000.

[27] L. Illusie. Grothendieck's Existence Theorem in formal geometry. In B. Fantechi, L. Goettsche, L. Illusie, S. Kleiman, N. Nitsure, and A. Vistoli, editors, *Fundamental Algebraic Geometry, Grothendieck's FGA explained*, Mathematical Surveys and Monographs. AMS, 2005.

[28] J. Lipman. The Picard group of a scheme over an Artin ring. *Publ. Math., Inst. Hautes Étud. Sci.*, 46:15–86, 1976.

[29] F. Loeser and J. Sebag. Motivic integration on smooth rigid varieties and invariants of degenerations. *Duke Math. J.*, 119:315–344, 2003.

[30] J. S. Milne. *Étale Cohomology*, volume 33 of *Princeton Mathematical Series*. Princeton University Press, 1980.

[31] J. Milnor. *Singular points of complex hypersurfaces*, volume 61 of *Annals of Math. Studies*. Princeton University Press, 1968.

[32] V. Navarro Aznar. Sur la théorie de Hodge-Deligne. *Invent. Math.*, 90:11–76, 1987.

[33] J. Nicaise. Formal and rigid geometry: an intuitive introduction and some applications. *Enseign. Math. (2)*, 54:1–37, 2008.

[34] J. Nicaise. A trace formula for rigid varieties, and motivic weil generating series for formal schemes. *Math. Ann.*, 343(2):285–349, 2009.

[35] J. Nicaise. Singular cohomology of the analytic Milnor fiber, and mixed Hodge structure on the nearby cohomology. *J. Algebraic Geom.*, 20:199–237, 2011.

[36] J. Nicaise. An introduction to *p*-adic and motivic zeta functions and the monodromy conjecture. In: G. Bhowmik, K. Matsumoto and H. Tsumura (eds.), *Algebraic and analytic aspects of zeta functions and L-functions*. Volume 21 of *MSJ Memoirs*, Mathematical Society of Japan, Tokyo, 2010, pages 115–140.

[37] J. Nicaise. A trace formula for varieties over a discretely valued field. *J. Reine Angew. Math.*, 650:193–238, 2011.

[38] J. Nicaise. Trace formula for component groups of Néron models. *preprint*, arXiv:0901.1809v2.

[39] J. Nicaise. Geometric criteria for tame ramification. *preprint*, arXiv:arXiv:0910.3812.

[40] J. Nicaise and J. Sebag. Motivic Serre invariants of curves. *Manuscr. Math.*, 123(2):105–132, 2007.

[41] J. Nicaise and J. Sebag. The motivic Serre invariant, ramification, and the analytic Milnor fiber. *Invent. Math.*, 168(1):133–173, 2007.

[42] J. Nicaise and J. Sebag. Motivic Serre invariants and Weil restriction. *J. Algebra*, 319(4):1585–1610, 2008.

[43] M. Rapoport and Th. Zink. *Period spaces for p-divisible groups*. Volume 141 of *Annals of Mathematics Studies*. Princeton University Press, Princeton, NJ, 1996.

[44] M. Raynaud. Géométrie analytique rigide d'après Tate, Kiehl, ... *Mémoires de la S.M.F.*, 39-40:319–327, 1974.

[45] B. Rodrigues and W. Veys. Poles of Zeta functions on normal surfaces. *Proc. London Math. Soc.*, 87(3):164–196, 2003.

[46] T. Saito. Vanishing cycles and geometry of curves over a discrete valuation ring. *Am. J. Math.*, 109:1043–1085, 1987.

[47] J. Sebag. Intégration motivique sur les schémas formels. *Bull. Soc. Math. France*, 132(1):1–54, 2004.

[48] J.-P. Serre. *Groupes algébriques et corps de classes*. Paris: Hermann & Cie, 1959.

[49] J.-P. Serre. *Corps locaux*. Paris: Hermann & Cie, 1962.

[50] J.-P. Serre. Classification des variétés analytiques p-adiques compactes. *Topology*, 3:409–412, 1965.

[51] J.H.M. Steenbrink. Mixed Hodge structure on the vanishing cohomology. In P. Holm, editor, *Real and complex Singularities, Proc. Nordic Summer Sch., Symp. Math., Oslo 1976*, pages 525–563. Alphen a.d. Rijn: Sijthoff & Noordhoff, 1977.

[52] M. Temkin. Desingularization of quasi-excellent schemes in characteristic zero. *Adv. Math.*, 219(2):488–522, 2008.

8

Motivic integration in mixed characteristic with bounded ramification: a summary

Raf Cluckers and François Loeser

1 Introduction

Though one can consider Motivic integration to have quite satisfactory foundations in residue characteristic zero after [9], [10] and [18], much remains to be done in positive residue characteristic. The aim of the present paper is to explain how one can extend the formalism and results from [9] to mixed characteristic.

Let us start with some motivation. Let K be a fixed finite extension of $\mathbb{Q}_p$ with residue field $\mathbb{F}_q$ and let K_d denote its unique unramified extension of degree d, for $d \geq 1$. Let us write by $\mathscr{O}_d$ the ring of integers of K_d and fix a polynomial $h \in \mathscr{O}_1[x_1, \cdots, x_m]$. For each d one can consider the Igusa local zeta function

$$Z_d(s) = \int_{\mathscr{O}_d^m} |h(x)|_d^s |dx|_d,$$

with $|\_|_d$ and $|dx|_d$ the corresponding norm and Haar measure such that the measure of $\mathscr{O}_d$ is 1 and such that $|a|_d$ for any $a \in K_d$ equals the measure of $a\mathscr{O}_d$. Meuser in [20] proved that there exist polynomials G and H in $\mathbb{Z}[T, X_1, \cdots, X_t]$ and complex numbers $\lambda_1, \cdots, \lambda_t$ such that, for all $d \geq 1$,

$$Z_d(s) = \frac{G(q^{-ds}, q^{d\lambda_1}, \cdots, q^{d\lambda_t})}{H(q^{-ds}, q^{d\lambda_1}, \cdots, q^{d\lambda_t})}.$$

Later Pas [22], [23] extended Meuser's result to more general integrals. In view of [15] and [16], it is thus natural to expect that there exists some

Motivic Integration and its Interactions with Model Theory and Non-Archimedean Geometry (Volume I), ed. Raf Cluckers, Johannes Nicaise, and Julien Sebag. Published by Cambridge University Press. © Cambridge University Press 2011.

motivic rational function $Z_{\text{mot}}(T)$ with coefficients in some localization of the Grothendieck ring $G_{\mathbb{F}_q}$ of definable sets over $\mathbb{F}_q$ such that, for every $d \geq 1$, $Z_d(s)$ is obtained from $Z_{\text{mot}}(T)$ by using the morphism $G_{\mathbb{F}_q} \to \mathbb{Z}$ counting rational points over $\mathbb{F}_{q^d}$ and letting T go to q^{-ds}. The theory presented here allows to prove such a result (more generally for h replaced by a definable function).

Another motivation for the present work lies in a joint project with J. Nicaise [12], where we prove some cases of a conjecture of Chai on the additivity of his base change conductor for semi-abelian varieties [2] and [3] by using Fubini's theorem for Motivic Integration from this paper for the mixed and from [9] for the equal characteristic zero case.

One of the achievements of motivic integration is the definition of measure and integrals on more general Henselian valued fields than just locally compact ones, for example on Laurent series fields over a characteristic zero field [17], [14], on complete discrete valuation rings with perfect residue field [19], [24], [21], and on algebraically closed valued fields [18]. A second achievement, since [16], and continued in [5] [9] [10], is that motivic integration can be used to interpolate p-adic integrals for all finite field extension of $\mathbb{Q}_p$ and integrals over $\mathbb{F}_q((t))$, uniformly in big primes p and its powers q. A third main achievement which goes together with the introduction of motivic additive characters and their motivic integrals, is the Transfer Principle of [10] which allows one to transfer equalities between integrals defined over $\mathbb{Q}_p$ to equalities of integrals defined over $\mathbb{F}_p((t))$ and vice versa. This is useful to change the characteristic in statements like the Fundamental Lemma in the Langlands program as is done in [4] [25]. In this paper we will focus on the first two mentioned achievements of motivic integration, in a mixed characteristic context. Firstly, for fixed prime p and integer $e > 0$, we will define the motivic measure and integrals on all Henselian discretely valued fields of mixed characteristic $(0, p)$ and ramification degree e, which will coincide with the standard measure in the case of p-adic fields. Secondly, our approach will be uniform in all unramified, Henselian field extensions, and hence, it will give an interpolation of p-adic integrals for all p-adic fields with ramification degree e.

Let us explain why the third mentioned achievement of motivic integration is not implemented. Of course it would be possible to implement a motivic additive character and motivic 'oscillating' integrals involving the character in our context, but this is left out since it is natural to do this as a next step. However, a Transfer Principle would not make sense in the present context. We will explain this in the most basic instance of the Transfer Principle, namely in a concrete setting falling under the Ax-Kochen Eršov principle. Suppose that one has a definable subassignment φ as in [9], namely, a definable subset in the equicharacteristic zero valued field context. Then it is explained in [10] that, for p big enough, φ gives a subset of some Cartesian power of $\mathbb{Q}_p$, and similarly of some Cartesian power of $\mathbb{F}_p((t))$, which is independent of the choice of the formulas which give φ. This independence can be easily obtained from Gödel's Completeness Theorem (together with the knowledge of the appropriate theory of Henselian valued fields as used in loc. cit.). Indeed, the fact that two formulas ψ_1 and ψ_2 both give the same definable subassignment φ is a first order sentence, and hence, by Gödel's result, can be deduced from finitely many axioms from the theory. A finite collection of axioms can at most specify that the residue field characteristic is different from some concrete primes, but it can never specify that the characteristic is zero. Hence, the fields $\mathbb{Q}_p$ for p big enough are also models of this finite collection of axioms and hence ψ_1 and ψ_2 both give the same definable set over $\mathbb{Q}_p$ (and similarly over $\mathbb{F}_p((t))$) for big p. Note that this is a basic instance of having some result in equicharacteristic 0, and deriving the analogous result in big enough residue field characteristic. More generally the Transfer Principle transfers results about integrals from a $\mathbb{Q}_p$ context to an $\mathbb{F}_p((t))$ context and vice versa, even when the results not necessarily hold motivically in equicharacteristic zero. It is clear that such techniques are not applicable to our context since the fact that a valued field has mixed characteristic $(0, p)$ and ramification degree e is completely expressible by a finite set of axioms. Hence, we cannot change the characteristic for any of the properties obtained for integrals on the mixed characteristic valued fields. On the other hand, what we gain (as opposed to the equicharacteristic zero context of [9] and [10]), is that any motivic relation, calculation, equality, and so on, will hold for all the p-adic fields of the correct ramification degree and correct residue characteristic (as opposed to for p big enough as in [9] and [10]).

A basic tool in our approach is to use higher order angular components maps $\overline{\mathrm{ac}}_n$ for integers $n \geq 1$, already used by Pas in [22], where $\overline{\mathrm{ac}}_n$ is a certain multiplicative map from the valued field K to the residue ring $\mathcal{O}_K/\mathcal{M}_K^n$ with $\mathcal{M}_K$ the maximal ideal of the valuation ring $\mathcal{O}_K$. We use several structure results about definable sets and definable functions in first order languages involving the $\overline{\mathrm{ac}}_n$, one of which is called cell decomposition and goes back to [22] and [5]. Note that the approach of this paper with the $\overline{\mathrm{ac}}_n$ would also work in equicharacteristic zero discretely valued Henselian fields, and it has the advantage of providing much more definable sets than with $\overline{ac} = \overline{ac}_1$ only, for instance all cylinders over definable sets are definable with the $\overline{ac}_n$, which is not the case if one uses only the usual angular component $\overline{ac}$. In mixed characteristic there is a basic interplay between the residue characteristic p, the ramification degree e, and the angular component maps $\overline{ac}_n$. Indeed, suppose for example that $p = 2$, and that we want to lift a root x_0 of a polynomial f by Hensel's Lemma. If it happens that $f'(x_0) = 2$ for some x_0, then we typically have to know that $f(x_0)$ is zero modulo 2^2 in order to uniquely lift x_0 to a zero of f. Hence, one should be able to speak about approximate roots modulo $\mathscr{M}_K^n$ with $n = 2e$. Such basic phenomena indicate the need of considering higher order residue rings in the setup, instead of only considering the residue field as in [9] [10].

Similarly as in [9] we consequently study families of motivic integrals, and we obtain many similar results as in [9]. However, we give a more direct approach to definitions and properties of the motivic measure and functions than in [9]: instead of the existence-uniqueness theorem of section 10 of [9], we explicitly define the motivic integrals and the integrability conditions and we do this step by step, as an iteration of more simple integrals. These explicit definitions give the same motivic measure and integrals as the ones that come from a direct image framework. Of course one has to be careful when translating conditions about integrability of an integral over a product space to conditions on the iterated integral. Since the value group is the group of integers, the origin of all integrability issues in our approach is summation on the integers, where one has the following Tonelli variant of Fubini's Theorem:

Suppose that $f : \mathbb{Z}^{k+\ell} \to \mathbb{R}$ is a function taking non-negative values in $\mathbb{R}$. Then one has in $\mathbb{R} \cup \{+\infty\}$ that (without any convergence conditions)

$$\sum_{(x,y)\in\mathbb{Z}^{k+\ell}} f(x,y) = \sum_{x\in\mathbb{Z}^k} g(x)$$

with $g : \mathbb{Z}^k \to \mathbb{R} \cup \{+\infty\} : x \mapsto \sum_{y\in\mathbb{Z}^\ell} f(x,y)$. Hence, f is summable over $\mathbb{Z}^{k+\ell}$ if and only if g is real valued and summable over $\mathbb{Z}^k$.

Hence, the Tonelli variant of Fubini's Theorem for non-negatively valued functions translates integrability issues to iterated, more simple integrals. The notion of non-negativity for motivic functions is defined as in [9], namely, by using semi-rings as value rings of the motivic functions (as opposed to rings). Note that in a semi-ring, every element is considered as nonnegative, like one does in the semi-ring $\mathbb{N}$.

One new feature that does not appear in [9], and which provides more flexibility in view of future applications, is the usage of the abstract notion of $\mathscr{T}$-fields, where $\mathscr{T}$ stands for a first order theory. The reader has the choice to work with some of the listed more concrete examples of $\mathscr{T}$-fields (which are close to the concrete semi-algebraic setup of [9] or the subanalytic setup of [5]) or with axiomatic, abstract $\mathscr{T}$-fields. Thus, $\mathscr{T}$-fields allow one to work with more general theories $\mathscr{T}$ than the theories in the original work by Pas. Note that also this feature of integration on $\mathscr{T}$-fields would work similarly in the equicharacteristic zero context, as a generalization of [9].

Some of the highlights of our formalism coincide with the highlights of [9], consisting of a general change of variables formula, a general Fubini Theorem, the ability to specialize to previously known theories of motivic integration (e.g. as in [19]), the ability to interpolate many p-adic integrals, avoidance of any completion process on the Grothendieck ring level, and, very importantly, the ability to work in parameter setups where the parameters can come from the valued field, the residue field, and the value group (this last property has been very useful in [4] and [25]).

To make our work more directly comparable and linked with [9], we write down in Section 11 how our more direct definitions of integrable constructible motivic functions lead naturally to a direct image formalism, analogous to the one in [9]. Let us indicate how [9] and this paper complement each other, by an example. Having an equality between two motivic integrals as in [9] implies that the analogous equality will hold over all p-adic fields for p big enough and all fields $\mathbb{F}_q((t))$ of big enough characteristic (the lower bound can be computed but is usually very bad). This leaves one with finitely 'small' primes p, say, primes which are less than N. For the fields $\mathbb{F}_q((t))$ of small characteristic, very little is known in general and one must embark on a case by case study. On the other hand, in mixed characteristic, one could use the framework of this paper finitely many times to obtain the equality for all p-adic fields with residue characteristic less than N and bounded ramification degree. Note that knowing an equality for a small prime p and all possible ramification degrees is more or less equivalent to knowing it in $\mathbb{F}_p((t))$, which as we mentioned can be very hard.

We end Section 11 with a comparison with work by J. Sebag and the second author on motivic integration in a smooth, rigid, mixed characteristic context. This comparison plays a role in [12].

This is a summary and contains almost no proofs. We refer to our paper in preparation [11] for complete proofs. Also in [11], we will implement both the mixed characteristic approach of this paper and the equicharacteristic zero approach, generalizing and complementing [9].

2 A concrete setting

2.1

A discretely valued field L is a field with a surjective group homomorphism ord : $L^\times \to \mathbb{Z}$, satisfying the usual axioms of a non-archimedean valuation. A ball in L is by definition a set of the form $\{x \in L \mid \gamma \leq \text{ord}(x - a)\}$, where $a \in L$ and $\gamma \in \mathbb{Z}$. The collection of balls in L forms a base for the so-called valuation topology on L. The valued field L is called Henselian if its valuation ring $\mathscr{O}_L$ is a Henselian ring. Write $\mathscr{M}_L$ for the maximal ideal of $\mathscr{O}_L$.

In the whole paper we will work with the notion of $\mathscr{T}$-fields, which is more specific than the notion of discretely valued field, but which can come with additional structure if one wants. The reader who wants to avoid the formalism of $\mathscr{T}$-fields may skip Section 3 and directly go to Section 4 and use the following concrete notion of $(0, p, e)$-fields instead of $\mathscr{T}$-fields.

Definition 2.1 Fix an integer $e > 0$ and a prime number p. A $(0, p, e)$-field is a Henselian, discretely valued field K of characteristic 0, residue field characteristic p, and ramification degree e, together with a chosen uniformizer π_K of the valuation ring $\mathscr{O}_K$ of K. That the ramification degree of K is e means that $\text{ord}\pi_K^e = \text{ord}p = e$.

Note that $\mathbb{Q}_p$ together with, for example, p as a uniformizer is a $(0, p, 1)$-field, as well as the algebraic closure of $\mathbb{Q}$ inside $\mathbb{Q}_p$, or any unramified, Henselian field extension of $\mathbb{Q}_p$. A $(0, p, e)$-field K comes with natural so-called higher order angular component maps for $n \geq 1$,

$$\overline{\text{ac}}_n : K^\times \to (\mathscr{O}_K \bmod \mathscr{M}_K^n) : x \mapsto \pi_K^{-\text{ord}x} x \bmod \mathscr{M}_K^n$$

extended by $\overline{\text{ac}}_n(0) = 0$. Sometimes one writes $\overline{\text{ac}}$ for $\overline{\text{ac}}_1$. Each map $\overline{\text{ac}}_n$ is multiplicative on K and coincides on $\mathscr{O}_K^\times$ with the natural projection $\mathscr{O}_K \to \mathscr{O}_K/\mathscr{M}_K^n$.

2.2

To describe sets in a field independent way, we will use first order languages, where the following algebraic one is inspired by languages of Denef and Pas. Its name comes from the usage of higher order angular component maps, namely modulo positive powers of the maximal ideal. Consider the following basic language $\mathscr{L}_{\text{high}}$ which has a sort for the valued field, a sort for the value group, and a sort for each residue ring of the valuation ring modulo π^n for integers $n > 0$. On the collection of these sorts, $\mathscr{L}_{\text{high}}$ consists of the language of rings for the valued field together with a symbol π for the uniformizer, the language of rings for each of the residue rings, the Presburger language $(+, -, 0, 1, \leq, \{\cdot \equiv \cdot \bmod n\}_{n>1})$ for the value group, a symbol ord for the valuation map, symbols $\overline{\text{ac}}_n$ for integers $n > 0$ for the angular component maps modulo the n-th power of the maximal ideal, and projection maps $p_{n,m}$ between the residue rings for $n \geq m$. On each $(0, p, e)$-field K, the language $\mathscr{L}_{\text{high}}$ has its natural meaning, where π stands for π_K, ord for the valuation $K^\times \to \mathbb{Z}$, $\overline{\text{ac}}_n$ for the angular component map $K \to \mathscr{O}_K/\mathscr{M}_K^n$, and $p_{n,m}$ for the natural projection map from $\mathscr{O}_K/\mathscr{M}_K^n$ to $\mathscr{O}_K/\mathscr{M}_K^m$.

Let $\mathscr{T}_{(0,p,e)}$ be the theory in the language $\mathscr{L}_{\text{high}}$ of sentences that are true in all $(0, p, e)$-fields. Thus, in particular, each $(0, p, e)$-field is a model of $\mathscr{T}_{(0,p,e)}$. In this concrete setting, we let $\mathscr{T}$ be $\mathscr{T}_{(0,p,e)}$ in the language $\mathscr{L}_{\text{high}}$, and $\mathscr{T}$-field means $(0, p, e)$-field. One can give a concrete list of axioms that imply the whole theory $\mathscr{T}_{(0,p,e)}$ (see [22]), but this is not relevant to this paper.

3 Theories on $(0, p, e)$-fields

In total we give three approaches to $\mathscr{T}$-fields in this paper, so that the reader can choose which one fits him best. The first one is the concrete setting of Section 2; the second one consists of a list of more general and more adaptable settings in Section 3.1, and the third approach is the axiomatic approach for theories and languages on $(0, p, e)$-fields in Section 3.2. Recall that for the first approach one takes $\mathscr{T} = \mathscr{T}_{(0,p,e)}$ in the language $\mathscr{L}_{\text{high}}$, and $\mathscr{T}$-field just means $(0, p, e)$-field.

3.1 A list of theories

In our second approach, we give a list of theories and corresponding languages which can be used throughout the whole paper.

1. Most closely related to the notion of $(0, p, e)$-fields is that of $(0, p, e)$-fields over a given ring R_0, for example a ring of integers, using the language $\mathscr{L}_{\text{high}}(R_0)$. Namely, for R_0 a subring of a $(0, p, e)$-field, let $\mathscr{L}_{\text{high}}(R_0)$ be the language $\mathscr{L}_{\text{high}}$ with coefficients (also called parameters) from R_0, and let $\mathscr{T}_{(0,p,e)}(R_0)$ be the theory of $(0, p, e)$-fields over R_0 in the language $\mathscr{L}_{\text{high}}(R_0)$. In this case one takes $\mathscr{T} = \mathscr{T}_{(0,p,e)}(R_0)$ with language $\mathscr{L}_{\text{high}}(R_0)$. By a $(0, p, e)$-field K over R_0 we mean in particular that the order and angular component maps on K extend the order and angular component maps on R_0.

2. In order to include analytic functions, let K be a $(0, p, e)$-field, and for each integer $n \geq 1$ let $K\{x_1, \ldots, x_n\}$ be the ring of those formal power series $\sum_{i \in \mathbb{N}^n} a_i x^i$ over K such that $\text{ord}(a_i)$ goes to $+\infty$ whenever $i_1 + \ldots + i_n$ goes to $+\infty$. Let $\mathscr{L}_K$ be the language $\mathscr{L}_{\text{high}}$ together with function symbols for all the elements of the rings $K\{x_1, \ldots, x_n\}$, for all $n > 0$. Each complete $(0, p, e)$-field L over K allows a natural interpretation of the language $\mathscr{L}_K$, where f in $K\{x_1, \ldots, x_n\}$ is interpreted naturally as a function from $\mathscr{O}_L^n$ to L. Let $\mathscr{T}_K$ be the theory in the language $\mathscr{L}_K$ of the collection of complete $(0, p, e)$-fields L over K. In this case one takes $\mathscr{T} = \mathscr{T}_K$ with language $\mathscr{L}_K$. For an explicit list of axioms that implies $\mathscr{T}_K$, see [5].

3. More generally than in the previous example, any of the analytic structures of [7] can be used for the language with corresponding theory $\mathscr{T}$, provided that $\mathscr{T}$ has at least one $(0, p, e)$-field as model.

4. For $\mathscr{T}_0$ and $\mathscr{L}_0$ as in any of the previous three items let $\mathscr{T}$ and $\mathscr{L}$ be any expansion of $\mathscr{T}_0$ and $\mathscr{L}_0$, which enriches $\mathscr{T}_0$ and $\mathscr{L}_0$ exclusively on the residue rings sorts. Suppose that $\mathscr{T}$ has at least one model which is a $(0, p, e)$-field.

For any of the listed theories in the corresponding languages, a $\mathscr{T}$-field is by definition a $(0, p, e)$-field that is a model for $\mathscr{T}$.

3.2 The axiomatic set-up

Our third approach to $\mathscr{T}$-fields consists of a list of axioms which should be fulfilled by an otherwise unspecified theory $\mathscr{T}$ in some language $\mathscr{L}$. The pairs of theories and languages for $(0, p, e)$-fields in the prior two approaches are examples of this axiomatic set-up by Proposition 3.8 (see Proposition 3.9 for more examples).

In our third approach, we start with a language $\mathscr{L}$ which contains $\mathscr{L}_{\text{high}}$ and has the same sorts as $\mathscr{L}_{\text{high}}$, and a theory $\mathscr{T}$ which contains

$\mathcal{T}_{(0,p,e)}$ and which is formulated in the language $\mathcal{L}$. The sort for the valued field is called the main sort, and each of the other sorts (namely the residue ring sorts and the value group sort) are called auxiliary. It is important that no extra sorts are created along the way.

The list of axioms will be about all models of $\mathcal{T}$, and not only about $(0, p, e)$-fields. Note that any model $\mathcal{K}$ of the theory $\mathcal{T}_{(0,p,e)}$ with valued field K carries an interpretation of all the symbols of $\mathcal{L}_{\text{high}}$ with the usual first order properties, even when K is not a $(0, p, e)$-field[1]. We will use the notation π_K, $\overline{\text{ac}}_n$ and so on for the meaning of the symbols π and $\overline{\text{ac}}_n$ of $\mathcal{L}_{\text{high}}$, as well as the notion of balls, and so on, for all models of $\mathcal{T}_{(0,p,e)}$. The axioms below will involve parameters, which together with typical model theoretic compactness arguments will yield all the family-versions of the results we will need for motivic integration. To see in detail how such axioms are exploited, we refer to [8]. By definable, resp. A-definable, we will mean $\mathcal{L}$-definable without parameters, resp. $\mathcal{L}$-definable allowing parameters from A, unless otherwise stated.

The following two Jacobian properties treat close-to-linear (local) behavior of definable functions in one variable.

Definition 3.1 (Jacobian property for a function)   Let K be the valued field of a model of $\mathcal{T}_{(0,p,e)}$. Let $F : B \to B'$ be a function with $B, B' \subset K$. We say that F has the Jacobian property if the following conditions hold all together:

(i) F is a bijection and B, B' are balls in K,
(ii) F is C^1 on B,
(iii)

$$\frac{\partial F}{\partial x} \text{ is nonvanishing and } \text{ord}\left(\frac{\partial F}{\partial x}\right) \text{ is constant on } B,$$

(iv) for all $x, y \in B$ with $x \neq y$, one has

$$\text{ord}\left(\frac{\partial F}{\partial x}\right) + \text{ord}(x - y) = \text{ord}(F(x) - F(y)).$$

If moreover $n > 0$ is an integer, we say that F has the n-Jacobian property if also the following hold

(v) $\overline{\text{ac}}_n\left(\frac{\partial F}{\partial x}\right)$ is constant on B,
(vi) for all $x, y \in B$ one has

$$\overline{\text{ac}}_n\left(\frac{\partial F}{\partial x}\right) \cdot \overline{\text{ac}}_n(x - y) = \overline{\text{ac}}_n(F(x) - F(y)).$$

---

[1] This can happen, for example, when K is not discretely valued.

Very often, the Jacobian property is used in families (with a model theoretic compactness argument), which explains our choice for the partial derivative notation in the above definition.

Definition 3.2 (Jacobian property for $\mathcal{T}$)   Say that the Jacobian property holds for the $\mathcal{L}$-theory $\mathcal{T}$ if for any model $\mathcal{K}$ with Henselian valued field K the following holds.

For any finite set A in $\mathcal{K}$ (serving as parameters in whichever sorts), any integer $n > 0$, and any A-definable function $F : K \to K$ there exists an A-definable function

$$f : K \to S$$

with S a Cartesian product of (the $\mathcal{K}$-universes of) sorts not involving K (these are also called auxiliary sorts), such that each infinite fiber $f^{-1}(s)$ is a ball on which F is either constant or has the n-Jacobian property (as in Definition 3.1).

The following notion of $\mathcal{T}$ being split is related to the model-theoretic notions of orthogonality and stable embeddedness.

Definition 3.3 (Split)   Call $\mathcal{T}$ split if the following conditions hold for any model $\mathcal{K}$ with Henselian valued field K, value group Γ and residue rings $\mathcal{O}_K/\mathcal{M}_K^n$

(i). Any $\mathcal{K}$-definable subset of Γ^r is Γ-definable in the language $(+, -, 0, <)$.

(ii). For any finite set A in $\mathcal{K}$, any A-definable subset $X \subset (\prod_{i=1}^s \mathcal{O}_K/\mathcal{M}_K^{m_i}) \times \Gamma^r$ is equal to a finite disjoint union of $Y_i \times Z_i$ where the Y_i are A-definable subsets of $\prod_{i=1}^s \mathcal{O}_K/\mathcal{M}_K^{m_i}$, and the Z_i are A-definable subsets of Γ^r.

The general notion of b-minimality is introduced in [8]. Here we work with a version which is more concretely adapted to the valued field setting.

Definition 3.4 (Finite b-minimality)   Call $\mathcal{T}$ finitely b-minimal if the following hold for any model $\mathcal{K}$ with Henselian valued field K. Each locally constant $\mathcal{K}$-definable function $g : K^\times \to K$ has finite image, and, for any finite set A in $\mathcal{K}$ (serving as parameters in whichever sorts) and any A-definable set $X \subset K$, there exist an integer n, an A-definable function

$$f : X \to S$$

with S a Cartesian product of (the $\mathscr{K}$-universes of) sorts not involving K (also called auxiliary sorts), and an A-definable function

$$c : S \to K$$

such that each nonempty fiber $f^{-1}(s)$ for $s \in S$ is either

1. equal to the singleton $\{c(s)\}$, or,
2. equal to the ball $\{x \in K \mid \overline{\mathrm{ac}}_n(x - c(s)) = \xi(s),\ \mathrm{ord}(x - c(s)) = z(s)\}$ for some $\xi(s)$ in $\mathscr{O}_K/\mathscr{M}_K^n$ and some $z(s) \in \Gamma$.

Note that in the above definition, the values $z(s)$ and $\xi(s)$ depend uniquely on s in the case that $f^{-1}(s)$ is a ball and can trivially be extended when $f^{-1}(s)$ is not a ball so that $s \mapsto z(s)$ and $s \mapsto \xi(s)$ can both be seen as A-definable functions on S.

Lemma 3.5 *For any model $\mathscr{K}$ with valued field K of a finitely b-minimal theory, any definable function from a Cartesian product of (the $\mathscr{K}$-universes of) auxiliary sorts to K has finite image, and so does any definable, locally constant function from any definable set $X \subset K^n$ to K.*

Corollary 3.6 *A finitely b-minimal theory is in particular b-minimal (as defined in [8]).*

Finally we come to the most general notion of $\mathscr{T}$-fields, namely the axiomatic one of our third approach.

Definition 3.7 Let $\mathscr{T}$ be a theory containing $\mathscr{T}_{(0,p,e)}$ in a language $\mathscr{L}$ with the same sorts as $\mathscr{L}_{\mathrm{high}}$, which is split, finitely b-minimal, has the Jacobian property, and has at least one $(0, p, e)$-field as model. Then by a $\mathscr{T}$-field we mean a $(0, p, e)$-field which is a model of $\mathscr{T}$.

We have the following variant of the cell decomposition statement and related structure results on definable sets and functions of [7] for our more concrete theories.

Theorem 3.8 ([7]) *The theory $\mathscr{T}_{(0,p,e)}$ as well as the listed theories in 3.1 satisfy the conditions of Definition 3.7.*

Finally we indicate how one can create new theories with properties as in Definition 3.7.

Proposition 3.9 *Let $\mathscr{T}$ be a theory that satisfies the conditions of Definition 3.7. Then so does the theory $\mathscr{T}(R)$ in the language $\mathscr{L}(R)$ for any ring R which is a subring of a $\mathscr{T}$-field, where $\mathscr{T}(R)$ is the theory* *of all $\mathscr{T}$-fields which are algebras over R (and which extend* ord *and the* $\overline{\mathrm{ac}}_n$ *on R).*

From now on we fix one of the notions of $\mathscr{T}$, $\mathscr{L}$, and $\mathscr{T}$-fields as in Definition 3.7 for the rest of the paper, which includes the possibility of $\mathscr{T}$ and $\mathscr{L}$ being as in Sections 2, 3.1, or as in Proposition 3.9. We will often write K for a $\mathscr{T}$-field instead of writing the pair K, π_K where π_K is a uniformizer of $\mathscr{O}_K$.

4 Definable subassignments and definable morphisms

4.1

We recall that definable means $\mathscr{L}$-definable without parameters[2]. For any integers $n, r, s \geq 0$ and for any tuple $m = (m_1, \ldots, m_s)$ of non-negative integers, denote by $h[n, m, r]$ the functor sending a $\mathscr{T}$-field K to

$$h[n, m, r](K) := K^n \times (\mathscr{O}_K/\mathscr{M}_K^{m_1}) \times \cdots \times (\mathscr{O}_K/\mathscr{M}_K^{m_s}) \times \mathbb{Z}^r.$$

The data of a subset X_K of $h[n, m, r](K)$ for each $\mathscr{T}$-field K is called a definable subasssignment (in model theory sometimes loosely called a definable set), if there exists an $\mathscr{L}$-formula φ in tuples of free variables of the corresponding lengths and in the corresponding sorts such that X_K equals $\varphi(K)$, the set of the points in $h[n, m, r](K)$ satisfying φ.

An example of a definable subassignment of $h[1, 0, 0]$ is the data of the subset $P_2(K) \subset K$ consisting of the nonzero squares in K for each $\mathscr{T}$-field K, which can be described by the formula $\exists y(y^2 = x \wedge x \neq 0)$ in one free variable x and one bounded variable y, both running over the valued field[3].

A definable morphism $f : X \to Y$ between definable subassignments X and Y is given by a definable subassignment G such that $G(K)$ is the graph of a function $X(K) \to Y(K)$ for any $\mathscr{T}$-field K. We usually write f for the definable morphism, Graph(f) for G, and f_K for the function $X(K) \to Y(K)$ with graph $G(K)$. A definable isomorphism is by definition a definable morphism which has an inverse.

---

[2] Note that parameters from, for example, a base ring can be used, see Section 3.1 and Proposition 3.9.

[3] Note that, as is standard, to determine $\varphi(K)$, each variable occurring in φ (thus also the variables which are bound by a quantifier and hence not free), runs over exactly one set out of K, $\mathbb{Z}$, or a residue ring $\mathscr{O}_K/\mathscr{M}_K^{\ell}$.

Denote by Def (or $\text{Def}(\mathcal{T})$) in full) the category of definable subassignments with the definable morphisms as morphisms. More generally, for Z a definable subassignment, denote by Def_Z the category of definable subassignments X with a specified definable morphism $X \to Z$ to Z, with as morphisms between X and Y the definable morphisms which make commutative diagrams with the specified $X \to Z$ and $Y \to Z$. We will often use the notation $X_{/Z}$ for X in Def_Z. In the prior publications [9] and [10], we consequently used the notation $X \to Z$ instead of the shorter $X_{/Z}$.

For every morphism $f : Z \to Z'$ in Def, composition with f defines a functor $f_! : \text{Def}_Z \to \text{Def}_{Z'}$, sending $X_{/Z}$ to $X_{/Z'}$. Also, fiber product defines a functor $f^* : \text{Def}_{Z'} \to \text{Def}_Z$, namely, by sending $Y_{/Z'}$ to $(Y \otimes_{Z'} Z)_{/Z}$, where for each $\mathcal{T}$-field K the set $(Y \otimes_{Z'} Z)(K)$ is the set-theoretical fiber product of $Y(K)$ with $Z(K)$ over $Z'(K)$ with the projection as specified function to $Z(K)$.

Let Y and Y' be in Def. We write $Y \times Y'$ for the subassignment corresponding to the Cartesian product and we write $Y[n, m, r]$ for $Y \times h[n, m, r]$. (We fix in the whole paper h to be the definable subassignment of the singleton $\{0\}$, that is, $h(K) = \{0\} = K^0$ for all K, so that $h[n, m, r]$, as previously defined, is compatible with the notation of $Y[n, m, r]$ for general Y.)

By a point on a definable subassignment X we mean a tuple $x = (x_0, K)$ where K is a $\mathcal{T}$-field and x_0 lies in $X(K)$. We denote $|X|$ for the collection of all points that lie on X.

4.2 Dimension

Since $\mathcal{T}$ is in particular b-minimal in the sense of [8] by Corollary 3.6, for each $\mathcal{T}$-field K and each definable subassignment φ we can take the dimension of $\varphi(K)$ to be as defined in [8], and use the dimension theory from [8]. In the context of finite b-minimality, for nonempty and definable $X \subset h[n, m, r](K)$, this dimension is defined by induction on n, where for $n = 0$ the dimension of X is defined to be zero, and, for $n = 1$, $\dim X = 1$ if and only if $p(X)$ contains a ball where $p : h[1, m, r](K) \to K$ is the coordinate projection, and one has $\dim X = 0$ otherwise. For general $n \geq 1$, the dimension of such X is the maximal number $r > 0$ such that for some coordinate projection $p : h[n, m, r](K) \to K^r$, $p(X)$ contains a Cartesian product of r balls if such r exists and the dimension is 0 otherwise. Note that a nonempty definable $X \subset h[n, m, r](K)$ has dimension zero if and only if it is a finite set.

The dimension of a definable subassignment φ itself is defined as the maximum of all $\varphi(K)$ when K runs over all $\mathscr{T}$-fields.

For $f : X \to Y$ a definable morphism and K a $\mathscr{T}$-field, the relative dimension of the set $X(K)$ over $Y(K)$ (of course along f_K) is the maximum of the dimensions of the fibers of f_K, and the relative dimension of the definable assignment X over Y (along f) is the maximum of these over all K.

One has all the properties of [8] for the dimensions of the sets $\varphi(K)$ and the related properties for the definable subassignments themselves, analogous to the properties of the so-called K-dimension of [9].

5 Summation over the value group

We consider a formal symbol $\mathbb{L}$ and the ring

$$\mathbb{A} := \mathbb{Z}\Big[\mathbb{L}, \mathbb{L}^{-1}, \Big(\frac{1}{1-\mathbb{L}^{-i}}\Big)_{i>0}\Big],$$

as subring of the ring of rational functions in $\mathbb{L}$ over $\mathbb{Q}$. Furthermore, for each real number $q > 1$, we consider the ring morphism

$$\theta_q : \mathbb{A} \to \mathbb{R} : r(\mathbb{L}) \mapsto r(q),$$

that is, one evaluates the rational function $r(\mathbb{L})$ in $\mathbb{L}$ at q.

Recall that $h[0,0,1]$ can be identified with $\mathbb{Z}$, since $h[0,0,1](K) = \mathbb{Z}$ for all $\mathscr{T}$-fields K. Let S be in Def, that is, let S be a definable subassignment. A definable morphism $\alpha : S \to h[0,0,1]$ gives rise to a function (also denoted by α) from $|S|$ to $\mathbb{Z}$ which sends a point (s, K) on S to $\alpha_K(s)$. Likewise, such α gives rise to the function $\mathbb{L}^\alpha$ from $|S|$ to $\mathbb{A}$ which sends a point (s, K) on S to $\mathbb{L}^{\alpha_K(s)}$.

We define the ring $\mathscr{P}(S)$ of constructible Presburger functions on S as the subring of the ring of functions $|S| \to \mathbb{A}$ generated by

1. all constant functions into $\mathbb{A}$,
2. all functions $\alpha : |S| \to \mathbb{Z}$ with $\alpha : S \to h[0,0,1]$ a definable morphism,
3. all functions of the form $\mathbb{L}^\beta$ with $\beta : S \to h[0,0,1]$ a definable morphism.

Note that a general element of $\mathscr{P}(S)$ is thus a finite sum of terms of the form $a\mathbb{L}^\beta \prod_{i=1}^{\ell} \alpha_i$ with $a \in \mathbb{A}$, and the β and α_i definable morphisms from S to $h[0,0,1] = \mathbb{Z}$.

For any $\mathscr{T}$-field K, any $q > 1$ in $\mathbb{R}$, and f in $\mathscr{P}(S)$ we write $\theta_{q,K}(f)$: $S(K) \to \mathbb{R}$ for the function sending $s \in S(K)$ to $\theta_q(f(s, K))$.

Define a partial ordering on $\mathscr{P}(S)$ by setting $f \geq 0$ if for every $q > 1$ in $\mathbb{R}$ and every s in $|S|$, $\theta_q(f(s)) \geq 0$. We denote by $\mathscr{P}(S)_+$ the set $\{f \in \mathscr{P}(S) \mid f \geq 0\}$. Write $f \geq g$ if $f - g$ is in $\mathscr{P}_+(S)$. Similarly, write $\mathbb{A}_+$ for the sub-semi-ring of $\mathbb{A}$ consisting of the non-negative elements of $\mathbb{A}$, namely those elements a with $\theta_q(a) \geq 0$ for all real $q > 1$.

Recall the notion of summable families in $\mathbb{R}$ or $\mathbb{C}$, cf. [1] VII.16. In particular, a family $(z_i)_{i \in I}$ of complex numbers is summable if and only if the family $(|z_i|)_{i \in I}$ is summable in $\mathbb{R}$.

We shall say a function φ in $\mathscr{P}(h[0, 0, r])$ is integrable if for each $\mathscr{T}$-field K and for each real $q > 1$, the family $(\theta_{q,K}(\varphi)(i))_{i \in \mathbb{Z}^r}$ is summable.

More generally we shall say a function φ in $\mathscr{P}(S[0, 0, r])$ is S-integrable if for each $\mathscr{T}$-field K, for each real $q > 1$, and for each $s \in S(K)$, the family $(\theta_{q,K}(\varphi)(s, i))_{i \in \mathbb{Z}^r}$ is summable. The latter notion of S-integrability is key to all integrability notions in this paper.

We denote by $\mathrm{I}_S\mathscr{P}(S[0, 0, r])$ the collection of S-integrable functions in $\mathscr{P}(S[0, 0, r])$. Likewise, we denote by $\mathrm{I}_S\mathscr{P}_+(S[0, 0, r])$ the collection of S-integrable functions in $\mathscr{P}_+(S[0, 0, r])$. Note that $\mathrm{I}_S\mathscr{P}(S[0, 0, r])$, resp. $\mathrm{I}_S\mathscr{P}_+(S[0, 0, r])$, is a $\mathscr{P}(S)$-module, resp. a $\mathscr{P}_+(S)$-semi-module.

The following is inspired by results in [13] and appears in this form in [9].

Theorem-Definition 5.1 *For each φ in $\mathrm{I}_S\mathscr{P}(S[0, 0, r])$ there exists a unique function $\psi = \mu_{/S}(\varphi)$ in $\mathscr{P}(S)$ such that for all $q > 1$, all $\mathscr{T}$-fields K, and all s in $S(K)$*

$$\theta_{q,K}(\psi)(s) = \sum_{i \in \mathbb{Z}^r} \theta_{q,K}(\varphi)(s, i). \tag{8.1}$$

Moreover, the mapping $\varphi \mapsto \mu_{/S}(\varphi)$ yields a morphism of $\mathscr{P}(S)$-modules

$$\mu_{/S} : \mathrm{I}_S\mathscr{P}(S \times \mathbb{Z}^r) \longrightarrow \mathscr{P}(S).$$

The proof of 5.1 is based on finite b-minimality, the fact that $\mathscr{T}$ is split, and explicit calculations, mainly of geometric series and their derivatives. Clearly, the above map $\mu_{/S}$ sends $\mathrm{I}_S\mathscr{P}_+(S \times \mathbb{Z}^r)$ to $\mathscr{P}_+(S)$. For Y a definable subassignment of S, we denote by $\mathbf{1}_Y$ the function in $\mathscr{P}(S)$ with value 1 on Y and zero on $S \backslash Y$. We shall denote by $\mathscr{P}^0(S)$ (resp. $\mathscr{P}^0_+(S)$) the subring (resp. sub-semi-ring) of $\mathscr{P}(S)$ (resp. $\mathscr{P}_+(S)$) generated by the functions $\mathbf{1}_Y$ for all definable subassignments Y of S and by the constant function $\mathbb{L} - 1$.

If $f : Z \to Y$ is a morphism in Def, composition with f yields natural pullback morphisms $f^* : \mathscr{P}(Y) \to \mathscr{P}(Z)$ and $f^* : \mathscr{P}_+(Y) \to \mathscr{P}_+(Z)$. These pullback morphisms and the subrings $\mathscr{P}^0(S)$ will play a role for the richer class of motivic constructible functions. First we turn our attention to another ingredient for motivic constructible functions, coming from the residue rings. Afterwards we will glue these two ingredients along the common subrings $\mathscr{P}^0_+(S)$ to define motivic constructible functions.

6 Integration over the residue rings

On the integers side we have defined rings of (nonnegative) constructible Presburger functions $\mathscr{P}_+(\cdot)$ and a summation procedure over subsets of $\mathbb{Z}^r$. On the residue rings side we will proceed differently.

Let Z be a definable subassignment in Def. Define the semi-group $\mathscr{Q}_+(Z)$ as the quotient of the free abelian semi-group over symbols $[Y]$ with $Y_{/Z}$ a subassignment of $Z[0, m, 0]$ for some $m = (m_1, \ldots, m_s)$ with $m_i \geq 0$ and $s \geq 0$, with as distinguished map from Y to Z the natural projection, by the following relations.

$$[\emptyset] = 0, \text{ where } \emptyset \text{ is the empty subassignment.} \tag{8.1}$$

$$[Y] = [Y'] \tag{8.2}$$

if there exists a definable isomorphism $Y \to Y'$ which commutes with the projections $Y \to Z$ and $Y' \to Z$.

$$[Y_1 \cup Y_2] + [Y_1 \cap Y_2] = [Y_1] + [Y_2] \tag{8.3}$$

for Y_1 and Y_2 definable subassignments of a common $Z[0, m, 0]$ for some m.

$$[Y[0, m', 0]] = [Y'] \tag{8.4}$$

if for the projection $p : Z[0, m + m', 0] \to Z[0, m, 0]$ one has $Y' = p^{-1}(Y)$ for some definable $Y \subset Z[0, m, 0]$ and some $m_i, m'_i \geq 0$.

We will still write $[Y]$ for the class of $[Y]$ in $\mathscr{Q}_+(Z)$ for $Y \subset Z[0, m, 0]$. In [9], the longer notation $SK_0(\mathrm{RDef}_Z)$ is used instead of $\mathscr{Q}_+(Z)$. Note that in [9] relation (8.4) is left out since it is redundant if one only uses $\overline{\mathrm{ac}}_1$ instead of all the $\overline{\mathrm{ac}}_n$. The semi-group $\mathscr{Q}_+(Z)$ carries a semi-ring structure with multiplication for $Y \subset Z[0, m, 0]$ and $Y' \subset Z[0, m', 0]$ given by

$$[Y] \cdot [Y'] := [Y \otimes_Z Y'],$$

where the fibre product is taken along the coordinate projections to Z. Similarly, for $f : Z_1 \to Z_2$ any morphism in Def, there is a natural pullback homomorphism of semi-rings $f^* : \mathcal{Q}_+(Z_2) \to \mathcal{Q}_+(Z_1)$ which sends $[Y]$ for some $Y \subset Z_2[0, m, 0]$ to $[Y \otimes_{Z_2} Z_1]$. Write $\mathbb{L}$ for the class of $Z[0, 1, 0]$ in $\mathcal{Q}_+(Z)$. Then, by relation (8.4), one has that the class of $Z[0, m, 0]$ in $\mathcal{Q}_+(Z)$ equals $\mathbb{L}^{|m|}$ with $m = (m_i)_i$ and $|m| = \sum_i m_i$. Clearly, for each $a \in \mathcal{Q}_+(Z)$, there exists a tuple m and a $Y \subset Z[0, m, 0]$ such that $a = [Y]$.

To preserve a maximum of information at the level of the residue rings, we will integrate functions in $\mathcal{Q}_+(\cdot)$ over residue ring variables in a formal way. Suppose that $Z = X[0, k, 0]$ for some tuple k, let a be in $\mathcal{Q}_+(Z)$ and write a as $[Y]$ for some $Y \subset Z[0, n, 0]$. We write $\mu_{/X}$ for the corresponding formal integral in the fibers of the coordinate projection $Z \to X$

$$\mu_{/X} : \mathcal{Q}_+(Z) \to \mathcal{Q}_+(X) : [Y] \to [Y]$$

where the class of Y is first taken in $\mathcal{Q}_+(Z)$ and then in $\mathcal{Q}_+(X)$. Note that this allows one to integrate functions from $\mathcal{Q}_+$ over residue ring variables, but of course not over valued field neither over value group variables. To integrate over any kind of variables, we will need to combine the value group part $\mathcal{P}_+$ and the residue rings part $\mathcal{Q}_+$.

7 Putting $\mathcal{P}_+$ and $\mathcal{Q}_+$ together to form $\mathcal{C}_+$

Many interesting functions on Henselian valued fields have a component that comes essentially from the value group and one that comes from residue rings. For Z in Def, we will glue the pieces $\mathcal{P}_+(Z)$ and $\mathcal{Q}_+(Z)$ together by means of the common sub-semi-ring $\mathcal{P}_+^0(Z)$. Recall that $\mathcal{P}_+^0(Z)$ is the sub-semi-ring of $\mathcal{P}_+(Z)$ generated by the characteristic functions $\mathbf{1}_Y$ for all definable subassignments $Y \subset Z$ and by the constant function $\mathbb{L} - 1$.

Using the canonical semi-ring morphism $\mathcal{P}_+^0(Z) \to \mathcal{Q}_+(Z)$, sending $\mathbf{1}_Y$ to $[Y]$ and $\mathbb{L} - 1$ to $\mathbb{L} - 1$, we define the semi-ring $\mathcal{C}_+(Z)$ as

$$\mathcal{P}_+(Z) \otimes_{\mathcal{P}_+^0(Z)} \mathcal{Q}_+(Z).$$

We call elements of $\mathcal{C}_+(Z)$ (nonnegative) constructible motivic functions on Z.

If $f : Z \to Y$ is a morphism in Def, we find natural pullback morphisms $f^* : \mathcal{C}_+(Y) \to \mathcal{C}_+(Z)$, by the tensor product definition of $\mathcal{C}_+(\cdot)$.

Namely, f^* maps $\sum_{i=1}^{r} a_i \otimes b_i$ to $\sum_i f^*(a_i) \otimes f^*(b_i)$, where $a_i \in \mathscr{P}_+(Y)$ and $b_i \in \mathscr{Q}_+(Y)$.

7.1 Interpretation in non-archimedean local fields

An important feature of our setting (as well as in the settings of [9] and [16]) is that the motivic constructible functions and their integrals interpolate actual functions and their integrals on non-archimedean local fields, and even more generally on $\mathscr{T}$-fields with finite residue field.

Let $X \subset h[n, m, r]$ be in Def, let φ be in $\mathscr{C}_+(X)$, and let K be a $\mathscr{T}$-field with finite residue field. In this case φ gives rise to an actual set-theoretic function φ_K from $X(K)$ to $\mathbb{Q}_{\geq 0}$, defined as follows:

For a in $\mathscr{P}_+(X)$, one gets $a_K : X(K) \to \mathbb{Q}_{\geq 0}$ by replacing $\mathbb{L}$ by q_K, the number of elements in the residue field of K.

For $b = [Y]$ with Y a subassignment of $X[0, m, 0]$ in $\mathscr{Q}_+(X)$, if one writes $p : Y(K) \to X(K)$ for the projection, one defines $b_K : X(K) \to \mathbb{Q}_{\geq 0}$ by sending $x \in X(K)$ to $\#(p^{-1}(x))$, that is, the number of points in $Y(K)$ that lie above $x \in X(K)$.

For our general φ in $\mathscr{C}_+(X)$, write φ as a finite sum $\sum_i a_i \otimes b_i$ with $a_i \in \mathscr{P}_+(X)$ and $b_i \in \mathscr{Q}_+(X)$. Our general definitions are such that the function

$$\varphi_K : X(K) \to \mathbb{Q}_{\geq 0} : x \mapsto \sum_i a_{iK}(x) \cdot b_{iK}(x)$$

does not depend on the choices made for a_i and b_i.

7.2 Integration over residue rings and value group

We have the following form of independence (or orthogonality) between the integer part and the residue rings part of $\mathscr{C}_+(\cdot)$.

Proposition 7.1 *Let S be in* Def. *The canonical morphism*

$$\mathscr{P}_+(S[0, 0, r]) \otimes_{\mathscr{P}_+^0(S)} \mathscr{Q}_+(S[0, m, 0]) \longrightarrow \mathscr{C}_+(S[0, m, r])$$

is an isomorphism of semi-rings, where the homomorphisms $p^* : \mathscr{P}_+^0(S) \to \mathscr{P}_+(S[0, 0, r])$ and $q^* : \mathscr{P}_+^0(S) \to \mathscr{Q}_+(S[0, m, 0])$ are induced by the pullback homomorphisms of the projections $p : S[0, 0, r] \to S$ and $q : S[0, m, 0] \to S$.

The mentioned canonical morphism of Proposition 7.1 sends $a \otimes b$ to $p_1^*(a) \otimes p_2^*(b)$, where $p_1 : S[0, m, r] \to S[0, 0, r]$ and $p_1 : S[0, m, r] \to S[0, m, 0]$ are the projections. Of course the proof of Proposition 7.1 uses that $\mathscr{T}$ is split. Recall that for a in $\mathscr{Q}_+(X)$, one can write $a = [Y]$ for some Y in Def_X, say, with specified morphism $f : Y \to X$. We shall write $\mathbf{1}_a := \mathbf{1}_{f(Y)}$ for the characteristic function of $f(Y)$, the "support" of a.

Lemma-Definition 7.2 *Let φ be in $\mathscr{C}_+(Z)$ and suppose that $Z = X[0, m, r]$ for some X in Def. Say that φ is X-integrable if one can write $\varphi = \sum_{i=1}^{\ell} a_i \otimes b_i$ with $a_i \in \mathscr{P}_+(X[0,0,r])$ and $b_i \in \mathscr{Q}_+(X[0, m, 0])$ as in Proposition 7.1 such that moreover the a_i lie in $\mathrm{I}_X \mathscr{P}_+(X[0,0,r])$ in the sense of Section 5. If this is the case, then*

$$\mu_{/X}(\varphi) := \sum_i \mu_{/X}(a_i) \otimes \mu_{/X}(b_i) \ \in \mathscr{C}_+(X)$$

does not depend on the choice of the a_i and b_i and is called the integral of φ in the fibers of the coordinate projection $Z \to X$.

The following lemma is a basic form of a projection formula which concerns pulling a factor out of the integral if the factor depends on other variables than the ones that one integrates over.

Lemma 7.3 *Let φ be in $\mathscr{C}_+(Z)$ such that φ is X-integrable, where $Z = X[0, m, r]$ for some X in Def. Let ψ be in $\mathscr{C}_+(X)$ and let $p : Z \to X$ be the projection. Then $p^*(\psi)\varphi$ is X-integrable and*

$$\mu_{/X}(p^*(\psi)\varphi) = \psi\mu_{/X}(\varphi)$$

holds in $\mathscr{C}_+(X)$.

Note that Lemma 7.3 is immediate when $m = 0$. Using the natural morphisms $\mathscr{P}_+(Z) \to \mathscr{C}_+(Z)$ which sends ψ to $\psi \otimes [Z]$, and $\mathscr{Q}_+(Z) \to \mathscr{C}_+(Z)$ which sends ν to $\mathbf{1}_Z \otimes \nu$, we can formulate the following. (Note that $\mathscr{P}_+(Z) \to \mathscr{C}_+(Z) : b \mapsto \mathbf{1}_Z \otimes b$ is not necessarily injective neither necessarily surjective.)

Lemma 7.4 *For any $\varphi \in \mathscr{C}_+(Z)$ there exist ψ in $\mathscr{P}_+(Z[0, m, 0])$ and ν in $\mathscr{Q}_+(Z[0, 0, r])$ for some m and r such that ν is Z-integrable and $\varphi = \mu_{/Z}(\psi) = \mu_{/Z}(\nu)$.*

Here is a first instance of the feature that relates integration of motivic functions with actual integration (or summation) on $\mathscr{T}$-fields with finite residue field.

Lemma 7.5 *Let φ be in $\mathscr{C}_+(Z)$ and suppose that $Z = X[0, m, r]$ for some X in Def. Let K be a $\mathscr{T}$-field with finite residue field and consider*

φ_K as in Section 7.1. If φ is X-integrable then, for each $x \in X(K)$, $\varphi_K(x, \cdot) : y \mapsto \varphi_K(x, y)$ is integrable against the counting measure, and if one writes ψ for $\mu_{/X}(\varphi)$, then

$$\psi_K(x) = \sum_y \varphi_K(x, y)$$

where the summation is over those y such that $(x, y) \in Z(K)$.

8 Integration over one valued field variable

For the moment let K be any discretely valued field. For a ball $B \subset K$ and for any real number $q > 1$, define $\theta_q(B)$ as the real number $q^{-\mathrm{ord}b}$, where $b \in K^\times$ is such that $B = a + b\mathcal{O}_K$ for some $a \in K$. We call $\theta_q(B)$ the q-volume of B.

Next we will define a naive and simple notion of step-function. Finite b-minimality will allow us to reduce part of the integration procedure to step-functions. A finite or countable collection of balls in K, each with different q-volume, is called a step-domain. We will identify a step-domain S with the union of the balls in S. This is harmless since one can recover the individual balls from their union since they all have different q-volume. Call a nonnegative real valued function $\varphi : K \to \mathbb{R}_{\geq 0}$ a step-function if there exists a unique step-domain S such that φ is constant and nonzero on each ball of S and zero outside $S \cup \{a\}$ for some $a \in K$. Note that uniqueness of the step-domain S for φ is automatic, except possibly when the residue field has two elements.

Let $q > 1$ be a real number. Say that a step-function $\varphi : K \to \mathbb{R}_{\geq 0}$ with step-domain S is q-integrable over K if and only if

$$\sum_{B \in S} \theta_q(B) \cdot \varphi(B) < \infty, \tag{8.1}$$

where one sums over the balls B in S, and then the expression (8.1) is called the q-integral of φ over K. Using Theorem 5.1 one proves the following.

Lemma-Definition 8.1 *Suppose that* $Z = X[1, 0, 0]$ *for some* X *in Def. Let* φ *be in* $\mathscr{P}_+(Z)$. *Call* φ *an* X-*integrable family of step-functions if for each* $\mathscr{T}$-*field* K, *for each* $x \in X(K)$, *and for each* $q > 1$, *the function*

$$\theta_{q,K}(\varphi)(x, \cdot) : K \to \mathbb{R}_{\geq 0} : t \mapsto \theta_{q,K}(\varphi)(x, t) \tag{8.2}$$

is a step-function which is q-integrable over K. If φ is such a family,
then there exists a unique function ψ in $\mathscr{P}_+(X)$ such that $\theta_{q,K}(\psi)(x)$
equals the q-integral over K of (8.2) for each $\mathscr{T}$-field K, each $x \in X(K)$,
and each $q > 1$. We then call φ X-integrable, we write

$$\mu_{/X}(\varphi) := \psi$$

and call $\mu_{/X}(\varphi)$ the integral of φ in the fibers of $Z \to X$.

Finally we define how to integrate a general motivic constructible function over one valued field variable, in families.

Lemma-Definition 8.2 *Let φ be in $\mathscr{C}_+(Z)$ and suppose that $Z = X[1,0,0]$. Say that φ is X-integrable if there exists ψ in $\mathscr{P}_+(Z[0,m,0])$ with $\mu_{/Z}(\psi) = \varphi$ as in Lemma 7.4 such that ψ is $X[0,m,0]$-integrable in the sense of Lemma-Definition 8.1 and then*

$$\mu_{/X}(\varphi) := \mu_{/X}(\mu_{/X[0,m,0]}(\psi)) \in \mathscr{C}_+(X)$$

is independent of the choices and is called the integral of φ in the fibers of $Z \to X$.

The proof of 8.2 is similar to the proofs in section 9 of [9].

9 General integration

In this section we define the motivic measure and the motivic integral of motivic constructible functions in general. For uniformity results and for applications it is important that we do this in families, namely, in the fibers of projections $X[n,m,r] \to X$ for X in Def. We define the integrals in the fibers of a general coordinate projection $X[n,m,r] \to X$ by induction on $n \geq 0$.

Lemma-Definition 9.1 *Let φ be in $\mathscr{C}_+(Z)$ and suppose that $Z = X[n,m,r]$ for some X in Def. Say that φ is X-integrable if there exist a definable subassignment $Z' \subset Z$ whose complement in Z has relative dimension $< n$ over X, and an ordering of the coordinates on $X[n,m,r]$ such that $\varphi' := 1_{Z'}\varphi$ is $X[n-1,m,r]$-integrable and $\mu_{/X[n-1,m,r]}(\varphi')$ is X-integrable. If this holds then*

$$\mu_{/X}(\varphi) := \mu_{/X}(\mu_{/X[n-1,m,r]}(\varphi')) \in \mathscr{C}_+(X)$$

does not depend on the choices and is called the integral of φ in the fibers of $Z \to X$, and is compatible with the definitions made in 8.2.

More generally, let φ be in $\mathscr{C}_+(Z)$ and suppose that $Z \subset X[n, m, r]$. Say that φ is X-integrable if the extension by zero of φ to a function $\widetilde{\varphi}$ in $\mathscr{C}_+(X[n, m, r])$ is X-integrable, and define $\mu_{/X}(\varphi)$ as $\mu_{/X}(\widetilde{\varphi})$. If X is $h[0, 0, 0]$ (which is a final object in Def), then we write μ instead of $\mu_{/X}$ and we call $\mu(\varphi)$ the integral of φ over Z.

One can prove 9.1 in two ways (both relying on all the properties of $\mathscr{T}$-fields of Definition 3.7): using more recent insights from [6] to reverse the order of the coordinates, or, using the slightly longer approach from [9] with a calculation on bi-cells.

One of the main features is a natural relation between motivic integrability and motivic integration on the one hand, and classical measure theoretic integrability and integration on local fields on the other hand:

Proposition 9.2 *Let φ be in $\mathscr{C}_+(X[n, m, r])$ for some X in Def. If φ is X-integrable, then, for each local field K which is a $\mathscr{T}$-field and for each $x \in X(K)$ one has that $\varphi_K(x, \cdot)$ is integrable (in the standard measure-theoretic sense). If one further writes ψ for $\mu_{/X}(\varphi)$, then, for each $x \in X(K)$,*

$$\psi_K(x) = \int_y \varphi_K(x, y),$$

where the integral is against the product measure of the Haar measure on K with the counting measure on $\mathbb{Z}$ and on the residue rings for y running over $h[n, m, r](K)$, and where the Haar measure gives $\mathscr{O}_K$ measure one.

10 Further properties

As mentioned before, the projection formula allows one to pull a factor out of the integral if that factor depends on other variables than the ones that one integrates over.

Proposition 10.1 (Projection formula) *Let φ be in $\mathscr{C}_+(Z)$ for some $Z \subset X[n, m, r]$ and some X in Def. Suppose that φ is X-integrable, let ψ be in $\mathscr{C}_+(X)$ and let $p : Z \to X$ be the projection. Then $p^*(\psi)\varphi$ is X-integrable and*

$$\mu_{/X}(p^*(\psi)\varphi) = \psi\mu_{/X}(\varphi)$$

holds in $\mathscr{C}_+(X)$.

In other words, if one would write $I_X\mathscr{C}_+(Z)$ for the X-integrable functions in $\mathscr{C}_+(Z)$, then

$$\mu_{/X} : I_X\mathscr{C}_+(Z) \to \mathscr{C}_+(Z) : \varphi \mapsto \mu_{/X}(\varphi)$$

is a morphism of $\mathscr{C}_+(X)$-semi-modules, where the semi-module structure on $I_X\mathscr{C}_+(Z)$ comes from the homomorphism $p^* : \mathscr{C}_+(X) \to \mathscr{C}_+(Z)$ of semi-rings, with $p : Z \to X$ the projection.

We will now explain Jacobians and relative Jacobians, first in a general, set-theoretic setting, and then for definable morphisms. This is done in several steps.

For any function $h : A \subset K^n \to K^n$ (in the set-theoretic sense of function) for some $\mathscr{T}$-field K and integer $n > 0$, let $\mathrm{Jac}\,h : A \to K$ be the determinant of the Jacobian matrix of h where this matrix is well-defined (on the interior of A) and let $\mathrm{Jac}\,h$ take the value 0 elsewhere in A.

In the relative case, consider a function $f : A \subset C \times K^n \to C \times K^n$ which makes a commutative diagram with the projections to C, with K a $\mathscr{T}$-field and with some set C. Write $\mathrm{Jac}_{/C}f : A \to K$ for the function satisfying for each $c \in C$ that $(\mathrm{Jac}_{/C}f)(c, z) = \mathrm{Jac}(f_c)(z)$ for each $c \in C$ and each $z \in K^n$ with $(c, z) \in A$, and where $f_c : A_c \to K^n$ is the function sending z to t with $f(c, z) = (c, t)$ and $(c, z) \in A$.

The existence of the relative Jacobian $\mathrm{Jac}\,g_{/X}$ in the following definable context is clear by the definability of the partial derivatives and continuity properties.

Lemma-Definition 10.2 *Consider a definable morphism $g : A \subset X[n, 0, 0] \to X[n, 0, 0]$ over X for some definable subassignment X. By $\mathrm{Jac}_{/X}g$ denote the unique definable morphism $A \to h[1, 0, 0]$ satisfying for each $\mathscr{T}$-field K that $(\mathrm{Jac}_{/X}g)_K = \mathrm{Jac}_{/X_K}(g_K)$ and call it the relative Jacobian of g over X.*

We can now formulate the change of variables formula, in a relative setting.

Theorem 10.3 (Change of variables) *Let $F : Z \subset X[n, 0, 0] \to Z' \subset X[n, 0, 0]$ be a definable isomorphism over X for some X in Def. Let φ be in $\mathscr{C}_+(Z)$. Then there exists a definable subassignment $Y \subset Z$ whose complement in Z has dimension $< n$ over X, and such that the relative Jacobian $\mathrm{Jac}_{/X}F$ of F over X is nonvanishing on Y. Moreover, if we take the unique φ' in $\mathscr{C}_+(Z')$ with $F^*(\varphi') = \varphi$, then $\varphi\mathbb{L}^{-\mathrm{ord}\,\mathrm{Jac}_{/X}F}$ is*

X-integrable if and only if φ' is X-integrable, and then

$$\mu_{/X}(\varphi \mathbb{L}^{-\text{ord Jac}_{/X} F}) = \mu_{/X}(\varphi')$$

in $\mathscr{C}_+(X)$, with the convention that $\mathbb{L}^{-\text{ord}(0)} = 0$.

The proof of Theorem 10.3 relies on the Jacobian property for $\mathcal{T}$. Finally we formulate a general Fubini Theorem, in the Tonelli variant for non-negatively valued functions.

Theorem 10.4 (Fubini-Tonelli) *Let φ be in $\mathscr{C}_+(Z)$ for some $Z \subset X[n, m, r]$ and some X in Def. Let $X[n, m, r] \to X[n - n', m - m', r - r']$ be a coordinate projection. Then φ is X-integrable if and only if there exists a definable subassignment Y of Z whose complement in Z has dimension $< n$ over X such that, if we put $\varphi' = 1_Y \varphi$, then φ' is $X[n - n', m - m', r - r']$-integrable and $\mu_{/X[n-n',m-m',r-r']}(\varphi')$ is X-integrable. If this holds, then*

$$\mu_{/X}(\mu_{/X[n-n',m-m',r-r']}(\varphi')) = \mu_{/X}(\varphi)$$

in $\mathscr{C}_+(X)$.

The proof of Theorem 10.4 relies essentially on Lemma-Definition 9.1.

11 Direct image formalism

Let Λ be in Def. From now on, all objects will be over Λ, where we continue to use the notation $\star_{/\Lambda}$ instead of $\star \to \Lambda$ to denote that some object $\star$ is considered over Λ.

Consider X in Def_Λ. For each integer $d \geq 0$, let $\mathscr{C}_+^{\leq d}(X_{/\Lambda})$ be the ideal of $\mathscr{C}_+(X)$ generated by characteristic functions 1_Z of $Z \subset X$ which have relative dimension $\leq d$ over Λ. Furthermore, we put $\mathscr{C}_+^{\leq -1}(X_{/\Lambda}) = \{0\}$.

For $d \geq 0$, define $C_+^d(X_{/\Lambda})$ as the quotient of semi-groups $\mathscr{C}_+^{\leq d}(X_{/\Lambda})/\mathscr{C}_+^{\leq d-1}(X_{/\Lambda})$; its nonzero elements can be seen as functions having support of dimension d and which are defined almost everywhere, that is, up to definable subassignments of dimension $< d$.

Finally, put

$$C_+(X_{/\Lambda}) := \bigoplus_{d \geq 0} C_+^d(X_{/\Lambda}),$$

which is actually a finite direct sum since $C_+^d(X_{/\Lambda}) = \{0\}$ for d larger than the relative dimension of X over Λ.

We introduce a notion of isometries for definable subassignments. This is some work since also residue ring and integer variables play a role.

Definition 11.1 (Isometries) Consider $\overline{\mathbb{Z}} := \mathbb{Z} \cup \{-\infty, +\infty\}$. Extend the natural order on $\mathbb{Z}$ to $\overline{\mathbb{Z}}$ so that $+\infty$ is the biggest element, and $-\infty$ the smallest.

Define $\overline{\text{ord}}$ on $h[1, 0, 0]$ as the extension of ord by $\overline{\text{ord}}(0) = +\infty$. Define $\overline{\text{ord}}$ on $h[0, m, r]$ by sending 0 to $+\infty$ and everything else to $-\infty$. Define $\overline{\text{ord}}$ on $h[n, m, r]$ by sending $x = (x_i)_i$ to $\inf_i \overline{\text{ord}}(x_i)$.

Call a definable isomorphism $f : Y \to Z$ between definable subassignments Y and Z an isometry if and only if

$$\overline{\text{ord}}(y - y') = \overline{\text{ord}}(f_K(y) - f_K(y'))$$

for all $\mathscr{T}$-fields K and all y and y' in $Y(K)$, where $y - y' = (y_i - y_i')_i$. In the relative setting, let $f : Y \to Z$ be a definable isomorphism over Λ. Call f an isometry over Λ if for all $\mathscr{T}$-fields K and for all $\lambda \in \Lambda(K)$, one has that $f_\lambda : Y_\lambda \to Z_\lambda$ is an isometry, where Y_λ is the set of elements in $Y(K)$ that map to λ, and f_λ is the restriction of f_K to Y_λ.

Definition 11.2 (Adding parameters) Let $f : Y \to Z$ and $f' : Y' \to Z'$ be morphisms in Def with $Y' \subset Y[0, m, r]$ and $Z' \subset Z[0, s, t]$ for some m, r, s, and t. Say that f' is obtained from f by adding parameters, if the natural projections $p : Y' \to Y$ and $r : Z' \to Z$ are definable isomorphisms and if moreover the composition $r \circ f'$ equals $f \circ p$.

We will now define the integrable functions (over Λ) inside $C_+(X_{/\Lambda})$, denoted by $\text{IC}_+(X_{/\Lambda})$, for any definable subassignment $X_{/\Lambda}$. The main idea here is that integrability conditions should not change under pullbacks along isometries and under maps obtained from the identity function by adding parameters. Consider φ in $\mathscr{C}_+^{\leq d}(X_{/\Lambda})$ and its image $\overline{\varphi}$ in $C_+^d(X_{/\Lambda})$ for some definable subassignment $X_{/\Lambda}$ over Λ. In general, one can write X as a disjoint union of definable subassignments X_1, X_2 such that there exists a definable morphism $f = f_2 \circ f_1 : Z \subset \Lambda[d, m, r] \to X_2$ for some m, r, $\mathbf{1}_{X_1}\varphi = 0$, f_2 is an isometry over Λ, and f_1 is obtained from the identity function $X_2 \to X_2$ by adding parameters. Call $\overline{\varphi}$ integrable if and only if $f^*(\varphi)$ is Λ-integrable as in Lemma-Definition 9.1. Note that this condition is independent of the choice of the X_i and f. This defines the grade d part $\text{IC}_+^d(X_{/\Lambda})$ of $\text{IC}_+(X_{/\Lambda})$, and one sets

$$\text{IC}_+(X_{/\Lambda}) := \sum_{d \geq 0} \text{IC}_+^d(X_{/\Lambda}).$$

The following theorem gives the existence and uniqueness of integration in the fibers relative over Λ (in all relative dimensions over Λ), in the form of a direct image formalism, by associating to any morphim $f : Y \to Z$ in Def_Λ a morphism of semi-groups $f_!$ from $\mathrm{IC}_+(Y_{/\Lambda})$ to $\mathrm{IC}_+(Z_{/\Lambda})$. This association happens to be a functor and the map $f_!$ sends a function to its integral in the fibers relative over Λ (in the correct relative dimensions over Λ). The underlying idea is that isometries, inclusions, and definable morphisms obtained by adding parameters from an identity map should yield a trivial $f_!$ coming from the inverse of the pullback f^*, and further there is a change of variables situation and a Fubini-Tonelli situation that should behave as in Section 10.

Theorem 11.3 *There exists a unique functor from Def_Λ to the category of semi-groups, which sends an object Z in Def_Λ to the semi-group $\mathrm{IC}_+(Z_{/\Lambda})$, and a definable morphism $f : Y \to Z$ to a semi-group homomorphism $f_! : \mathrm{IC}_+(Y_{/\Lambda}) \to \mathrm{IC}_+(Z_{/\Lambda})$, such that, for φ in $\mathrm{IC}_+^d(Y_{/\Lambda})$ and a representative φ^0 in $\mathscr{C}_+^{\leq d}(Y_{/\Lambda})$ of φ one has:*

M1 (Basic maps):

If f is either an isometry or is obtained from an identity map $C \to C$ for some C in Def by adding parameters, then $f_!(\varphi)$ is the class in $\mathrm{IC}_+^d(Z_{/\Lambda})$ of $(f^{-1})^*(\varphi^0)$.

M2 (Inclusions):

If $Y \subset Z$ and f is the inclusion function, then $f_!(\varphi)$ is the class in $\mathrm{IC}_+^d(Z_{/\Lambda})$ of the unique ψ in $\mathscr{C}_+^{\leq d}(Z_{/\Lambda})$ with $f^*(\psi) = \varphi^0$ and $\psi \mathbf{1}_Y = \psi$.

M3 (Fubini-Tonelli):

If $f : Y = \Lambda[d, m, r] \to Z = \Lambda[d - d', m - m', r - r']$ is a coordinate projection, then φ^0 can be taken by Theorem 10.4 such that it is $\Lambda[d - d', m - m', r - r']$-integrable and then $f_!(\varphi)$ is the class in $\mathrm{IC}_+^{d-d'}(\Lambda[d - d', m - m', r - r']_{/\Lambda})$ of

$$\mu_{/\Lambda[d-d', m-m', r-r']}(\varphi^0).$$

M4 (Change of variables):

If f is a definable isomorphism over $\Lambda[0, m, r]$ with $Y \subset \Lambda[d, m, r]$ and $Z \subset \Lambda[d, m, r]$ then φ^0 and a $\psi \in \mathscr{C}_+^{\leq d}(Y_{/\Lambda})$ can be taken by Theorem 10.3 such that $\varphi^0 = \psi \mathbb{L}^{-\mathrm{ordJac}_{/\Lambda[0,m,r]}f}$, and then $f_!(\varphi)$ is the class in $\mathrm{IC}_+^d(Z_{/\Lambda})$ of $(f^{-1})^*(\psi)$.

Theorem 11.3 thus yields a functor from the category Def_Λ to the category with objects $\mathrm{IC}_+(Z/\Lambda)$ and with homomorphisms of semi-groups (or even of semi-modules over $\mathscr{C}_+(\Lambda)$) as morphisms. This functor is an embedding (that is, injective on objects and on morphisms). The functoriality property $(g \circ f)_! = g_! \circ f_!$ is a flexible form of a Fubini Theorem.

Note that $C_+(X_{/\Lambda})$ is a graded $\mathscr{C}_+(X)$-semi-module (but not so for $\mathrm{IC}_+(X_{/\Lambda})$ which is just a graded $\mathscr{C}_+(\Lambda)$-semi-module). Using this module structure, we can formulate the following form of the projection formula.

Proposition 11.4   *For every morphism $f : Y \to Z$ in Def_Λ, and every α in $\mathscr{C}_+(Z)$ and β in $\mathrm{IC}_+(Y_{/\Lambda})$, $\alpha f_!(\beta)$ belongs to $\mathrm{IC}_+(Z_{/\Lambda})$ if and only if $f^*(\alpha)\beta$ is in $\mathrm{IC}_+(Y_{/\Lambda})$. If these conditions are verified, then $f_!(f^*(\alpha)\beta) = \alpha f_!(\beta)$.*

The analogy with the direct image formalism of Theorem 14.1.1 of [9] with $S = \Lambda$ is now complete. An important ingredient of the proof of Theorem 11.3 is that general definable morphisms can be factored, at least piecewise, into definable morphisms of the specified simple types falling under M1 up to M4.

11.1 Comparison with [19]

We end the paper by comparing our motivic integrals with the ones in [19] by means of specialization. Let R be a complete discrete valuation ring with fraction field K and perfect residue field k. In [19], motivic integration on smooth rigid varieties over K was developed. We shall now compare it with the approach in the present paper in unequal characteristic (although it works similarly in equal characteristic zero).

Let X be a smooth quasi-compact and separated rigid variety over K endowed with a gauge form ω. Assume that K has unequal characteristic. Note that, if we write p for the residue field characteristic of K and e for the ramification degree of K, one can naturally consider $X(L)$ for any complete $(0, p, e)$-field L. Consider the analytic theory $\mathscr{T} = \mathscr{T}_K$ with language $\mathscr{L}_K$, as in 2 of Section 3.1. Then we can look at X as a definable subassignment (by using affine charts). One may define $\int_X |\omega|$ using definable morphisms as charts and as transition functions, say for X of dimension n. For each chart in a finite disjoint covering of X by (definable) charts, one takes the pullback of ω on that chart to get a volume form on an open definable subassignment O of $h[n, 0, 0]$. One then expresses the latter as a multiple f of $dx_1 \wedge \ldots \wedge dx_n$, where clearly f is a definable morphism to $h[1,0,0]$. Finally one integrates $\mathbb{L}^{-\mathrm{ord}f}$ on O as is defined in section 9, with the convention that $\mathbb{L}^{-\mathrm{ord}(0)} = 0$, and one takes the sum for all the charts, where integrability follows from the quasi-compactness assumption. This is well defined thanks to the Change of Variables Theorem 10.3.

We will now link this integral $\int_X |\omega|$ to the integral as defined in [19]. Let $K_0(\mathrm{Var}_k)$ be the Grothendieck ring of varieties over k, moded out by the extra relations $[X] = [f(X)]$, for $f : X \to Y$ radical. Note that radical means that for every algebraically closed field ℓ over k the induced map $X(\ell) \to Y(\ell)$ is injective, and that $[f(X)]$ is an abbreviation for the class of the constructible set given as the image of f by Chevalley's Theorem. (In the case that the characteristic of k is zero, the extra relations would be redundant.) In any case, this ring is isomorphic to the Grothendieck ring of definable sets with coefficients from k for the theory of algebraically closed fields in the language of rings.

Similarly as the morphism γ as in Section 16.3 of [9], there is a canonical morphism $\delta : \mathscr{C}_+(\mathrm{point}) \to K_0(\mathrm{Var}_k) \otimes \mathbb{A}$. Indeed, we may note that $\mathscr{C}_+(\mathrm{point})$ is isomorphic to $\mathbb{A}_+ \otimes_{\mathbb{Z}} \mathscr{Q}_+(\mathrm{point})$, and for $x = \sum_{i=1}^r a_i \otimes b_i$ with $a_i \in \mathbb{A}_+$ and $b_i \in \mathscr{Q}_+(\mathrm{point})$, we may set $\delta(x) = \sum_i [b_i] \otimes a_i$, where $[b_i]$ is the class in $K_0(\mathrm{Var}_k)$ of the constructible set obtained from b_i by elimination of quantifiers for the theory of algebraically closed fields in the language of rings (Chevalley's Theorem).

In [19] an integral $\int_X^{LS} |\omega|$ in the localization of $K_0(\mathrm{Var}_k)$ with respect to the class of the affine line is defined. Hence we may consider the image of $\int_X^{LS} |\omega|$ in the further localization $K_0(\mathrm{Var}_k) \otimes \mathbb{A}$.

Proposition 11.5 *Let X be a smooth separated rigid variety over K endowed with a gauge form ω. Assume that K has unequal characteristic and that X is quasi-compact. Then, with the above notation, $\delta(\int_X |\omega|)$ is equal to the image of $\int_X^{LS} |\omega|$ in $K_0(\mathrm{Var}_k) \otimes \mathbb{A}$.*

References

[1] N. Bourbaki, *Topologie générale*, Chapitres 5 à 10, Hermann 1974.

[2] C. L. Chai, *Néron models for semiabelian varieties: congruence and change of base field*, Asian J. Math. **4** (2000), 715–736.

[3] C. L. Chai, J. K. Yu, *Congruences of Néron models for tori and the Artin conductor* (with an appendix by E. de Shalit), Ann. Math. **154** (2001), 347–382.

[4] R. Cluckers, T. Hales, F. Loeser, *Transfer Principle for the Fundamental Lemma*, to appear in a book edited by M. Harris on "Stabilisation de la formule des traces, variétés de Shimura, et applications arithmétiques", available at www.institut.math.jussieu.fr/projets/fa/bp0.html and arXiv:0712.0708.

[5] R. Cluckers, L. Lipshitz, Z. Robinson, *Analytic cell decomposition and analytic motivic integration*, Ann. Sci. École Norm. Sup., **39** (2006), 535–568.

[6] R. Cluckers, G. Comte, F. Loeser, *Lipschitz continuity properties for p-adic semi-algebraic functions and subanalytic functions*, Geom. Funct. Anal. **20** (2010), 68–87.

[7] R. Cluckers, L. Lipshitz, *Fields with analytic structure*, to appear in J. Eur. Math. Soc. (JEMS), **13** (2011), 1147–1223, arXiv:0908.2376.

[8] R. Cluckers, F. Loeser, *b-minimality*, Journal of Mathematical Logic, Vol. 7, No. 2, 195 - 227 (2007)

[9] R. Cluckers, F. Loeser, *Constructible motivic functions and motivic integration*, Invent. Math., Vol. 173, No. 1, 23–121 (2008).

[10] R. Cluckers, F. Loeser, *Constructible exponential functions, motivic Fourier transform and transfer principle*, Ann. Math. **171** (2010) 1011–1065.

[11] R. Cluckers, F. Loeser, *Motivic integration in mixed characteristic with bounded ramification*, preprint, arXiv:1102.3832.

[12] R. Cluckers, F. Loeser, J. Nicaise, *Chai's conjecture and Fubini properties of dimensional motivic integration*, preprint, arXiv:1102.5653.

[13] J. Denef, *On the evaluation of certain p-adic integrals*, Séminaire de théorie des nombres, Paris 1983–84, 25–47, Progr. Math., **59**, Birkhäuser Boston, Boston, MA, 1985.

[14] J. Denef, F. Loeser, *Germs of arcs on singular algebraic varieties and motivic integration*, Invent. Math. **135** (1999), 201–232.

[15] J. Denef, F. Loeser, *Motivic Igusa zeta functions*, J. Algebraic Geom. **7** (1998), 505–537.

[16] J. Denef, F. Loeser, *Definable sets, motives and p-adic integrals*, J. Amer. Math. Soc., **14** (2001), 429–469.

[17] M. Kontsevich, Lecture at Orsay (December 7, 1995).

[18] E. Hrushovski, D. Kazhdan, *Integration in valued fields*, in Algebraic geometry and number theory, Progress in Mathematics 253, 261–405 (2006), Birkhäuser.

[19] F. Loeser, J. Sebag, *Motivic integration on smooth rigid varieties and invariants of degenerations*, Duke Math. J. **119** (2003), 315–344.

[20] D. Meuser, *The meromorphic continuation of a zeta function of Weil and Igusa Type*, Invent. math. **85** (1986), 493–514.

[21] J. Nicaise, *A trace formula for rigid varieties, and motivic Weil generating series for formal schemes* Math. Ann. **343** no. 2 (2009) 285–349.

[22] J. Pas, *Cell decomposition and local zeta functions in a tower of unramified extensions of a p-adic field*, Proc. London Math. Soc. **60** (1990), 37–67.

[23] J. Pas, *Local zeta functions and Meuser's invariant functions*, J. Number Theory **38** (1991), 287–299.

[24] J. Sebag, *Intégration motivique sur les schémas formels*, Bull. Soc. Math. France, **132** (2004), 1–54.

[25] Zhiwei Yun, with appendix by J. Gordon, *The fundamental lemma of Jacquet-Rallis*, Duke Math. J. **156**, no. 2 (2011), 167–227.

Printed in the United States
by Baker & Taylor Publisher Services